Steel Fibre Reinforced Concrete

Steel fiber reinforced concrete (SFRC) for structures dates back to 1925, yet it remains largely unknown to many engineers and contractors. Essential information is often buried in codes, reports, and standards. This book presents the essential information required by engineers to utilize SFRC effectively, achieving superior results across a range of applications.

The focus is on its most common use in flat, horizontal cast in situ applications, such as ground-bearing slabs, ground-suspended slabs, suspended elevated slabs, and raft slab foundations for buildings, which present more opportunities than three-dimensional structures. Prefabrication and paving are discussed where they intersect with these main applications. Material and structural testing are addressed to guide the design for these typical applications, with attention also given to shrinkage, cracking, fire resistance of slabs and walls, and environmental benefits, alongside future opportunities and challenges.

- This unique book provides simple practical guidance for designing and placing SFRC.
- It is especially helpful for design-and-build contractors and professional engineers.
- It offers simple tools and methods to launch a project.

Steel Fibre Reinforced Concrete

A Practitioner's Guide to SFRC Slabs

Xavier Destrée

CRC Press
Taylor & Francis Group
Boca Raton London New York

CRC Press is an imprint of the
Taylor & Francis Group, an **informa** business

Designed cover image: by courtesy of Alpha-Con, Calgary, Canada

First edition published 2025
by CRC Press
2385 NW Executive Center Drive, Suite 320, Boca Raton FL 33431

and by CRC Press
4 Park Square, Milton Park, Abingdon, Oxon, OX14 4RN

CRC Press is an imprint of Taylor & Francis Group, LLC

ISBN: 978-1-032-03641-0 (hbk)
ISBN: 978-1-032-03642-7 (pbk)
ISBN: 978-1-003-18831-5 (ebk)

DOI: 10.1201/9781003188315

Typeset in Sabon
by SPi Technologies India Pvt Ltd (Straive)

Contents

4   SFRC pile supported slabs (G-SFRS)   150

7 SFRC foundations · 249

8 Shrinkage of slabs · 293

Note to Readers

Dear Reader,

I hope and I'm convinced that you'll read this book to find how Steel Fiber Reinforced Concrete (SFRC) is not only for simple slabs on grade but also rather for slab structures. It is simple to understand and design. You'll see that the steel fibers are not only used to enhance plain concrete.

This book is not academic since I'm not an academic writer but a practical engineer who has been involved in his work for over 45 years of experience, in steel fiber reinforcing, regarding inventions such as fiber types and applications, manufacturing, design, testing, selling, and more.

I have learned for over 45 years that steel fibers can be used as a structural principal reinforcing and, as you will see, I have written this book to share my experience with you.

Currently, there are many applications of SFRC but the book's major focus is on horizontal structure applications where two dimensions are predominant over the third, like for slabs and rafts or footings. The book is mainly over slabs and rafts on grade or pile supported when the ground is too soft or too deformable with time. Statically indeterminate free suspended elevated SFRC slabs are also reviewed here as well as ground retaining walls. Applications of SFRC in beams and columns are very limited as the structural stresses are of high intensity and are very concentrated in the section of the structure. Indeed steel fibers are not concentrated in some parts of the section like rebars are, but rather distributed over the whole section of the structure. However, some positive SFRC experience has been gained in foundation piles. Some standards are available today with regard to the fiber nature and geometry, the performances of the steel fibers in the concrete material, and also the design of structures. The fiber performance standards are limited to the analysis of small statically determinate prismatic specimens in flexion. It results today that actual SFRC design standards limit severely the use of SFRC in structures. We'll see that the small size of the standard beams tested to define the performances does not represent a statically indeterminate structure by far.

Steel fibers are known to have a great positive influence on concrete cracking, an advantage gained by their small size, their high number, the good bonding of the steel fibers, and the ductility in concrete.

In my 45 years of experience, I observed that concrete shrinkage is the main culprit of cracking in slab and raft applications. The traditional reinforced concrete, in reality, doesn't control the shrinkage cracking practically as expected according to the standard calculations. Therefore the book devotes a chapter to SFRC shrinkage and cracking of slabs. You will see also that SFRC slabs and walls are quite resistant under fire conditions. A full chapter is devoted to full scale-tests of SFRC slabs and walls under fire, indeed a world premiere at that time. This book is not a handbook about the steel fiber reinforcing of concrete. It is not a repetition of what has already been written in several thousand academic papers on SFRC. It doesn't cover important applications like in the prefabrication or in the tunneling techniques, nor the shotcrete applications. It doesn't replace all the excellent publications of the American Concrete Institute Steel Fiber Reinforced Concrete Committee 544 and of its subcommittees for structural applications, testing methods, design in general, durability, production and application, shotcrete, design and construction of suspended elevated SFRC slabs…

I recommend strongly however, the reader to make sure they read all these ACI 544 publications. All are achieved by a multidisciplinary team of motivated professional engineers in materials and structures, academics, architects, steel fiber producers, and suppliers of related specialty materials, prepared under a unique complete contradictory method in writing, including vetos followed by a cumbersome balloting process.

In general, there are no documents better than ACI publications to discover, learn, understand, specify, and use all sorts of concretes including the SFRC.

This book shows what is my understanding of SFRC, indeed gained over more than 45 years of SFRC experiences in production and application, testings, design, and its installation on-site and shrinkage.

I would like now to summarize here the book content by Chapters:
Preface is a narration of my engineering history and milestones

Chapter 1 is about the basics of understanding steel fibers, including its brief history.
Chapter 2 deals with some material properties and some important standard material properties.
Chapter 3 outlines the series of 15 full-scale SFRC slab tests I have been involved with from 1983 to 2020.
Chapter 4 is devoted to G-SFRS, indeed the SFRC ground suspended slabs that rest on a grid of piles.
Chapter 5 is to explain the E-SFRS, the SFRC suspended elevated slabs
Chapter 6 is about fire resistance of SFRC elevated suspended slabs (denoted as S-SFRS) and of the SFRC bearing walls.

Chapter 7 summarizes the SFRC experiences in raft foundations and shallow footings of buildings.

Chapter 8 is the analysis of the shrinkage of SFRC slabs from the constituent materials of concrete to its installation on site.

Chapter 9 summarizes how to overcome the shrinkage in slabs by taking advantage of the chemical process of the slab post-tensioning. It is indeed a comparison between SFRC and self-stressing SFRC.

Chapter 10 shows a comparison of various design standards for slabs and structures.

Chapter 11 is about SFRC metal decking slabs.

Chapter 12 is about the sustainability of the SFRC.

My sincere wish is that this book will not only help engineers to start their practical experience of SFRC in general, or extend their current experience into more efficient and innovative designs of structural applications, but also to understand the importance of SFRC mix design and installation to minimize shrinkage.

Steel fibers don't make good concrete from a bad one.

My wish is that you'll immediately start a productive reading of this book to more efficiently use the SFRC.

You'll see that SFRC can be used in most slabs and rafts, or footings, instead of the cumbersome rebar reinforcing in design, drafting, estimating, handling, and installation.

Good Luck and Best Regards,
Xavier Destrée, ir. FACI

A tribute to many

Throughout my 47 years of experience in fiber reinforcement, I have never been alone. I have learned, observed, and progressed significantly, thanks to the many questions, observations, and conditions raised by others. These questions—whether theoretical, practical, experimental, economical, social, or environmental—could only be addressed with clear, simple, and understandable answers. This process has been a constant source of growth and understanding for me.

All stakeholders in fiber and traditional reinforced concrete have contributed to my development. A lot of thanks to all of them.

Special thanks to:

- Prof. Dr. Barzin Mobasher of Arizona State University, former President of the ACI 544 Fiber Concrete Committee, for his vast and shared experience in composite materials and fiber-reinforced concrete. His work on constitutive laws for structural applications, design solutions, and application methods has been invaluable. We have worked together as the lead authors of the ACI 544 6R15 and I must admit, I couldn't have done it without him.
- Prof. Dr. Nemy Banthia of the University of British Columbia, whom I first met in 1986 at the University of Quebec. His persistent encouragement in 2004 convinced me—and my wife—that I should join the ACI 544 committee as a voting member to prepare a report on SFRC structural slabs, which eventually became the ACI 544-6R15 report.
- Dr. Corina Aldea, Director of Materials at WSP Engineers and a prominent member of ACI, for her support with SFRC-related Special Publications and seminars. Her assistance helped me articulate my vision for SFRC within and beyond ACI. These successful efforts led me to become an ACI Fellow in 2019.
- Mrs. Ann Ellis, a past President of ACI, for opening doors within ACI to support the progress and circulation of SFRC knowledge and reputation within the ACI and beyond.

M.A. Lazzari, my first boss 45 years ago and who was willing to quickly implement my novel ideas in his businesses.

M. Dan Baker, a past President of ACI and founder of Baker Concrete, for welcoming my designs and the SFRC BakerBuilt joint-free slabs into his company in the 1980s.

Prof. Dr.ès. Sciences Wolfgang Jalil, Technical Manager of Socotec, the international control and guarantee company, who helped me and my company with the first certified SFRC slab application in France in the 1980s, as well as later with SFRC pile-supported ground slabs and anti-seismic Twincone CFA piles.

Mr. Pierre Weicherding, General Manager of ARBED steel wire plants, for his early and significant investment in producing high-quality, low-cost Eurosteel and Twincone steel fibers.

M. Roland Jung, the first CEO of the ArcelorMittal group, welcomed me to Arcelor as an independent R&D consultant in SFRC, and wanted me to remain at the same position for the next 24 years until retirement, which I indeed achieved for them for 26 years.

All concrete contractors on five continents who have promoted and installed SFRC jointfree floors, piled slabs and rafts.

Janis Oslejs, CEO and Founder of Primekss concrete contractors in the Scan-Baltic Region and Overseas who, though licensing partners, has implemented my ideas regarding ultimate slab design with zero-shrinkage, chemically post-tensioned jointfree SFRC slabs on grade and piled slabs, rafts or even elevated suspended slabs. He has purposefully and aggressively directed his Primekss R&D department to improve sustainability by drastically reducing the slab carbon footprint by up to 65%.

I must deeply thank my wife Marie-France for our 47 years together, for her patience, attention, and love. She has been a firm believer and supporter of these fibers, especially through the inevitable crisis events and periods of doubts. It was natural for her to build our 250 m³ SFRC family house in 1990, out of steel fiber reinforced concrete using prototype structural Twincone fiber reinforced concrete for the raft, elevated slabs, columns, and walls.

Thanks to all these people. No doubt I would relive a second life for the same.

I am also grateful to the CRC team, Mr. Tony Moore and Ms. Aimee Wragg who invited me to prepare this book, for their patience, encouragement, understanding, and continuous support throughout the process.

Xavier Destrée,
La Hulpe, Belgium

Foreword

It is with great pleasure and anticipation that I introduce you to the remarkable journey chronicled within the pages of *Steel Fiber Reinforced Concrete: A Practitioner's Guide to SFRC Slabs* by Mr. Xavier Destree. In this captivating narrative, Mr. Destree shares his profound insights and experiences, unveiling the fascinating evolution of Steel Fiber Reinforced Concrete (SFRC) design throughout his illustrious career.

Xavier's career has meant a real innovation on how we consider our built world, with a dedicated focus on development, quality, efficiency, and sustainability. His pioneering spirit has left an indelible mark on the realm of SFRC design, shaping the way we approach construction and infrastructure projects.

What sets this book apart is not merely the technical expertise it imparts but also the person anecdotes and reflections that provide a glimpse into the passionate journey of a dedicated professional. Through Mr. Destree's lens, we witness the fusion of knowledge and passion that has driven the advancements in SFRC design.

Xavier has many significant milestones in his career, but for me, he is my father in SFRC design. Xavier took me under his wing and took me from a non-engineering background to having a high level of understanding. Xavier also taught me to think out of the box without the limitations ingrained into all engineers who are governed by outdated codes. The knowledge I have gained has allowed me to understand SFRC as a material and seek innovation that matters.

As we delve into the world SFRC and design, we are granted access to Mr. Destree's unparallel expertise and contributions to this specialized field of concrete technology. With each turn of the page, we gain valuable glimpses into the challenges, breakthroughs, and innovations that have shaped the landscape of SFRC design.

As you embark on this enlightening journey through the pages of Mr. Destree's book, I am confident that you, too, will recognize the profound influence he has had on the world of SFRC. *Steel Fiber Reinforced Concrete:*

A Practitioner's Guide to SFRC Slabs is not only a comprehensive guide but also an inspiring tribute to a true pioneer in the field.

In conclusion, let us contemplate the essence of knowledge. Its true purpose emerges when it is shared and applied in the real world. This is Xavier's true legacy.

Brett Meadway,
Technical sales/Project Manager,
Kingdom of Saudi Arabia, UAE, South Africa.

Preface

Early in my student life, back in the late 1970s at the Polytechnic School of the University of Brussels, I chose the cycle on composite materials as an option. I was immediately motivated to delve deeply into the concept, finding it both innovative and exciting. In contrast, the study of traditional reinforced concrete felt dull to me, as it was already an old and established field at the time.

For my master's thesis, I proposed investigating the flexural behavior of flat beams made of mortar, gypsum, and micro-concrete, reinforced with continuous glass roving fibers (E- or A-glass) arranged parallel at the bottom of the form without twist. Early in my research, I discovered the bad effect of alkali attack on glass fibers and the necessity of protecting them with a cost-effective coating.

The solution I developed involved soaking the fibers in a bath of Venezuela hard bitumen in a water-based emulsion. The protection was total, and I discovered that the bituminous coating had a positive effect on the bonding since the debonding, starting at the tip of a crack, became controlled, and even more, was reversible. The crack could close back when the loading was removed.

So I had to learn about the debonding of smooth fibers, in general, based on the Boussinesq model indeed no longer applicable with coating as a viscous relative displacement was possible between the fibers and the matrix.

While the beams demonstrated ductility in flexion, they retained a major disadvantage: sudden failure under shear forces, like all brittle materials. This was due to the use of oriented continuous fibers. While randomly distributed fiber reinforcement could have addressed this limitation, it was outside the scope of my thesis.

I developed a design method for these glass roving-reinforced beams and accurately back-calculated my test results. My thesis, concise at only 42 pages compared to my peers' 200-page submissions, earned me the prestigious "Jacques Verdeyen Prize" in 1978, a rare honor awarded by the university. My family was so proud of this achievement that they included it on our wedding announcement card beneath my name: "Civil engineer—Jacques Verdeyen Prize award—1978."

Afterward, I sought to further develop the concept within the Belgian concrete and contracting industry. I met with leading companies, but the results were disappointing. The feedback was always the same: everything was already known in the construction industry, and the industry was in crisis, with no money or resources to invest in unproven innovations.

I was hired by a general contractor of Brussels, where I wanted to deepen my knowledge of fibers, and more precisely of randomly distributed short steel fibers.

I found the ACI 544 fiber-reinforced concrete committee literature developed in the 1960s and 1970s and even a "WIRAND" handbook on SFRC concrete floor design written by the Battelle Development Company in the 1960s.

So, I was hired by the Silidur concrete contractor, which operated in Belgium, Holland, France, and Germany, to develop SFRC slabs on grade from scratch. The owner and president, Mr. A. Lazzari, gave me the opportunity to establish EUROSTEEL S.A., a new company dedicated to steel fiber activities, where I eventually became a stockholder as an incentive.

Mr. Lazzari completely understood the limitations imposed by engineers relying on outdated standards—a concept he viewed as a pleonasm. With his support, I had no restrictions on developing fibers and applications, provided the return on investment was swift and substantial, and the applications proved successful. Lazzari decided to grow his company through the innovative use of steel fiber reinforcement. My mindset and education have always been to do a lot with minimal resources.

Over the 14 years I have worked for Silidur, it grew by a factor of 50! It was the result of an SFRC design-built slab strategy. Mr. Angelo Lazzari retired at 65 years of age in 1993 and sold the company to a prominent Cement Company, which ruined it in some years with a different strategy. I was fired by them in 1995.

In 1982, on a job site road, I stumbled upon a 250 mm long piece of hard-crimped steel wire. I immediately realized that, if cut to 60 mm, it could become the ideal steel fiber. This wire turned out to be a "lost hair" from a rotating broom machine used to clean pavements and roads. I visited the machine manufacturing plant in Germany and learned that the crimped wire was produced by a small plant located in a townhouse near Dortmund, originally for musical instrument steel wires. Although the plant was uninterested in producing steel fibers, they agreed to provide me with a 250 kg sample for testing. These fibers were 0.8 mm in diameter, 60 mm long, and had 7 waves of 1 mm depth along their centerline.

We tested the fibers at a site in Luxembourg and observed significant challenges. Introducing the fibers into the truck mixer was difficult, and numerous dry fiber balls formed, even at a low dosage rate of 20 kg/m³.

The owner of the musical instrument wire plant referred me to Mr. Willy Krampe, who had recently started a steel fiber production activity from scratch. Mr. Krampe, a former executive at a German wire plant, had launched a small-scale production in his garage for short steel fibers used in refractory concrete, benefiting from SFRC's ductility at high temperatures ("Feuerfestigkeit" in German).

Mr. Krampe, a highly dynamic and hardworking person, agreed immediately to start producing my undulating fiber. Based on the initial site experience, I decided to increase the fiber diameter to 1 mm. This adjustment significantly improved the separation, introduction, and mixing of the fibers in a truck mixer.

With minimal precaution regarding speed, mix design, and mixing duration, the balling effect disappeared completely at a dosage rate of 20 kg/m^3. The Eurosteel fibers I invented were protected by a robust international patent, covering both Europe and North America. Willy Krampe became the first licensed producer of the patented fiber and remained a high-quality supplier to Eurosteel and Silidur until Silidur's collapse in 2000, 5 years after I was fired in 1995. Willy Krampe, indeed actually the second generation, has become a leading producer of steel fibers today, almost on a global scale.

The first samples of the Eurosteel fibers (1 mm × 60 mm, 1200 MPa tensile strength) were presented for the 100,000 m^2 NATO Slab Project in Luxembourg. This opportunity came through an introduction by the management of ARBED, a global steel company based in Luxembourg. Over time, ARBED's name evolved to ARCELOR and later to ArcelorMittal, as it is known today.

In the chapter on full-scale tests, I will provide more details about this large project, which we eventually secured at a dosage rate of 20 kg/m^3. This project marked a significant milestone, acting as a green light for the rapid expansion of the SFRC flooring market in Western Europe. I owe this success to Mr. Pierre-Marie Dubois, the principal engineer at the SECO control and guarantee office, whose approval was based on successful full-scale tests.

ARBED became the second licensee of our patent and consistently supplied us from its Ghent plant in Belgium and Bissen plant in Luxembourg. Over time, ARBED expanded its fiber production to plants in the Czech Republic, the UK, Poland, and Bosnia.

After 12 months of site experience, I observed that sawn cuts (shrinkage joints) at 6 m intervals did not open when the dosage exceeded 30 kg/m^3. This phenomenon was initially more common in Holland than in Belgium, due to the superior concrete quality in Holland, thanks to the clean sands dredged from the Rhine and Maas rivers.

Soon I had to invent a machine to introduce the steel fibers into the truck mixer in order to be more productive than by hand, safer on the job site or at the batching plant, and more precise, to reduce as much as possible the variations caused by mixing in a truck mixer. I discovered through my home vacuum cleaner, despite the negative opinion of a specialist company in airflow transport, that the EUROSTEEL undulating steel fibers could take off in a flow of air at high speed.

The so-called fiber blast machine was invented in 1983. It was mounted on a trailer and consisted of a hopper with a strong vibrating grid to separate the fibers and rain them down after a venture-like reduced pipe section, fed with high-velocity air powered by a powerful centrifugal ventilator. The fibers were ejected at 30 m/s velocity straight in the truck mixer. Later the machine was improved as the fibers were loaded in a bucket (the workers called it "the coffin" of similar shape) at the ground level, to load a rotating cage where the fibers were separated and rained down to fall further than from the venturi. At the end, it was a great progress so that these machines are built today by more companies and are sold across four continents. The machine has been reported in the ENR magazine in the USA in 1986.

The machine was first used for SFRC slabs in Canada (Toronto) in 1984 at Plastiglide Ltd, installed by Tricon Concrete Finishing for the Inducon developer in Toronto. Earlier, in 1983, we had completed the first successful SFRC joint-free floor at the Keystone Valves plant in Breda-Noord, Holland. This project covered 1,800 m² with a thickness of 150 mm over a compacted sand bed.

Eurosteel SA applied a patent on how a jointfree floors of up to 3000 m² area could be obtained by the application of minimum 30 kg/m³ dosage rate in a ready mixed concrete with a C 25-30 low "shrinkage content," laid on a rather smooth sandy grade with a minimum k-value of 30 MPa/m.

In 1984, the first superflat FF100 SFRC jointfree floor was installed in Toronto at American Hospital Supplies by Tricon Concrete Finishing in the Toronto area.

The patent was extended beyond Europe to North America with a number of licensees to exploit it. I applied "Conductil" as the brand name of Eurosteel SA SFRC jointfree floor. It took more time to introduce the SFRC jointfree floor technique in Germany, France, and the UK. Nevertheless, the 20 kg/m³ dosage rate with saw cuttings shrinkage joints became very popular immediately as a result of the NATO testing in full-scale and the successfully completed slab.

In the laboratory of Professors Dehousse and Sahloul at the University of Génie Civil (Liège, Belgium), initial small beam flexion tests showed disappointing results with significant scatter, making predictions unreliable. I shifted the focus to statically indeterminate square plates (600 mm × 600 mm × 100 mm), supported on a fresh plaster perimeter to avoid torsion, and subjected to center point loading with a 100 mm × 100 mm footprint.

This approach eliminated scatter and consistently showed multiple cracking, increasing with the dosage rate. The 20 kg/m³ dosage rate of the EUROSTEEL (of 1 mm x 60 mm) showed a perfect elastoplastic diagram with a deflection of up to 10 mm.

It became obvious to me that the SFRC slabs should not be designed like a statically determinate beam, especially when the reference beam is so small.

I presented at the 1986 University of Birmingham UK fiber concrete conference, ancestor of the BEFIB conferences, on slab testing as well as the need to carefully consider the mix design for steel fiber concrete. The main attention was on the fiber concrete ductility instead of its compressive strength. German customers quickly recognized the advantages of SFRC slabs over conventional 200 mm thick slabs with top and bottom wire meshes, with huge economic benefits. Silidur sold millions of square meters of this slab type, marking an immense success. The SFRC patented jointfree floor activity expanded in Germany and France in the second half of the 1980s. In France, therefore, I developed from 1983, a specification with the help of Dr. Es Sciences. Wolgang Jalil, technical manager of the SOCOTEC control and guarantee company, professor at Ecole Nationale des Ponts et Chaussées (ENPC).

I invited Mr. Jalil and his colleague to visit completed or ongoing SFRC jointfree floors in Holland. Mr. Jalil, a brilliant engineer, sought effective, simple, and reliable solutions. The SOCOTEC-guaranteed applications covered design, concrete mix, constituent materials, on-site installation, and stringent quality control.

Prof. Dr. Jalil later presented a paper at a Vienna conference on Twincone fiber-reinforced anti-seismic piles and authored a highly regarded seismic design handbook.

The first "CONDUCTIL" SFRC jointfree floor in France was built in 1984 at the Flodor potato chips plant in Péronnes, covering 2,000 m² with a thickness of 150 mm at 30 kg/m³ dosage rate of 1 mm diameter, 1200 MPa steel wire tensile strength undulating fibers under the EUROSTEEL product name.

This success was followed by a number of applications across PEUGEOT plants in France, featuring 150 mm thick floors and even up to 400 mm rafts under the first robotic units. Mr. Schey, Mr. Boilleau together with SOCOTEC have been promoting the concept within the PEUGEOT industries and later the PSA group...

To supply the market in France, "Trefileries de Conflandey" became a licensee to produce the Eurosteel fibers, which were then seen as a truly French product—always a positive point in France.

The top managers of Conflandey, including the Lemaire family and Mr. Gache, invested significantly to supply us, but not as much or as intensely as ARBED. As they said, "*we are so depending on you that we don't know what could happen if you were crushed under a bus in Paris when you come to visit us.*"

At the end of the 1980s, we had no fewer than five licensed steel fiber manufacturers supplying us on an exclusive basis. We never failed to supply a job anywhere. Our purchasing costs for steel fibers were under control, as was the quality. I think this is the best approach to obtaining the budgeted quantity for the year, at the right cost and with consistent quality. My advice, based on experience, is to never restrict yourself to a single supplier.

The first SFRC jointless slab in the UK was installed at a Brook Bond Foods plant in 1989: four bays of 1,200 m² each, with a thickness of 180 mm, placed by Stuarts Flooring Ltd, as explained in Reference 1.

Later, EUROSTEEL fibers were also produced at the Sheffield plant of ARBED. The SFRC jointfree slab was branded under the Conductil trade name as it was the abbreviation of continuous and ductile.

It was very frustrating at that time to see our competitors attempting to out-compete us with low dosage rates of steel fibers and saw cuts—solutions *that became plagued by severe joint curling and uneven saw-cut openings by the 1990s.*

The extension of the sawcuts distance to 12 m together with low dosage rates of steel fibers proved to be a disaster in general. Our competitors were selling 20–25 kg/m³ low dosage rate of steel fibers to recommend saw cuts distance of up to 10 m apart. Some saw cuts, every two or three saw cuts, opened significantly and showed curling so that they invented the advanced concept of "dominant joints" a bit like the scelerate wave of the Ocean!

The Eurosteel SFRC jointfree patent was infringed several times but stopped through appropriate court actions. Today, it is very satisfying for me to see that the SFRC joinffree floor, just as I patented it, has become common in large logistics applications across many European countries as well as in the USA and Canada. It was clear to me at the time that steel fibers could effectively control cracking.

When I discovered in Holland around 1986 that many industrial slabs were built on piles and reinforced with a large percentage of traditional top and bottom reinforcement, I immediately recognized this as the next massive application for steel fiber reinforcing.

Therefore, I invented and developed the TWINCONE steel fiber, applied its brand name, and protected it with an international patent in Europe, the Americas, and further in East Asia.

The TWINCONE fiber was a straight 1 mm diameter fiber, designed with an anchoring cone at each end featuring a 60° summit angle to lock the head into the concrete matrix. This design provided a unique and highly effective total anchorage for steel fiber.

All flexion tests of Twincone-reinforced small beams showed a steep strengthening after the limit of proportionality, up to the maximum load, followed by a decrease in loading intensity caused by the progressive rupture of fibers in the bottom section. A significant number of Twincone fibers exhibited rupture by striction!

The Twincone fibers proved to be the best crack arresters or crack eliminators for piled slabs. The current testing methodology for SFRC flexion tests relies primarily on EN 14651 f_{r1} values for SLS and the relatively low value of f_{r4} for ULS. It is regrettable, as the curvature at f_{r4} doesn't even exist in any SFRC slab structure, even at ULS, according to all full-scale test observations summarized in Chapter 3. Instead of a single large crack or hinge, as shown in the EN 14651 beam tests, SFRC slabs show a larger number of smaller, controlled cracks that develop into more controlled yield lines.

The Twincone fiber production was a real nightmare to develop but was eventually achieved with the strong support and help of Mr. Pierre Weicherding, head of the steel wire plants at ARBED.

The production process involved high-speed stamping (1,500 per minute) of two wires cut precisely to length by a WAFIOS machine.

WAFIOS, a leader in wire processing and manufacturing machines located in the Stuttgart area of Germany, was an impressive company to visit, including its school for training highly specialized personnel.

Mr. Weicherding ordered up to 40 of these machines, capable of producing 10,000 tons of Twincone per year. The manufacturing process was so quick that the movements within the machine and its output were invisible to the naked eye! Mr. P. Weicherding also promoted the parallel dense packing of steel fibers in 25 kg compact boxes to ease handling on-site. He was a firm believer in the future of steel fibers and continuously reduced the direct costs of steel fiber production. He developed the continuous steel wire drawing process from a 6 mm diameter coil wire to a 1 mm EUROSTEEL diameter wire, achieving speeds of 50 m per second. When production started at the ARBED Ghent plant in 1983, the drawing speed was limited to 13 m/sec. Thanks to him and his management until 1995, the rapid and intense reduction of direct costs in fiber production became a reality. We owe much of our success to him. I have always believed that to succeed in an open competitive market, you need to minimize the direct costs of your business.

The Twincone fibers were used as the sole reinforcement in the first two jobs during 1993: a 15,000 m² slab on piles for the Ziegler company in the Rotterdam Seaport area, followed by a 40,000 m² piled slab for the logistics center of the Belgian mail in Brussels.

By the end of 1995, we had already completed half a million m² of Twincone-suspended reinforced slabs through Silidur concrete contractors. Foundation rafts were also built using Twincone-reinforced concrete, the first being 1 m thick for AIDE, a Belgian government agency for water treatment plants, in 1993. Hundreds followed across Europe and North America. Many clad rack, structure rafts indeed, for up to 35 m height racking automated warehouses, were built using SFRC across Europe, the UK, Canada, and the USA. I can say that my ideas quick-started the structural application market for steel fibers. The technique proved to be very competitive and was ready to take off.

At the same time, I became an independent R&D SFRC consultant for ARBED and continued in this role until 2002, then for ARCELOR until

2007, and finally for ARCELORMITTAL until 2022, retiring at the age of 67 two years more than the official limit.

Sadly, our competitors, having been left behind, continued to bad-mouth our SFRC structural applications, claiming their intellectual impossibility. I must admit they were correct if designs are limited by the flexion strength derived from small prismatic specimens in bending tests. However, such a method leads to gross overdesign and a waste of resources.

You will see in Chapter 3, that we, at Sildur company, organized a full-scale test of an SFRC piled slab in elevated conditions at the Ternat site in 1994.

At the University of Nord-Pas de Calais laboratory in France (Béthunes), a number of Round Indeterminate Panels (150 mm thick with a 1.50 m clear span) were tested in flexion under center-point loading in the early 1990s. This was done to derive the plastic post-cracking flexion strength of SFRC and eliminate the high artificial scatter associated with small prism specimens testing in flexion.

Later, competitors claimed credit for inventing the hybrid design of SFRC slabs on piles, which included a beam-type reinforcement from pile to pile and top meshes over each pile.

The same Ternat test was repeated in 2000 at James Cook University in Australia, producing the same conclusions as the original Ternat test. Prof. Ginger concluded that Twincone-reinforced concrete was fit for suspended slab reinforcement.

The hybrid reinforcing of steel fibers and rebars piled slab patent was canceled in 2004 at the court at the European level in Munchen, Germany. After the court decision had been given the competitor lawyers came to congratulate me with a firm shake-hands, indeed true gentlemen.

At the end of the 1990s, I introduced the yield line method to determine the design moment. This method was fully developed as a theory by Johansen in the early 1950s. A first attempt to define it was made in Russia in the 1920s, and a Dutch standard publication appeared in the early 1960s under the name CUR 26 A, B, and C. However, it was never widely used or taught at universities, likely because it was considered too revolutionary.

For example, in the case of a simple UDL between four columns, the design moment (kNm/m) is proportional to the total load intensity only, not the span. In the case of a simple point load, the design moment depends on the span, unlike the square of the span in linear elastic methods. This approach suggests significant cost savings.

I implemented the yield line slab theory, already known for over 50 years but unused in slab design, in my paper presented at BEFIB 2000 in Lyon. The resisting moment was derived from the round panel indeterminate slab test in flexion.

However, to introduce it into the reality of the market, I have been supported by the Belgian SECO control and guarantee company, same as at the time of SFRC slabs on grade introduction to the NATO slab in 1984. This time, Mr. Ely Schmit, the principal engineer at SECO, contributed significantly to the success of the plastic method design otherwise completely unknown by most engineers. M.Ely Schmit later became a professor at the University of Louvain.

When Ely had to approve the method and oversee a first real slab to installation, I invited him to Holland to visit already completed and in-service a number of SFRC piled slabs. The feedback was great but he noticed some form of hairline, and non-detrimental cracking due to shrinkage. I decided then to finish the day with a visit to a TNT traditional reinforced concrete piled slab with 140 kg/m³ rebars, next to Brussels airport.

The traditional slab was 1 year old and under full-service conditions. I knew the slab because a regular SECO engineer at the time of the possible order refused to consider our method, although it was quite cheaper than the traditional method.

I also had this understanding that the concrete supply for the installation of this traditional slab had been very difficult due to super-dense traffic jams on the Ring Road of Brussels during weeks. Moreover, the concrete was supplied from a town 60 km away and included, as constituent aggregates, crushed recycled demolition bricks to still meet the C 30 standard strength. The visited slab was horrible, looking like an old slab after 30 years of heavy-duty use with lots of wide cracks, spalling, heavy crazing, chipping out of cracks and joints, significant abrasion, and a black, very dirty surface. In total opposition to the SFRC piled slabs we had visited earlier in Holland.

Ely Schmit gave final approval for SECO, admitting he could not decide otherwise after seeing the stark difference between the SFRC slabs and the traditional reinforced slab concrete design approved earlier by SECO.

With experienced, educated engineers, it is possible to discuss and innovate but with rigid standards, it is impossible.

In my opinion, standards freeze the past to solidify it for the future. Many engineers believe traditional reinforced concrete provides good crack control. Nothing is more wrong. It enjoys a crack control of up to 0.3 or 0.4 mm by standard calculations. The design standards are wrong and these engineers can't be blamed. Traditional reinforced concrete cracks are everywhere and have shown openings of more than 0.3 mm.

A few years later, yield line-based design moment equations were included in TR 34 4th Edition, with significant contributions from Prof. D. Beckett at Greenwich University.

We never patented our SFRC-only piled slab application because we believed the Man of Art could adopt it by simply increasing the steel fiber dosage of slabs on grade and improving the fiber type.

Our structural slab applications grew rapidly in Belgium, Holland, the UK, and later in other continents such as North America and East Asia.

We continued testing numerous round indeterminate slabs (1.50 m and 2.0 m clear spans) at universities in Béthunes Nord-Pas de Calais, Brussels, Montreal, Birmingham, Sheffield, Bilbao, and Pretoria.

Given the success of SFRC piled slabs, I decided to start a new R&D program on SFRC suspended elevated slabs. The span-to-depth ratio of approx. 20 (maximum for piled slabs), which is increased to 30 (for suspended elevated slabs).

Initial testing was planned at the Polytechnic of Montréal with Prof. B. Massicotte.

We decided to include the anti-progressive collapse (APC) rebars in any suspended elevated application, from column to column at the bottom to ensure the slab would never detach from the columns under any possible circumstances, such as accidents or terrorist explosions, as specified in the Canadian Reinforced Concrete Code of 1992 for all types of reinforced or post-tensioned slabs.

In 2004, a public full-scale test of an SFRC-suspended elevated slab was organized at the Bissen Plant of Arcelor in Luxembourg. The slab measured 18 m x 18 m, 200 mm thick, with a center-to-center span of 6 m.

Over 300 attendees were invited. The entire test was supervised by Prof. U. Gossla of the Fachhochschule Aachen (Germany).

The test concluded brilliantly, with the loading intensity and ductility far exceeding the expectations of most attendees.

I remember a lottery was organized to guess the collapse load at the center-point loading. Most replies were in the range of 200 kN. The winner was Louis Crépeau, a prominent structural engineer in Montréal, who predicted 500 kN, the exact number. He won four magnum bottles of excellent Bordeaux red wine from the best millesimal at that time.

The engineer from our main competitor, whom we also invited, guessed a collapse load of 30 kN, the lowest guess by far among the dozens received—17 times less than the reality. No wonder they didn't believe in structural steel fiber-only reinforcement.

At the ARBED company, there was a strong tradition of impeccable food and service for invited guests. When I was invited by the top management of ARBED in the 1980s and 1990s, it was at their old palace and castle in Luxembourg's downtown center Its starred, elegant, and distinguished restaurant offered incredible wines, champagne, and French gastronomy. We had amicable and productive discussions there. Sadly, it was only an ARBED tradition, and the castle was later sold.

ARBED was established in the early 1900s by its founder, Belgian engineer M. Mayerich de Saint Hubert, stood for "Aciéries Réunies de Burbach Et Differdange." It later became a world-leading steel company, almost a State within the Grand Duchy of Luxembourg to employ 70,000 personnel in a country with a population of 400,000. I'm happy and proud to have known it, from the bottom to the very top.

At the Bissen plant test day in October 2004, there were guests from three continents who enjoyed a lavish reception, true to tradition, with different booths for each food type. Table clothes, linen napkins, and excellent wines and foods were served during the whole day.

It has been a way for me to enjoy the business we did—thanks to the great opportunities to meet and understand the people, food, habits, countries, cities, towns, and villages of those I met. The human factor is often a game-changer!

The same Bissen test was repeated in 2007 in Estonia with the ArcelorMittal Scandinavian branch and the Rudus concrete and cement companies, supervised by the University of Tallinn and Prof. J. Pelloo.

By 2005, Prof. N. Banthia of the University of British Columbia invited me to become an active voting member of the American Concrete Institute 544 fiber-reinforced committee. I was asked to prepare a report on my steel fiber experience in design and construction of structural applications.

I met Prof. Barzin Mobasher of Arizona State University, a wealth of knowledge in composite material reinforcement and steel fiber-reinforced concrete design theories and applications. Together, we became the lead authors of the famous ACI 544-6R 15 Report on design and construction of elevated suspended steel fiber reinforced concrete slabs, issued in 2015. Prof. Joaquim Barros of the University of Guimaraes, Portugal, joined us to prepare the yield-line-based design chapter and annexes.

I discovered the unparalleled ACI procedure for document preparation—a multidisciplinary, open, fully contradictory process in writing, with ballots and vetoes. It took us 8 years to finalize the ACI issues and circulations, which included 256 vetoes, five ballots, and two reviews by the ACI TAC committee.

From 2009 to 2011, I worked with ArcelorMittal on the design of the Friends Arena, a 65,000-seat stadium built on a piled raft foundation of 17,000 m³ SFRC. The design followed the Johansen yield-line method, eliminating almost all 175 kg/m³ traditional rebars and replacing them with the dosage rate of 45–50 kg/m³ of HE 1/60 fibers with 1500 MPa tensile strength.

The project was so successful that the local contractors association requested the Swedish Standards Association to establish a committee for an SFRC STRUCTURAL design standard. I was invited to become member which I accepted. The SS 812310 was issued in 2014.

This document introduced the structural ductility concept through a statical indeterminacy factor for SFRC structures.

Year 2015 was quite a successful one, as it marked the release of two critically important design documents regarding SFRC suspended slabs, both far ahead of existing standards and the Annex L of the Eurocode 2 draft, which is expected to circulate from 2027. Annex L, in comparison, is arguably a significant step backward.

The SS 812310 structural indeterminacy factor was set at 2 for two-way slabs, as I proposed in my BEFIB 2004 paper. This factor multiplies the strength derived from EN 14651 by 2 in case of a 2-way slab, which has been proved many times by testing indeterminate slabs vs. EN 14651 small beams.

It has been very frustrating that the main and leading competitor insisted at the Swedish committee that the future Swedish standard could simply be a translation of the German DafstB-DIN 1045 standard, which I opposed on all occasions since it could have stopped all SFRC structural activities in Scandinavia. In Germany, practically, the steel fiber are typically used in structural slabs at about 35 to 40% content, with rebars making up 60 to 65%. As a result, steel fibers are perceived merely as an enhancement to plain concrete, aimed at reducing the number and size of cracks in theory.

I demonstrated it once more in my paper of BEFIB 2008 in India, so that Professor Lucie Vandewalle of the University of Leuven (Belgium) invited me to become member of the FIB ModelCode steel fiber structural subcommittee.

We owe it also to Prof. Johan Silfwerbrand of K.T.H Stockholm who was the lead author of the SS812310. We knew each other already a long time in the past as we had prepared papers together For FIB 2012 Congress in Stockholm regarding the Friends Arena SFRC raft design regarding the differences between the Swedish method and the Belgian method as defined by Johan Silfwerbrand.

It was amazing that Johan named it Belgian Method, as the HE+1/60 type of fibers were manufactured by Arcelor Luwembourg and that I was their R. and Design consultant.

In 2011, I participated actively in the TU Eindhoven (Holland) full-scale test of a 180 mm thick, one-way continuous slab consisting of three consecutive spans of 5.4 m center-to-center, reinforced with SFRC at 50 kg/m³ of HE 0.9/60 hooked end fibers supplied by Arcelor Mittal and Bekaert. The test showed that such a slab could collapse under almost five times the Dutch standard residential UDL of 2.5 kN/m². It also revealed that cracking in flexion exhibited a multiple pattern, with crack openings of less than 0.5 mm at 6 kN/m²—more than twice the residential loading intensity.

Unlike our recommendation, the slabs were not provided with APC rebars in the bottom, as it was argued that these rebars could prevent observers from evaluating the contribution of the steel fibers. The goal should have been to develop a new construction process where the slabs were installed on tunnel forms, completing the full floor level, including the walls, in SFRC.

The relieving of the tunnel form started 16 hours after placement, when the concrete reached 18 N/mm² crushing strength.

At BEFIB 2012, I presented a paper written together with Mrs. Ann Lambrechts of the Bekaert company, detailing how to design the Eindhoven test slab using the FIB ModelCode provisions for steel fibers.

In 2019, a full-scale test of the same slab size was organized by Primekss concrete contractor under Euro standard test setups at the ISO Fire in Poprad Slovakia at the FIRE official laboratory. Fire resistance and insulation up to 200 minutes were achieved, which was well above expectations.

These slabs included APC rebars in accordance with the ACI 544 6R15 report.

The test was a world premiere since SFRC had previously been investigated only as a material but not as a structure. All SFRC material tests concluded with the very beneficial effects of steel fibers in concrete under fire conditions.

In 2020, I prepared two papers for the FIB Shanghai and BEFIB 2020 Valencia (Spain) conferences to summarize and circulate the fire test results.

My collaboration with Primekss, which began in 2002 with Janis Oslejs, CEO, for SFRC jointfree floors and later for piled slabs, deepened in 2005 to continue my ever-lasting conquest against the challenges of concrete shrinkage. This effort led to the first step toward a zero-shrinkage concrete slab method in 2009, which combined SFRC slab together with the use of expansive binders mixed on-site in the truck mixer.

The R&D team at Primekss, under my advice, dedicated significant effort to demonstrate chemical post-tensioning achieved through restrained chemical expansion and the internal friction provided by fiber reinforcement. I presented these findings in my paper at the ACCTA 2016 conference in Tanzania, held at the Dar Es Salaam White Sands Palace.

After the presentation, I was asked by local team of concrete experts for my opinion on Tanzanian standards, specifically whether they should follow the Eurocode and other Euro standards. I candidly stated that if the intention is to build like Germany, then it is a good idea. However, I added that doing so would lead to the loss of local traditional knowledge in technologies such as vegetal or animal fiber reinforcement of clays and other slightly hardened raw materials. They would need to build a structure similar to Germany with lots of waste, overdesign, a lack of long-term durability, and the loss of local resources.

I participated in additional full-scale tests, such as the one conducted in 2009 at the University of Vaasa, Finland, with a SFRC by Primekss. In this test, an SFRC slab with a 35 kg/m³ dosage rate of HE 75/50 steel fibers was installed on EPS 200 insulation (Expanded Polystyrene) with a thickness of 100 mm. The slab was so strong that the hydraulic jack under the portal frame collapsed at 270 kN point loading, yet the slab did not rupture. It was clear to me that most industrial slabs without shrinkage could be designed at 100 mm thickness in most case of loadings like 70 kN back-to-back legs.

From that milestone, Primekss began to grow significantly, achieving a leading position in the Scan-Baltic region. This success was partly due to the region's open-minded mentality, which allowed innovations to thrive without being hindered by the kind of local bureaucracy prevalent in France and Germany.

Janis Oslejs, the CEO, and I gradually built a network of a dozen licensees in the USA and Canada to successfully expand the PrimX jointfree slab system in those markets. The results far exceeded all prescriptions and specifications of the ACI 223 Committee for expansive cements in slabs.

I continued to explore the implications of shrinkage on cracking in SFRC slabs. At BEFIB 2016, I presented a paper on predicting crack openings in jointfree SFRC floors based on the shrinkage content of plain concrete, the type and dosage rate of steel fibers, and the influence of the grade or piles beneath the slab.

I was also involved with Primekss in two additional full-scale tests of piled slabs: the first test was conducted in Lithuania under elevated slab conditions with a thickness of 210 mm over a 4 m x 4 m pile grid. The slab was loaded at 30 kN/m^2 for three months and showed less than 1 mm total deflection. According to the TR34, 4th Edition design standards, a minimum thickness of 300 mm would have been required. The second test involved a 300 m^2 slab, built in Göteborg for this purpose. It was a collaboration between Primekss SIA, the NCC Company, and ArcelorMittal to verify the slab's 40 kN/m^2 UDL SLS capacity.

The CBI (Cement och Betong Institutet) laboratory in Stockholm also tested a Primekss piled slab in a Tingstad warehouse under service conditions. The test confirmed a 44 kN/m^2 UDL SLS limit, exceeding the 40 kN/m^2 specified in the contract.

Professor Plizzari of Brescia University (Italy) contributed to the definition of the Göteborg test, particularly regarding the size and footprint of the most critical UDL for flexure, shear, and torsion, as well as the ULS preliminary design. The analysis concluded with a 1.49 loading factor, nearly 1.5.

The SFRC slab by Primekss, with a 45 kg/m^3 dosage rate of HE 1/60 fibers at 1500 MPa tensile strength, far exceeded the requirements of all tests

I contributed to the design process of the test slab together with Prof. Barzin Mobasher of Arizona State University and Prof. Giovanni Plizzari of the University of Brescia (Italy). The Primekss traditional SFRC-only slab met the expectations at SLS and exceeded them at ULS and further to the final collapse.

The slab, calculated for a 40 kN/m^2 UDL capacity, was able to carry up to 78 kN/m^2 just before the final collapse.

Following the Fire Slab test at the Slovakian Laboratory with Primekss concrete, SFRC walls of 200 mm thickness were subjected to a 330 kN vertical loading intensity per meter run while exposed to the ISO standard fire. This caused significant thermal bending due to the thermal gradient between the heated and cold faces. Despite this massive thermal aggression, the 3 m-high SFRC walls successfully resisted the fire for 4 hours.

It was a world première of which we are very proud. Both the slabs and walls tested demonstrated that SFRC statically determinate full-scale structures, with the appropriate mix design, remain strong under fire conditions for over 3 hours and are suitable for use even in critical buildings.

Ultimately, as it is ongoing now, we at Primekss decided to organize four large full-scale tests on suspended elevated slabs of 300 m^2 surface each during 2023 and 2024. These tests aim to observe and understand the SLS, ULS, and the final collapse under critical patterns of uniformly distributed loading and point loadings for SFRC, compared to Primekss self-stressing SFRC.

This is a unique research effort and a significant R&D investment that no other entity—whether a concrete contractor, general contractor, steel fiber manufacturer, large international engineering or design-build firm, official research institution, or owner—has undertaken. Only Primekss does it.

Over the past 15+ years, I have supported Primekss in growing and maintaining a permanent R&D effort with a budget of up to 2% of gross income, a percentage reached in 2023, which is far above the usual in the construction industry.

A specialty laboratory has been developed in Riga, Latvia, fully equipped with top-tier human resources and scientific equipment devoted to fiber-reinforcing solutions, anti-shrinkage technologies, and sustainability. Today, we are proud to rely on a brilliant team of seven material, chemical, and structural engineers, led by Dr. Prof. Rolands Cepuritis and Dr. Gita Sakale.

I have always promoted the concept of full-scale testing of SFRC slabs to free the design and understanding of SFRC from the limitations imposed by standard methods to, what is said to be, the SFRC material properties, for us in an attempt to know material intrinsic ductility together with the structural ductility.

I attended no fewer than 18 full-scale tests, among which 15 were devoted to suspended slab SFRC applications. There is no better way to observe and understand the structural ductility of these applications.

As you read this book, you'll agree that material and structural ductility are fundamental for the lean and sustainable design of SFRC structures.

I have been supported and helped in this concept and its understanding for years by Dr. Prof. B. Mobasher of Arizona State University and Prof. Dr. G. Fischer of TU Denmark. This has been, and still is, a world-unique effort made by a concrete contractor like Primekss. Generally, specialty and general contractors do very little R&D, if they do anything at all.

Since I met him in 2002 while he was passing through Brussels, Janis Oslejs, founder and CEO of Primekss, has dreamed of creating the ideal concrete that will not age rapidly due to shrinkage or reliance on standard strengths. Janis Oslejs has always been ready to wage the battle, as hard and risky as it might have initially appeared.

He decided that his company would grow quickly so he could finance a record R&D effort for a long time. I often told colleagues that our way is the hard way.

As Yves Malier, president of the ITBTP (Paris-France), once said when as speaker at a Conference of the Institution, "They did it because they didn't know it was impossible," citing Mark Twain.

At the end of the day, what matters most is seeing customers return for repeated orders over the years and ensuring that every stakeholder—from owners to workers and society as a whole—benefits from it. I also told colleagues that keeping a customer for repeated orders is far more effective than continually finding new ones.

The product of R&D is to invent and develop for the future. This cannot be achieved by simply following standards or asking customers how they would make it. It is our relentless job.

We know we will always encounter the market and standard barriers that delay results and make innovations seem too risky. The academic world, in general, needs to face the reality of construction sites; otherwise, it risks ignoring the very real dangers to owners, workers, and society. Without this connection, it can become excessively risk-averse, focused solely on the past.

This is a frustrating reality, as the academic world is reputed within the building industry for its concentration of knowledge and is a reference point for the industry when it comes to standards or public questions.

Being the most knowledgeable group, academics should act ahead of the standards, setting the catalogue of knowledge and limits for the man of art and the public.

The man of art, by definition, is one who can improve anything incrementally by 10–15%—or perhaps a bit more—while innovators and inventors change the picture completely at once. I'm happy and proud that I never changed ideas or business in all these 46 years.

The bottom line for the future is to understand the reality and to understand what people and stakeholders think about it. At that point and under those conditions, your job becomes clear.

REFERENCE

1 R. Tilden Smith, "Jointless floors slabs-Breaking from the mould." *Concrete. (UK)* February 1991.

Author

Xavier Destrée is a consulting engineer and former R&D consultant of ARBED and later of ARCELORMITTAL, a worldwide steel company. He is a member of numerous technical committees issuing SFRC standards, guidelines, and reports in several countries, a Fellow of the American Concrete Institute, and arguably a world's leading authority on SFRC.

Chapter 1

Steel fiber reinforced concrete basics

1.1 GENERALITIES

Steel fiber reinforced concrete (SFRC) is often regarded as a composite material consisting of two phases: the concrete matrix and the steel fiber reinforcement. The volume fraction of the matrix typically ranges from 99.875% to 99%, depending on the application, while the steel fiber reinforcement ranges from 0.125% to 1%.

The concrete matrix is brittle, as it can't deform significantly before rupturing when subjected to tensile stresses.

Crack propagation in a brittle matrix doesn't need any mechanical energy, allowing the crack to grow continuously until it encounters a crack arrester, such as steel fibers, which absorb mechanical energy to stop the crack growth.

The brittle matrix is by a high number of microcracks—so minute they are invisible—that grow uncontrolled over time due to local tensile stresses caused by deformations from humidity and thermal variations, static, or dynamic forces.

Restrained hygral or thermal shrinkage variations can cause cracking in any concrete body as it begins to harden. A hardened concrete structure or body becomes riddled with cracks and is no longer purely elastic soon after hardening. A crack-free concrete does not exist.

If a crack-free slab is desired, concrete is not the right material—consider using something else. All concrete cracks.

It is expected that professionals understand and accept that plain concrete, as well as reinforced concrete, will often develop cracks over time, sometimes in completely unpredictable ways.

Concrete cracking is acceptable as long as the concrete structure remains serviceable.

Even when the concrete matrix functions as a static filling mass, subjected to relatively small mechanical stresses, it remains prone to uncontrolled cracking due to restrained hygral and thermal shrinkage.

The addition of a steel fiber phase is an effort to practically control cracking under all possible conditions. The aim is to mitigate the brittleness of the

DOI: 10.1201/9781003188315-1

concrete matrix by utilizing the mechanical energy available when a micro-crack propagates and encounters a crack-arresting steel fiber.

The steel fiber's bond strength, tensile strength, and modulus of elasticity—approximately ten times greater than that of concrete—play a critical role in controlling cracking.

The ability of the fiber phase in a given matrix to control composite cracking, not only instantly but also over time—spanning months and years—depends largely on the ratio between the modulus of elasticity of the reinforcing material and the matrix, along with the quality of the bonding.

When steel reinforcement is used in concrete, this ratio is generally around 10 but can vary from 7 to 20, depending on the duration of the load application on the structure and the extent of cracking in the matrix.

A high modulus ratio is a reason why stresses in concrete are "deviated" and attracted to the steel reinforcement, relieving the matrix of excessive stress.

When the modulus ratio drops below 1, stresses remain within the concrete and tend to bypass any inclusion of such a low-modulus fiber.

The modulus of elasticity of most synthetic fibers is smaller than that of the concrete matrix, which is why synthetic fibers do not reinforce concrete. However, synthetic fibers can significantly enhance the ductility of plain concrete.

Synthetic fibers in concrete are indeed used as a kind of enhancement material of concrete. Some applications of synthetic fibers are very popular, such as slabs on grade, and have been in use for decades.

Thanks to the high value of the modulus ratios, concrete stresses are driven into steel fibers, relieving the intergranular mortar from high tensile stress. Hence, microcracking can be prevented or delayed by closely spaced steel fibers.

From this observation, we understand the importance of two critical questions:

1. How far apart are two adjacent steel fibers?
2. How strong is the bond between the steel fiber and the concrete matrix?

Hopefully, everybody understands that without a strong bond to the matrix, it is impossible to take full advantage of the high modulus of elasticity of the steel fiber reinforcement phase.

Thus, without bonding to the matrix, there is no reinforcement.

Steel fiber reinforcement becomes effective when a minimal dosage of suitable fibers is added to the mix.

Since 1905, a significant number of patents related to different steel fiber shapes, sizes, bonding methods, or types of steel have been filed.

Regarding potential applications, the most notable SFRC patent is the French N° 587.169, dated 14 December 1923, by Maison Fougerolles S.A., which was quite ahead of its time and remains relevant even in the Eurocode 2024 draft on steel fiber concrete.

Fougerolles discloses that steel wire pieces with a diameter of 0.5 mm and a length of 30 mm can be homogeneously mixed into concrete at a rate of 1% by volume (78 kg/m³ dosage rate), or adjusted depending on the wire type and the application.

As a result, shrinkage or swelling cracks are eliminated in ordinary concrete, and this enhanced concrete can be used in columns, foundations, elevated suspended slabs, and more. It can also be used alongside reinforcing bars when necessary.

The "Maison Fougerolles" patent is shown in Figure 1.1

Finally, the steel fibers must "fit" within the skeleton of the concrete, specifically the array of large aggregates in the concrete matrix.

The objective is to transform brittle plain concrete into a ductile composite concrete capable of absorbing a significant amount of energy before rupture. The resulting composite material is safer and more predictable in terms of cracking and rupture compared to plain concrete, as well as many forms of traditional reinforced concrete.

The volume percentage or dosage rate are different metrics used to define the fiber concentration in concrete, as shown in Table 1.1.

1.2 HOW FAR APART ARE TWO ADJACENT STEEL FIBERS?

Although careful visual observation of a sample of SFRC gives quickly a good indication, it is useful to refer to Romualdi and Mandel (1), Krenchel (2), Souroushian and Lee (3, 4). Indeed, the literature about the composite material theory offers a simple expression of the distance S between two adjacent steel fibers:

From the literature we know:

$$S = 13,8 * d_f / (100 * V_f)1/2$$

where d_f is the fiber diameter and V_f is the volume fraction percentage.

Given: $V_f = 0,5\%$, $d_f = 1$ mm, $S = 19,5$ mm

In commercial units, the volume percentage is not used and replaced by V_m, the dosage rate in kg/m³ so that the expression becomes:

$$S = 122 \times d_f / (V_m)^{1/2}$$

Example: 1 mm diameter, 40 kg/m² dosage rate gives S = 122 x 1/ 40 ½ **= 19,29 mm**

RÉPUBLIQUE FRANÇAISE.

MINISTÈRE DU COMMERCE ET DE L'INDUSTRIE.

DIRECTION DE LA PROPRIÉTÉ INDUSTRIELLE.

BREVET D'INVENTION.

VII. — Construction, travaux publics et privés.

1. — Matériaux et outillage.

N° 587.169

Perfectionnement aux bétons.

Société anonyme : MAISON FOUGEROLLE Frères résidant en France (Seine).

Demandé le 14 décembre 1923, à 14ʰ 45ᵐ, à Paris.

Délivré le 13 janvier 1925. — Publié le 14 avril 1925.

[Brevet d'invention dont la délivrance a été ajournée en exécution de l'art. 11 § 7 de la loi du 5 juillet 1844 modifiée par la loi du 7 avril 1902.]

La présente invention a pour objet un béton perfectionné se distinguant des bétons usuels à base de cailloux, de gravillon, de sable, etc., et de ciment ou autre agglomérant, en ce qu'il renferme, en proportion variable suivant l'usage en vue, de courts tronçons de fils métalliques de nature, de diamètre et de longueur convenables, par exemple des bouts de fils d'acier de 5 dixièmes de millimètre de diamètre et de 3 centimètres de longueur, ces tronçons de fil étant mélangés aussi bien que possible avec les éléments usuels du béton afin que le tout forme masse homogène.

Les dimensions indiquées peuvent varier en plus et en moins. La forme de la section des fils peut aussi varier : ronde, carrée, plate, etc.

Comme proportion à employer, une partie environ de fils d'acier pour 100 parties de béton (en volume) convient dans l'exemple susvisé, mais il va sans dire que cette proportion peut varier entre de larges limites sans qu'on s'écarte de l'invention, suivant la nature des fils et les usages en vue.

Dans tous les cas, le produit obtenu a l'avantage sur les bétons ordinaires d'être plus résistant à la compression et à la traction. En outre, la tendance à la fissuration, due au retrait ou à la dilatation, se trouve très réduite, sinon supprimée.

L'emploi de ce béton perfectionné est indiqué plus particulièrement pour tous les usages où l'on recherche une grande résistance aux efforts mécaniques ou à l'usure, colonnes, bâtis de machines, fondations, voussoirs, poutres, planchers, chaussées, dallages, enduits, etc.

Bien entendu, le béton suivant l'invention pourra, si on le juge utile, être armé au moyen de tiges métalliques disposées dans la masse de toutes manières convenables.

RÉSUMÉ.

Béton dont la composition comprend, outre les éléments usuels, cailloux, gravillon, sable, etc. et ciment ou autre agglomérant, une petite proportion de fils métalliques en tronçons relativement courts.

Société anonyme : MAISON FOUGEROLLE Frères.

Par procuration :

Brévet.

Prix du fascicule : 2 francs.

Pour la vente des fascicules, s'adresser à l'Imprimerie Nationale, 27, rue de la Convention, Paris (15ᵉ).

Figure 1.1 A remarkable patent very ahead of its time and of today as well.

Table 1.1 Steel fiber concentration

V_m (kg/m³)	78	60	50	40	30	20	10
Volume percentage (%)	1	0.77	0.64	0.5	0.38	0.25	0.13
V_m (lbs/cu.yd)	132	102	85	66	51	33	17

Table 1.2 Average fiber spacing S function of the fiber diameter d_f and the dosage rate V_m

D_f, V_m, S	10 (kg/m³)	20	30	40	50	60
0,6 mm	S = 23 mm	*16 (mm)*	*13*	*12*	*10*	9
0,7	27	**19**	*16*	*13*	*12*	*11*
0,8	31	**22**	**18**	*15*	*14*	*13*
0,9	34	24	**20**	**17**	*16*	*14*
1,00	39	27	**22**	**19**	**17**	*16*
1,25	48	34	28	24	**22**	**20 (mm)**

If d_f is in inch and V_m is in pounds per cubic yard, the expression becomes:

$$S = 158 \times d_f / (V_m)^{1/2}$$

Example: 1 mm diameter, 40 kg/m³ dosage rate gives S = 122 x 0.04/ 66 ½ = 0,78 in

When the fiber volume proportion ρ_f is used instead of V_m, as the dosage rate, then:

$$S = (\pi\, d^2_f L_f / 4\, \rho_f)^{1/3}$$

Example: ρ_f = 0.005 (0.5% or 39.5kg/m³), d_f = 1 mm, L_f = 50 mm: S = 19.87 mm

Table 1.2 shows typical values of the predicted fiber spacing in function of the dosage rate and the fiber diameter.

The mix is practicable and useful when the fiber length is between two and three times the nominal aggregate size.

In the vast majority of cast-in-situ slabs, the ready-mixed concrete used typically contains aggregates with a maximum size of 20 mm (D), though in some cases, 25 mm or 16 mm aggregates are used.

Figures in italic indicate an increased difficulty in achieving homogeneous mixing as the aggregate size decreases from 16 mm to 9 mm, while the underline represents the opposite trend as the size increases from 22 mm to 48 mm.

As the ratio of S to D decreases, the introduction and mixing of steel fibers become increasingly challenging.

A difficult mix will require significantly more energy for mixing, placing, and compacting, as fibers that cannot fit within the aggregates will appear on the surface of the concrete slab.

The purpose of Table 1.2 is to clearly summarize when mixing, placing, and finishing SFRC slabs is likely feasible and when difficulties may arise.

The goal is to control the cracking at its origin, specifically in the intergranular mortar of the concrete. In Table 1.2, all cases marked in italic indicate practical difficulties on a real jobsite. However, this does not mean these issues cannot be addressed with attention to detail and proper care.

The L_f/d_f ratio, also known as the fiber length-to-diameter ratio or fiber aspect ratio, is a critical parameter that strongly influences the conditions for introduction and mixing, as well as the final strength of the composite concrete. This ratio directly affects the effectiveness of fiber reinforcement in the matrix.

When the aspect ratio is less than 45, steel fibers are generally user-friendly and easy to introduce and mix into the concrete.

For aspect ratios above 45, introducing and mixing remain feasible, albeit with increased care and attention, up to a ratio of 60.

Beyond an aspect ratio of 60, specific techniques are typically employed, such as using glued fibers that dissolve and separate during mixing. Deploying steel fibers at high speed with specialized machines, often referred to as "fiber blast machines" or "steel blower," is also a convenient way to achieve fiber mixing.

We will explore these processes in greater detail in a later chapter.

For example, a mix of a 0.7 mm fiber diameter and 50 mm length at a 40 kg/m³ dosage rate, combined with a 25 mm maximum aggregate size, would be difficult to mix, transport, install, and finish. Such a mix would be prone to fiber exposure. Reducing the aggregate size to 16 mm can help mitigate these challenges. Additionally, enhanced finishing skills for the slab surface can also aid in resolving these difficulties.

Extra skills in finishing the surface of the slab can also help.

On the other hand, the engineer will expect effective cracking control and should exclude all cases marked with a blue number in Table 1.2, which indicates fibers spaced too far apart.

A mix with a 0.7 mm fiber diameter at a 10 kg/m³ dosage rate and a 20 mm maximum aggregate size will not provide effective cracking control.

As a guideline, practical and useful applications are represented by the diagonal of black figures in Table 1.2, running from the top left to the bottom right.

If the fiber length is smaller than the maximum aggregate size, mixing difficulty is eliminated because the fibers can easily fit between the aggregates. However, as we'll explore in the second question, this results in the loss of any reinforcing effect.

Smaller fibers, which can fit into the intergranular mortar in large numbers, will inhibit and delay the growth of the smallest cracks, which are invisible to the naked eye. Over time, the phenomenon of cracking coalescence occurs, where the smallest adjacent microcracks join together to form larger cracks. This process continues until macro-cracks with much larger openings develop.

At this stage, the fiber length must be significantly larger than the maximum aggregate size to bridge the intergranular mortar where cracking originates.

For standard slab mix designs, the fiber length should be between 1.5 and 3 times the maximum aggregate size. The optimum length for slabs is typically between 50 mm and 60 mm, depending on the fiber diameter.

1.3 EXAMPLE OF TYPICAL COMMERCIAL FIBERS GEOMETRY AVAILABLE

1.3.1 Hooked end anchoring: denoted HE

- With HE1/50 (1 mm diameter x 50 mm length) steel fibers, from 20 to 60 kg/m³ dosage rates of steel fibers dosage rate indeed, quite user-friendly to mix and place the fiber concrete, limited slump loss
- With HE1/60 (1 mm x 60 mm length) steel fibers almost like HE1/50at the usage
- With HE75/50 (0.75 mm x 50 mm length) steel fibers from 20 to 40 kg/m³, need of equipment to introduce and mix. Fiber showing more likely to happen in the finished surface
- With HE75/60 (0.75 mm x 60 mm length, L_f/d_f = 80) from 20 to 40 kg/m³, the introducing and mixing equipment is very needed. Glued fiber ease the introduction in the truck mixer.

Depending on the manufacturer, the glue that is supposed to dissolve, does it too early or too late. Questionable fiber glue can entrain air and generate bubbles in the concrete.

1.3.2 Undulated fibers (also called crimped fibers)

The undulation depth at the centerline is approximately 0.80 times the diameter, with a wavelength of 6 to 8 times the diameter.

For L/d < 40, the fibers are very user-friendly to introduce and mix.

Above L/d = 50, careful attention is required, and special equipment, such as a mechanical blower, becomes necessary.

When L/d > 55, the use of special equipment for introduction and mixing becomes absolutely essential.

1.4 WILL THE STEEL FIBERS REINFORCE THE MATRIX?

Consider the β factor in Table 1.3.

The $\beta = V_m \times L_f/d_f$ factor is not only an important index of difficulty to introduce and mix the fibers in the concrete but also a first relative indication of the efficiency of fiber reinforcing obtained.

For a slab that does not break like a piece of glass, the factor β should exceed a value of 1000.

To ensure easy introduction and mixing for a workable mix in slabs, β should remain below a value of 3000.

When special methods for introducing and mixing fibers are employed, combined with an optimized mix design, β can reach as high as 4500, providing optimal reinforcement and smooth operations on site.

When β is less than 1000, it can no longer be considered reinforced concrete. However, this does not mean it cannot replace traditional reinforcement technology in cases of low geometrical percentages. Geometrical percentages as low as 0.10% are occasionally used in ground beams, shallow footings under light buildings, or slabs on grade.

1.5 HOW STRONG IS THE BOND OF THE STEEL FIBER TO THE CONCRETE MATRIX?

The length of the fiber is a critical parameter for composite concrete, as the tensile stresses in the matrix are transferred to the steel fiber through its total lateral surface.

The natural bond strength of the steel fiber surface is practically limited by the upper tensile strength of the fine mortar.

For a typical mortar tensile strength of 3 N/mm², which depends on the concrete's strength, a fiber with a diameter of 0.8 mm and a length of 60 mm would have a maximum pull-out force calculated as: 60 mm/2 * π * d_f * 3 N/mm² = 226 N. This pull-out force would generate a steel tensile stress of 226 N/ π * $(d_f/2)^2$ = 449 N/mm². This value is significantly smaller than the tensile strength of the steel wire, which is typically between 1000 N/mm² and 2000 N/mm², depending on the type of fiber used.

Mathematically, the fiber pull-out force P can, in theory, be considered a point load P located along the z-axis and applied to the external surface of a semi-infinite elastic body. The resulting shear stress τ_{rz} at a distance r from the z-axis can be calculated using the Timoshenko formula:

$$\tau_{rz} = \left(3 * P / 2\pi\right) * r * z^2 / \left(r^2 + z^2\right)^{5/2}$$

so that its maximum value, after all calculations, is $\tau_{rz} = 0.278 * \sigma_f$, attained at a depth z = 0.82 r. It shows that in the case of a natural bond along

Table 1.3 $\beta = V_m \times L_f/d_f$ as a fiber introduction and mixing parameter

$d_f V_{m\,Length}\, V_m\, L_f/d_f$	10 kg/m³		20 kg/m³		30 kg/m³		40 kg/m³		50 kg/m³		60 kg/m³	
	50mm	60mm	50mm	60mm	50mm	60mm	50mm	60mm	50mm	60mm	50mm	60mm
0.6 mm	833	1000	1666	2000	2499	3000	3332	4000	4165	5000	5000	6000
0.7 mm	714	857	1428	1713	2142	2570	2856	3427	3570	4284	4284	5140
0.8 mm	625	750	1350	1620	1875	1944	2500	3000	3125	3750	3750	4500
0.9 mm	555	666	1110	1332	1665	2000	2220	2400	2775	3330	3330	4000
1.0 mm	500	600	1000	1200	1500	1800	2000	2400	2500	3000	3000	3600
1.25 mm	400	480	800	960	1200	1440	1600	1920	2000	2400	2400	2880

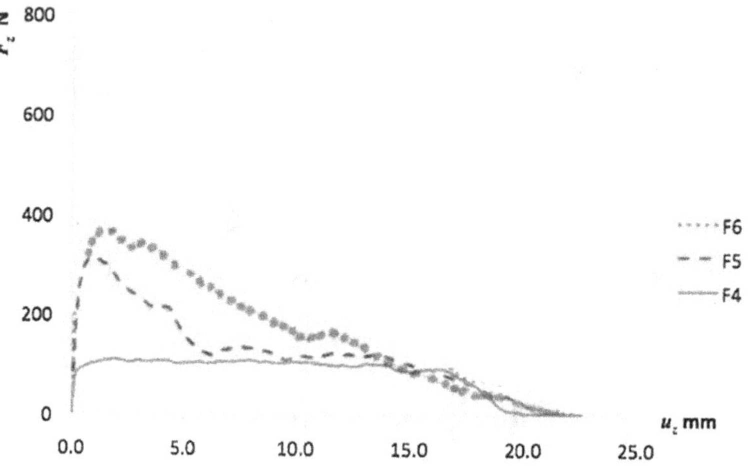

Figure 1.2 Pull-out diagram of 0.75 mm diameter x 50 mm fiber length for undulated, hooked ends and smooth steel fibers.

a smooth fiber, the fiber stress can't exceed ca. 4 times the mortar tensile strength!

For the steel fiber to function as an efficient reinforcement, it must be stressed as close as possible to its yielding stress. Therefore, a specific deformed shape distributed along the fiber length or a deformed anchoring at the ends is necessary.

Many different steel fiber shapes have been developed and are still available today, including indented fibers, undulated and corrugated fibers, flat-ended fibers (like a swallow tail), conical ends, hooked ends, or double and triple hook ends.

The anchoring shape of a deformed fiber is designed to "lock" the fiber into the matrix, allowing both tensile and compressive stresses to develop as the fiber is pulled out. The shape must be optimized based on the fiber size, the strength of the concrete matrix, and the tensile strength of the constituent steel wire.

If the anchoring shape is too large, the steel fibers may tangle and ball together during introduction into the mixer and during mixing. Conversely, if the anchoring shape is too small, the limited natural bond strength of the fiber will fail to prevent complete debonding.

Most commercial steel wire fibers with specific anchoring shapes can be pulled out gradually, allowing for up to 10 mm of fiber-matrix relative displacement under a relatively constant or increasing pull-out load. As a result, the steel wire stress typically reaches 35% to 80% of its tensile strength, depending on factors such as the fiber's length, diameter, shape, and anchoring type.

In some cases, with optimal mortar strength, the ideal anchoring can bring the fiber to the point of steel wire striction at rupture. This is often

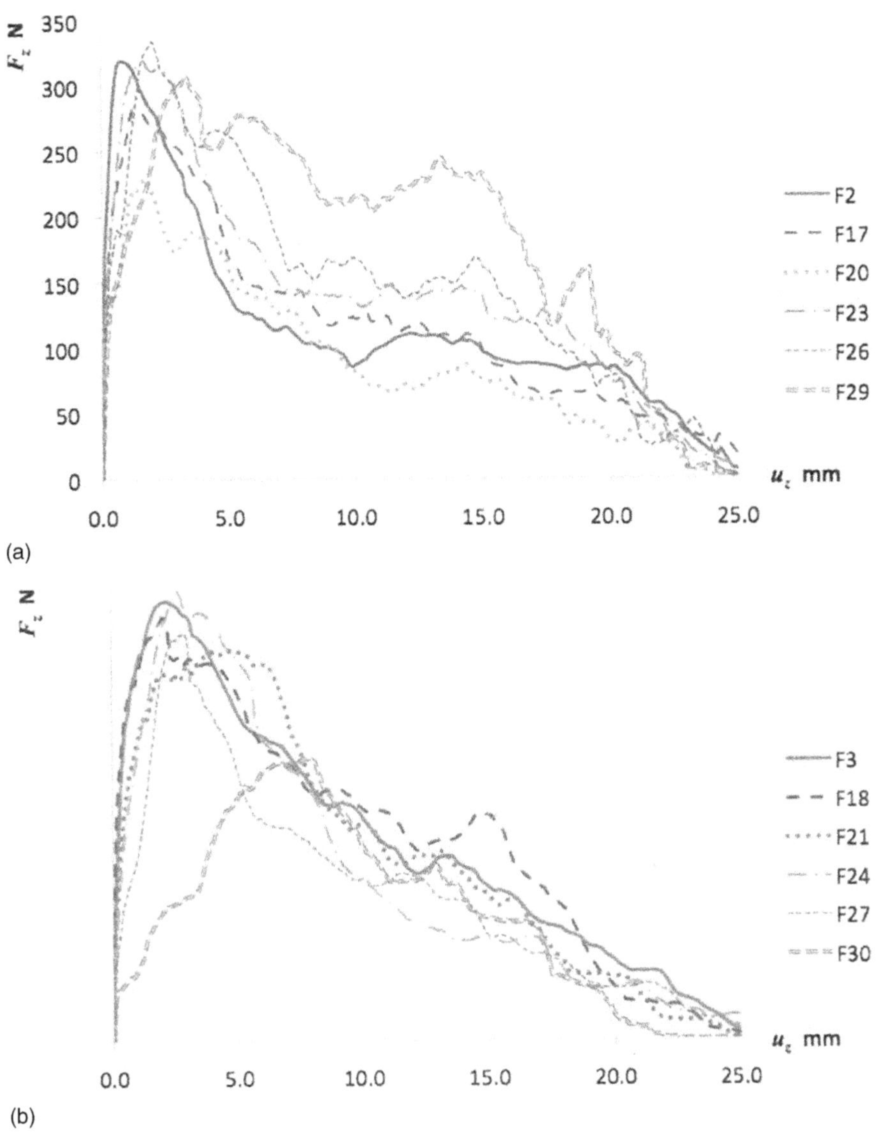

Figure 1.3 (a) Inclined pull-out diagram of hooked end fiber, the same as in Figure 1.1. (b) Inclined pull-out diagram of undulated fiber, the same as in Figure 1.3.

observed with Twincone fibers, which feature conical ends and provide a perfect lock within the matrix.

Inclined fiber pull-out tests, as opposed to uniaxial tests, typically show fibers being extracted over a longer length but under a slightly smaller pull-out force. However, the mechanical energy dissipated during an inclined fiber pull-out test is not less than that in a uniaxial pull-out test.

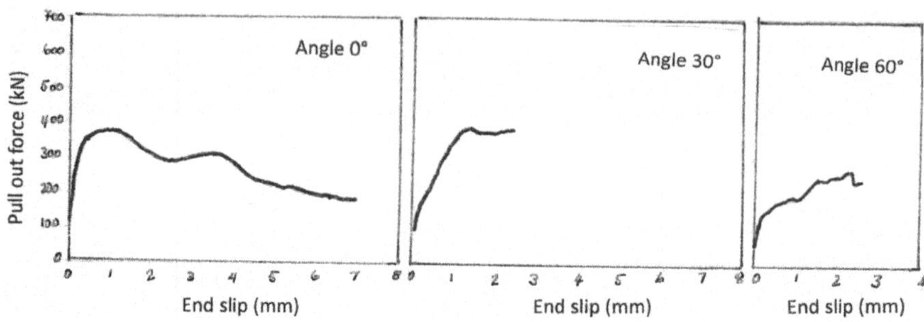

Figure 1.4 Experimental average pullout load curves.

Steel fibers are randomly distributed within the composite, aligning in all directions. This randomness ensures that fibers in various orientations contribute to reinforcing the composite and dissipating mechanical energy, from the ultimate state to the final collapse of the structure.

Some literature mistakenly suggests that fibers should be oriented preferentially in the main direction to maximize reinforcement efficiency. However, in slabs, concrete stresses occur in multiple directions due to flexion, shear, shrinkage, differential shrinkage, and dynamic actions. Therefore, a composite material solution with reinforcement in all directions is more practical and effective.

In contrast, for structures like tension legs or statically determinate beams, where stresses are concentrated and uniaxial, a uniaxial steel rod or bottom rebar is significantly more effective for reinforcement.

It is important not to rely solely on steel fiber reinforcement in statically determinate structures.

Figures 1.2, 1.3, and 1.4 show the average pull-out force versus displacement diagram of a single steel fiber.

Figure 1.2 illustrates the uniaxial pull-out behavior of fibers with a 0.75 mm diameter and 50 mm length, each of different shapes, when half their length is embedded in mortar. An undulated fiber is depicted in F6 (red dotted line), hooked-end fibers in F5, and a straight, smooth fiber in the F4 case.

Figure 1.3 presents the average pull-out force of the same hooked-end fibers as a function of fiber displacement, based on the angle of inclination. The angles and corresponding cases are as follows: 0° (uniaxial) in F2, 10° in F17, 20° in F20, 30° in F23, 45° in F26, and 60° in F29.

In the case of Figure 1.4, in reference (5), the same as in Figure 1.2 but for undulated fibers, where 0° is in F3; 10° in F18; 20° in F 21; 30° in F24; 45* in F27, and 60° in F30.

The main observations are as follows:

- For fibers of the same size, both undulated and hooked-end types exhibit almost the same ultimate pull-out force.

- Beyond a displacement of 10 mm (for an initial embedded length of 25 mm), deformed fibers show the same residual pull-out force as straight fibers without any anchoring.
- Both types of fibers demonstrate a similar residual pull-out force, nearly equal to the maximum pull-out force, up to a displacement of 5 mm.
- Both types of fibers exhibit a similar residual pull-out force, regardless of the inclination angle.
- Both types of fibers show a high potential for crack bridging up to a displacement of 10 mm.

It is evident that inclined fibers, relative to the principal stress direction, contribute significantly to the deformation process, leading up to the final ductile rupture of the composite concrete.

However, there is currently no theory that directly links individual pull-out behavior to the final mechanical properties of the composite. As a result, a fiber pull-out diagram does not provide sufficient information to design a concrete structure.

Figure 1.4 (referred to as Figure 4 in the cited article) illustrates pull-out forces vs. the angle of inclination, along with the article author's comments.

For some, it was once a dream to focus solely on the pull-out test to design structures, as it could have greatly simplified the development of innovative, superior fibers and significantly reduced the effort required to create different structural design methods.

Simple average assumptions can be made based on Table 1.1 and a typical residual pull-out stress of 50–70% of the wire strength. For example, 40 kg/m^3 of fibers with a 1 mm diameter and 60 mm length, made from steel with a tensile strength of 1200 N/mm^2 and equipped with suitable hooked ends, can be gradually extracted at 60% of the steel's yield strength.

At 40 kg/m^3 of fibers with a 1 mm diameter and 50 mm length, resulting in a 19 mm spacing between fibers, the local pull-out force could reach as high as N = 0,70 x 1^2/4 x 3.14 x 1200 N/mm^2 = 660 N. This corresponds to an average intergranular concrete mortar tensile strength of 660 N/19 *19 mm^2 = 1,83 N/mm^2.

This kind of strength is present in all directions within the composite material, providing the concrete with ductile tensile strength in all directions, unlike traditional reinforced concrete.

The pull-out test of a single fiber is a classical method used to evaluate the crack-bridging potential of a fiber type. While there is a vast number of academic papers on this subject, no structural design method based solely on the pull-out performance of fibers has ever been established.

A final conclusion on the issue has been given by A. Orbe whom we cite herein (6):

Figure 1.5 Deformed pulled-out hooked end steel fibers after a flexion test.

Figure 1.6 SFRC slab section after the final collapse of an SFRC elevated suspended slab (E-SFRS).

Even though information about the influence of matrix strength, fiber geometry, orientation and embedded length in the debonding process and pull-out mechanism of individual fibers can be obtained, it is not suitable for design purposes in structural elements. The randomness of the position and inclination of fibers in a real case, requires regarding the material as a composite, characterized as such.

Nevertheless, numerous formally "valid" steel fiber patents exist for recent, sophisticated double or triple-hooked-end fiber-reinforced concrete structures. These patents are based on the superior uniaxial pull-out behavior of the invented fibers, claiming the ability to design and reinforce concrete structures.

However, these patents fail to explain how to extrapolate a resisting moment value for an SFRC structure from a pull-out test. The "man of art" is left without guidance, as the patents provide no explanation on how to achieve this. As a result, these patents appear to lack practical utility.

All the geometric properties and parameters of steel fibers influence their efficiency in reinforcing concrete, making testing essential to determine the potential offered by each fiber type.

The importance of the flexural pull-out behavior of steel fibers is illustrated in Figures 1.5 and 1.6, where hooked-end fibers are visible following the complete rupture and failure of an SFRC section. Short straight fibers or brittle steel fibers, which do not release any plastic ductile energy, as seen in these figures, are excluded from use as reinforcement in structural applications.

1.6 INFLUENCE OF SIX PARAMETERS

Figure 1.7, taken from a private ArcelorMittal presentation, summarizes the influence of the dosage rate, fiber length, fiber shape, aspect ratio, steel wire strength, and concrete matrix on flexion resistance in general.

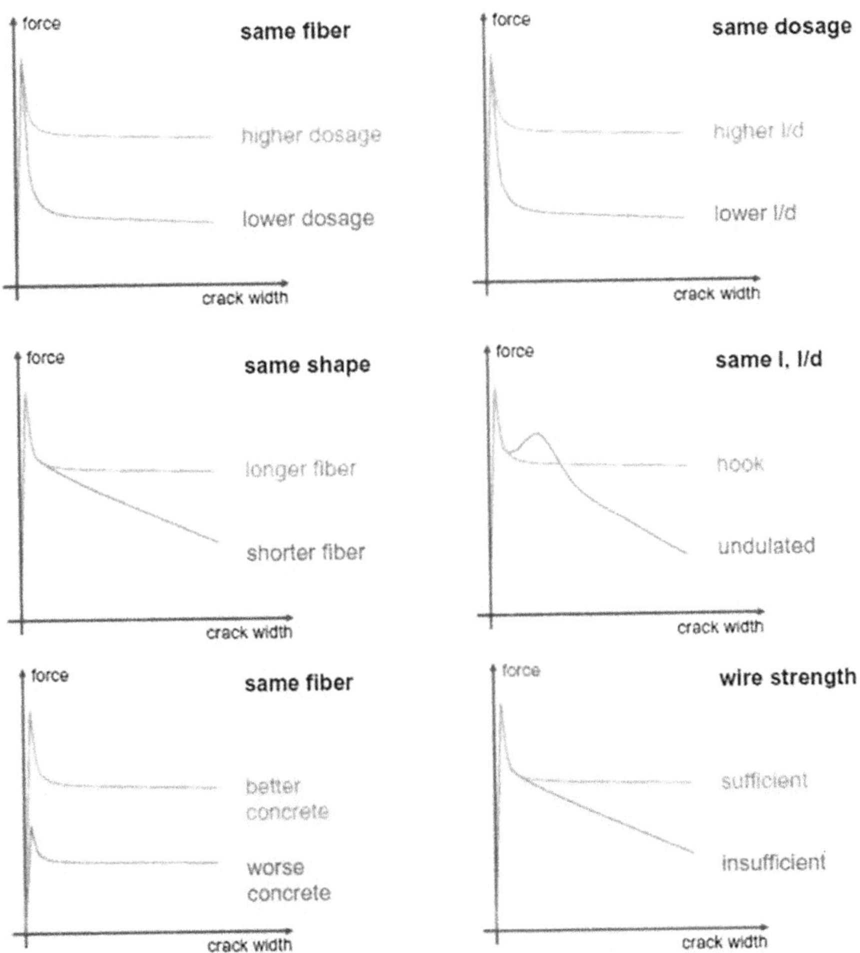

Figure 1.7 Influence of the steel fiber parameters vs. the crack width.

1.7 INFLUENCE OF THE STEEL WIRE TENSILE STRENGTH

The steel wire strength significantly influences performance, provided the fiber has a shape specifically adapted to the wire strength to anchor it effectively in the matrix.

Higher wire's strength can stiffen the anchoring shape, requiring more tensile stress to plastically deform the anchorage as it is pulled out of a crack. A stiff anchorage reduces deformation and relative displacement between the fiber and the matrix, resulting in a higher EN 14651 f_{r1} flexion strength compared to softer anchorage.

If the anchorage starts stiff but softens gradually, longer relative displacement becomes possible, leading to a higher EN 14651 f_{r3} flexion strength.

The behavior of the fiber in its matrix also heavily depends on the nature and strength of the matrix. Using high-strength fibers in a weak matrix is ineffective.

For example, the Twincone fiber, which featured true conical heads (though no longer manufactured), demonstrated very high f_{r1} flexion strength. These values were so high that many Twincone fibers fractured in striction across cracks with minimal opening. As a result, f_{r3} values were much lower, often only 40% of f_{r1}. Figure 1.8 shows Twincone fibers after experiencing striction instead of pull-out.

As we will see later, the modern standard design (refer to the FIB Model Code 2010) of a section is primarily based on the f_{r3} value. Consequently, interest in Twincone fibers has diminished, partly due to their higher production cost compared to deformed fibers.

A Dutch concrete contractor once told me that he had never seen better crack control as with Twincone-reinforced concrete.

In contrast, traditional single hooked-end fibers exhibit a plateau in post-cracking behavior between f_{r1} and f_{r3} as defined in EN 14651.

Figure 1.8 Striction rupture of Twincone steel fibers.

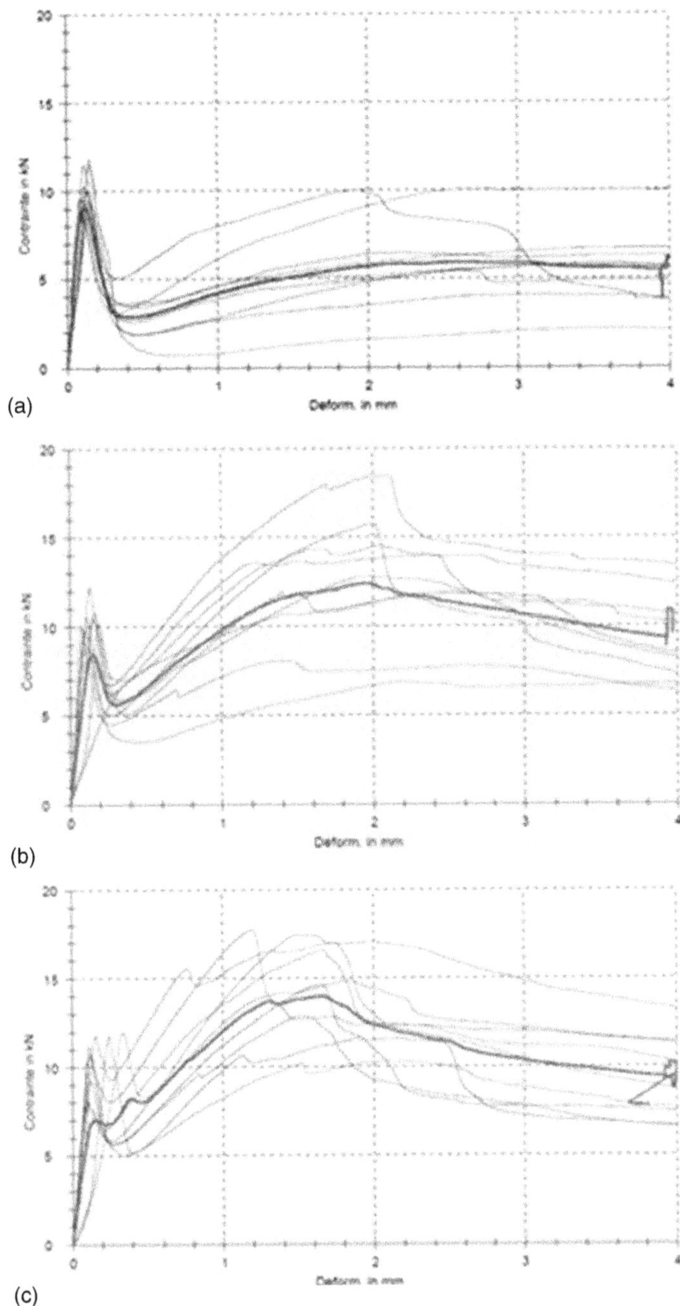

Figure 1.9 C25-30 EN14651 flexion diagrams of Triple hook ends fibers at different dosage rates: (a) 20 kg/m³; (b) 30 kg/m³; (c) 40 kg/m³.

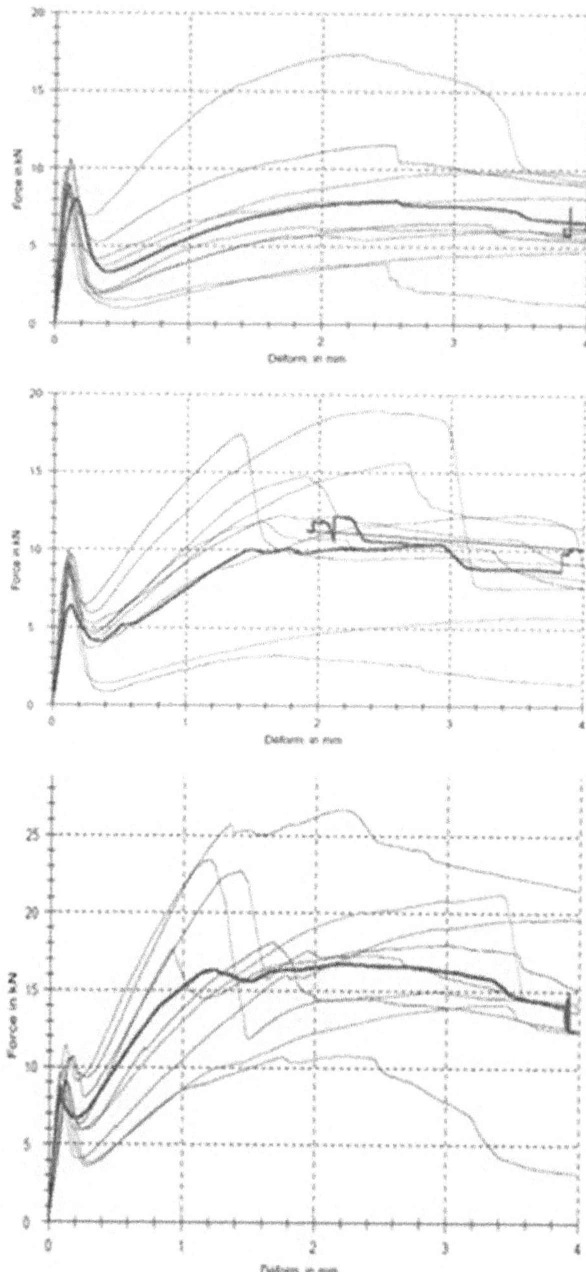

Figure 1.10 EN 14651 SFRC flexion diagrams with Triple Hook ends, 2100 MPa tensile strength steel wire fibers in a C40-50 mix.

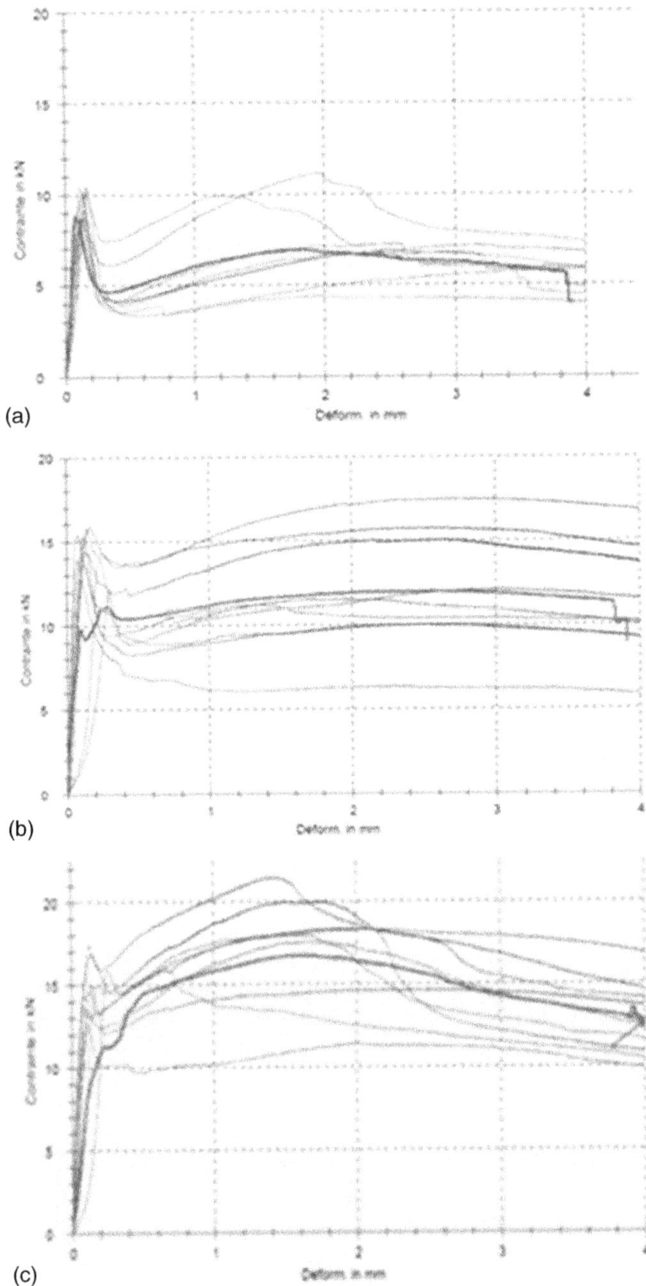

Figure 1.11 EN 14651 SFRC flexion diagrams with Double Hook ends, 1500 MPa tensile strength steel wire fibers in a C25-30 mix. (a) 20 kg/m³; (b) 30kg/m³; (c) 40 kg/m³.

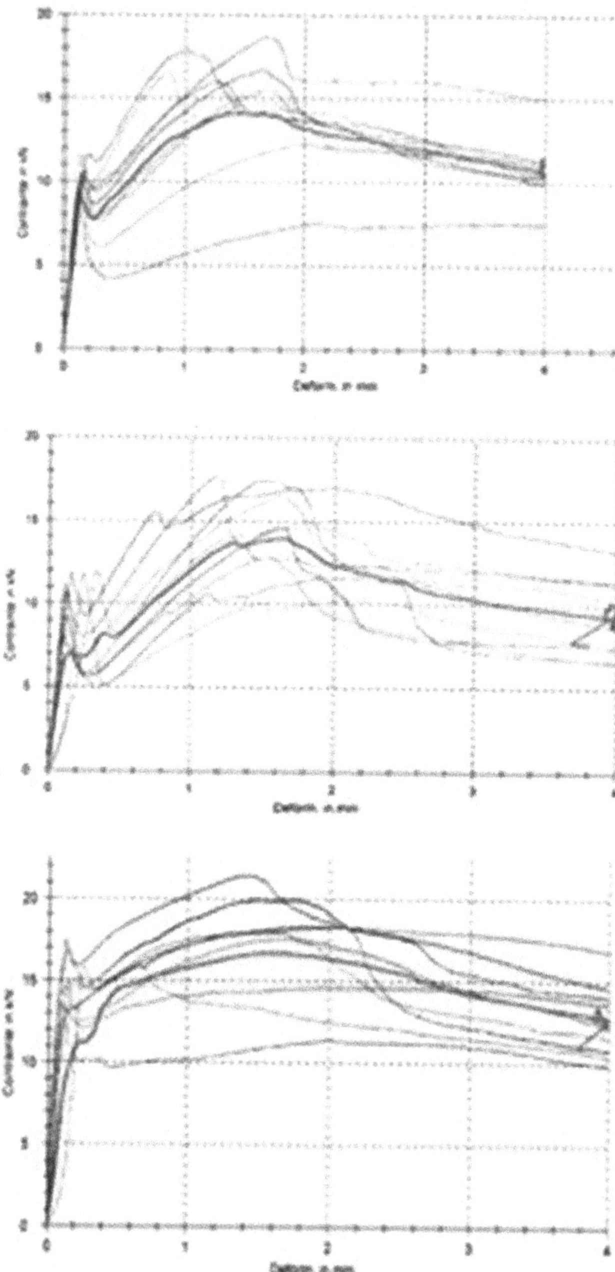

Figure 1.12 At 40 kg/m³ dosage, EN 14651 flexion test diagrams in a C40 − 50 in case of using, DHE-1500 MPa; THE-2100 MPa; HE+1/60-1500 MPa (SHE).

Table 1.4 EN 14651 f_{r1} and f_{r3} values in laboratory

kg/m3		DHE *1500 MPa*	THE *2100 MPa*	SHE *1500 MPa*	f_{Ri}
20	C25-30	4 MPa	3 MPa	6 MPa	i = 1
20	C25-30	6 MPa	5 MPa	7 MPa	i = 3
30	C25-30	7 MPa	6 MPa	10 MPa	i = 1
30	C25-30	10 MPa	11 MPa	11 MPa	i = 3
40	C25-30	9 MPa	8 MPa	14 MPa	i = 1
40	C25-30	12 MPa	12 MPa	14 MPa	i = 3
40	C40-50	6 MPa	8 MPa	10 MPa *2000 MPa wire*	i = 1
40	C40-50	13 MPa	16 MPa	13 MPa *2000 MPa wire*	i = 3

Note: DHE: double hooked ends; THE: triple hooked ends; SHE: single hooked end.

Figure 1.9 shows typical EN 14651 flexion diagrams for Triple Hooked-End 2100 MPa steel fibers C25-30, W/C = 0.52 - 0.55

Figure 1.10 shows C40-50 EN14651 THE (Triple Hook Ends) fibers at different dosage rates

Figure 1.11 shows EN 14651 typical Double Hooked Ends 1500 MPa steel fibers flexion diagrams C25-30, W/C = 0.52 - 0.55

Figure 1.12 shows that in a C40-50 concrete, the HE+1/60 5 (SHE shape) is not at all outperformed by the DHE type-1500MPa and THE type-2100 MPa.

These images show that low dosage rates (20 kg/m³ and below, and likely even 25 kg/m³) do not perform as reinforced concrete sections. It can also be observed that there is a significant drop in strength after the first crack, resulting in a low f_{r1} value, which, in standard theory, indicates poor performance under service conditions (SLS).

The drop of strength at f_{r1} is not observed here at 40 kg/m³ and higher dosage rates.

This is summarized in Table 1.4:

The EN 14651 f_{ri} values here are average values and depend strongly on the mix design used. Do not use these values to design a structure as it is given only as trend values.

It can be observed that double hooked-end (DHE) and triple hooked-end (THE) fibers typically exhibit quite high average f_{r1} values, while Single Hooked-End (SHE) fibers appear similarly good, if not slightly superior, in terms of f_{r1}.

Additionally, super high-strength steel fibers do not seem to significantly improve the results, although this conclusion could change with the use of a very specific mix design.

From over 30 years of experience in structural design and applications, a practical dosage of 40 kg/m³ with 1 mm diameter fibers achieves matrix saturation, where the average distance between two adjacent fibers matches the maximum aggregate size.

1.8 SECTION DESIGN EXAMPLE

As an example, let's consider a case with f_{r3} = 12 MPa (SHE - 1500 MPa steel fibers, 1 mm diameter, 60 mm length), resulting in a characteristic value of 0.7 x 12 = 8.4 MPa.

For a two-way piled slab, a design strength can be calculated as follows:

Given, γ_m = 1.5, as the material coefficient η_{det} = 2, a structural indeterminacy factor, as per the EN-SS 812310 Swedish standard (Design of Fiber Reinforced Concrete Structures), typically applicable for piled slabs (ground suspended slabs) (7)

The flexion design strength is:

$$f_{Rd} = 0.85 \text{ x } 8.4 \text{ x } 2 \text{ } /1.5 = 9.52 \text{ N}/\text{mm}^2$$

A 200 mm slab thickness can offer M_{Rd}= 9.52 x 200²/6 = 63,46 kNm/m as a resisting moment. The rebar equivalent section of a 200 mm slab is given as:

$$S = 63460(\text{Nmm}/\text{m})/ \text{ } 0.9 \text{ x } 160 \text{ mm x } 435 \text{ N}/\text{mm}^2 = 1013 \text{ mm}^2/\text{m or}$$
$$5 \text{ rebars of diameter } 16 \text{ mm}/\text{m}$$

For top and bottom steel rebars of 16 mm diameter at 200 mm spacing, the weight of re-steel is 20 m/m² x 1.6 kg/m = 32 kg/m² or 160 kg/m³. This shows that 40 kg/m³ of steel fibers could replace up to 160 kg/m³ steel rebars. This ratio has been verified in the past for numerous projects including, several significant ones.

Based on general experience, as a rule of thumb, for SHE steel fibers with a tensile strength of 1500 MPa, the rate of rebars per cubic meter of concrete can be divided by a factor of 2.2 to determine the fiber dosage rate for slabs and foundation raft applications.

1.9 1100 MPA OR 1500 MPA WIRE STRENGTH?

In Table 1.5, we can compare C35-45 concrete type, as in Table 1.4, the effect of the wire strength increase from 1100 MPa to 1500 MPa on the same shape and geometry steel fiber: the HE1/60 – 1100 MPa (Simple Hooked

Table 1.5 Influence of the constituent wire strength on f_{ri} EN 14651 flexion strength

HE 1/60

Residual Flexural Strengths f_r according EN 14651 - C35/45

Average Values	30kg	35kg	40kg	45kg	50kg	55kg	60kg	65kg	70kg
fLm	6,07	5,90	5,73	5,92	6,10	6,11	6,12	6,13	6,14
fr,1m	4,77	5,89	7,01	7,17	7,33	7,55	7,78	8,00	8,23
fr,2m	5,03	6,20	7,37	7,61	7,85	8,08	8,32	8,55	8,79
fr,3m	4,59	5,75	6,91	7,15	7,39	7,58	7,78	7,97	8,17
fr,4m	4,06	5,03	6,00	6,34	6,68	6,89	7,09	7,30	7,51

HE+ 1/60

Residual Flexural Strengths f_r according EN 14651 – with C35/45

Average Values	30kg	35kg	40kg	45kg	50kg	55kg	60kg	65kg	70kg
fLm	5,66	5,85	6,03	5,97	5,91	6,00	6,09	6,19	6,28
fr,1m	5,09	6,23	7,37	7,42	7,48	7,73	7,98	8,23	8,48
fr,2m	5,71	6,75	7,79	8,07	8,34	8,57	8,80	9,03	9,26
fr,3m	5,63	6,42	7,21	7,38	7,55	7,72	7,89	8,06	8,23
fr.4m	5,35	6,01	6,67	6,83	6,99,	7,08	7,17	7,26	7,34

ends,1 mm diameter – 60 mm long, 1100 MPa tensile strength) against the HE + 1/60 (as HE 1/60 but of 1500 MPa wire tensile strength-SHE).

The mix design is not the same as in Table 1.4, so that we see once more how much the mix design matters when the EN 14651 flexion test method is used.

We observe a significant increase in the f_{ri} strength from 30 kg/m³ to 40 kg/m³ dosage rate.

The increase is surprising as 40 kg/m³ dosage rate is the composite is at matrix saturation, indeed when the average fiber spacing "s", is similar to the largest aggregate size of the matrix.

Indeed, s = (122 x fiber diameter) / √(dosage rate), where units are in mm and kg/m³.

Both types don't show an appreciable difference between f_{r1} and f_{r3} above 40 kg/m³ so that the flexion diagrams are rather flat or "plateau" like to reflect an ideal "elasto-plastic" behavior under EN 14651 test method. In numbers, this is shown by the f_{r3}/f_{r1} ratio of 0.98 to 1.01 with 1.1 as one single exception as shown in Table 1.6 here below.

fr3/fr1 <1 is flexion softening while fr3/fr1 > 1 is flexion hardening as defined in the Mode Code 2010. (8, 9)

We will, however, see in Table 1.6 in the Full-Scale tests chapter that real SFRC slab structures clearly demonstrate flexion hardening due to the combination of the material and the structural ductilities.

There is a small f_{r3} difference of 7% and 5% between 40 kg/m³ and 50 kg/m³, and approximately 3% and 2% between HE 1/60 and HE+1/60 at 40 kg/m³ and 50 kg/m³, respectively.

In full-scale and quasi-full-scale test slabs, as we will see in Chapter 3, which focuses on Full-Scale Tests, a radical difference is observed. Specifically, with HE 1/60 or +1/60 fibers, there is no load drop after the first crack.

Multiple flexural cracks form in significant numbers, and the number of cracks increases with a higher dosage rate, providing a completely different perspective on SFRC behavior.

In the EN 14651 flexion test, the strength difference between 70 kg/m³ and 50 kg/m³ dosage rates is only 7–9%, despite the dosage rate varying by 40%.

To place and compact SFRC in the standard EN 14651 test form, smaller aggregates and more mortar—inevitably requiring more cement—are needed. This, however, leads to increased shrinkage. Such a standard mix would not typically be used in a real structure due to its higher cost and the increased shrinkage, which, over time, results in shrinkage cracks.

The risk with the EN 14651 flexion test is that new steel fibers may be developed solely to achieve better f_{r3}/f_{r1} ratios, without considering the actual cracking behavior or the impact of a smaller f_{r1} value to obtain better ratios and qualify for a higher class.

As we will explore in a later chapter, in some standard section designs, the f_{r1} value has little influence and can almost be disregarded compared to the f_{r3} value.

Table 1.6 EN 14651 fr1, fr3, fr3/fr1 for HE 1/60 and HE+1/60 steel fibers

Dosage rates	30 kg/m³		40 kg/m³		50 kg/m³		70 kg/m³		
HE 1/60	4.67		7.01		7.33		8.23		fr1
HE 1/60	4.59	0.98	6.91	0.98	7.39	1.01	8.17	0.99	fr3
HE+ 1/60	5.09		7.37		7.48		8.48		fr1
HE+ 1/60	5.63	1.10	7.21	0.98	7.55	1.01	8.23	0.97	fr3
		fr3/fr1		fr3/fr1		fr3/fr1		fr3/fr1	

1.10 THE FIB MODEL CODE 2010 CLASSIFICATION OF SFRC

In the FIB Model Code in Reference 8 (Vol. 1, p. 239), the f_{r3}/f_{r1} ratio ranges from 0.5 to 1.3, or higher, is used to define five important categories, from class *a* to class *e*, for specifying the use of SFRC.

Does the f_{r3}/f_{r1} ratio of the FIB Model Code 2010 significantly matter, given the design flexion strength depends 95% on f_{r3} and no more than 5% of f_{r1}? Not significantly, indeed.

According to the FIB Model Code 2010 and the FIB Bulletin no. 105, rebars substitution becomes possible if $f_{r1}/f_L > 0.4$ and $f_{r3}/f_{r1} > 0.5$ (Class and above), ensuring prevention of brittleness in structural members.

Hence, these two conditions combined are equivalent to $f_{r1}/f_L > 0.5 \times 0.4 = 0.2$, indicating a minimum value of one-fifth of the EN 14651 standard f_L first flexion crack strength, which is the highest nominal stress value within the 0.0 to 0.05 mm CMOD-interval.

Accordingly a concrete slab of 200 mm such that $f_L = 4.0$ N/mm² could be "structural" acceptable with an SFRC of $f_{r3} = 0.8$ N/mm², of design resisting moment of 0.8 x 200 ²/6 = 5.33 kNm/m.

$f_{r3} = 0.8$ N/mm² is typical of a 10 kg/m³ HE steel fiber dosage rate, indeed a very small dosage that I wouldn't use to reinforce any structural member!

The equivalent traditional reinforcing section is of ca. S = 5.33 10⁶ (N.mm) / 0.9 x 160 (mm) x 435(N/mm²) = 86 mm² /m width of slab! Thus a 4 mm diameter x 150 mm x 150 mm wire mesh size or a 0.043 % of rebar reinforcing! Indeed a very small percentage for a structural member! We can't understand such a light percentage in a structural application

What fiber types do I recommend to use in slabs, rafts, and shallow foundations?

These preferred steel fibers must be user-friendly, ensuring they are smoothly introduced into the concrete mix. They should not significantly alter the mix's workability, visibly appear on the surface, or adversely affect the pumping of the concrete. Additionally, they must provide high post-cracking strength in standard flexion tests for C25-30, C 30-37, and C 35-45 mixes.

Our preference is the: SHE(single hook end) 0.9 mm x 60 mm @ 1200 MPa for slabs on grade and slabs on pile.

Other type of fibers are also quite suitable and practical

- SHE 1 mm x 60 mm @ 1200 MPa for slabs on grade and rafts.
- SHE 0.9 mm x 60 mm @ 1200 and 1500 MPa for slabs on pile and elevated suspended slabs.
- SHE 1 mm x 60 mm @ 1500 MPa for slabs on piles and elevated suspended slabs and rafts.
- SHE 1 mm x 50 mm @ 1200 MPa for slabs on grade with sawn joints, shallow foundations like isolated footings, line footings, ground beams, as these fibers are very user-friendly and easily used by low-tech concrete plants to supply SFRC.

All these types are loose, unglued steel fibers, which are widely available at competitive rates and can be introduced into the mix on-site as needed.

At a higher unit cost, DHE and THE fibers are not recommended here, as their advantages in post-cracking performance, if any, are not significant. Additionally, they are not generally available at competitive rates.

1.11 PHOTO OF EXCEPTION

The Figure 1.13 shows an unnotched standard-sized small beam from the Bissen full-scale test in 2004, conducted under the German four-point bending method, exhibited ideal double cracking, almost directly beneath each load. Capturing such a result in a photo I took, was indeed rather exceptional.

The mix design was with a 100 kg/m³ dosage rate of the undulating type TABIX fibers of 1.3 mm diameter by 50 mm length with an 850 MPa wire tensile strength. TABIX is a registered name of ArcelorMittal fibers.

1.12 SUMMARY

- Steel fiber-reinforced concrete is a composite material where steel fibers, added at a suitable dosage rate, are randomly distributed in all directions. This provides the concrete with tensile strength in all directions. The steel fibers do not need to be oriented and, in fact, perform better when randomly oriented.

Figure 1.13 Double cracking of a German standard flexion test beam.

- Steel fibers are characterized by their length, diameter, shape, and the tensile strength of the steel wire, which helps anchor them to the matrix.
- The geometrical properties of the steel fibers significantly influence their introduction and mixing into the concrete.
- Steel fibers must fit between the large and smaller aggregates of the concrete matrix to ensure proper mixing, transportation, placement, and finishing of the slab.
- The purpose of steel fiber reinforcement is to strengthen the matrix and control cracking, regardless of its cause, to produce a ductile composite using suitable, ductile, anchored steel fibers.
- Very low dosage rates of steel fibers are used to replace very low geometrical percentages of traditional reinforcement.
- The Model Code 2010 defines classes of SFRC, ranging from class a to class e, based on a specific ratio for structural design. However, it is noted that this ratio is not considered in the calculation of section resistance.
  - The EN 14651 standard flexion test is the reference test method for the Model Code 2010.

REFERENCES

1. Romualdi, J. P., and J. A. Mandel "Tensile strength of concrete affected by uniformly distributed and closely spaced short length of wire reinforcement." *ACI Journal*, 61(38), pp 657–670 June 1964.
2. H. Krenchel: *Fiber spacing and specific fiber surface*, Technical University of Denmark.
3. P. Souroushian and C.D. Lee. "Tensile strength of steel fiber reinforced concrete: correlation with some measures of fiber spacing." *ACI Materials Journal*, 87 541–546, Title n° 87-M157, pp 541-546 Nov.-Dec. 1990.
4. A. Pupurs, A. Krasnikovs, T.U. Riga, 2011. "Load bearing capacity prediction of steel fiber reinforced concrete elements subjected to bending loads" Dissertation thesis.

5. V. Cunha, J. M. Serna-Cruz, J. Barros. "Impact of the manufacturing quality on the fiber Pull Out performance".
6. A. Orbe, J. Cuadrado, R. Losada, E. Roji "Framework for the design and analysis of steel fiber reinforced self-compacting concrete structures." In *Construction and building materials* N°35(2012) pp. 676–686, Elsevier.
7. EN-SS 812310, a Swedish standard: Fiber concrete-Design of fiber concrete structures.
8. The FIB Bulletin 65, Model Code 2010, volume 1, 5.6.3 p.239 classification.
9. B. Lancini, *Structural behavior of fiber reinforced concrete slabs with class-L reinforcement*. James Cook University, Australia, Sept. 2010.

Chapter 2

SFRC material properties and standard hardened material properties

In general, the steel fiber reinforced concrete (SFRC) structure will not crack in flexion, shear, or torsion under service conditions; however, cracking may occur due to both hydraulic and thermal shrinkage, or imposed displacements. The post-cracking ductility of the SFRC provides material ductility to the structure material.

2.1 HISTORY OF TESTING SFRC

The first significant reference to SFRC in structural applications dates back to 1923, with a patent N° 587.169 by Maison Fougerolles (Paris, France)as already explained in the Chapter 1.

Despite the patent, no significant developments followed until SFRC was reintroduced in 1960 by Battelle Development Corporation in the USA. Battelle focused on small fibers at high dosage rates to create a new material, that is, a composite concrete.

The testing method for this material was initially adapted from mortar testing, using very small prismatic specimens in flexion.

In 1964, the ACI 544 committee was formed to describe, analyze, and review fiber-reinforced concrete. The committee defined a prismatic specimen size of 400 mm x 100 mm x 100 mm for determining flexural strength.

As steel fibers began to be used in regular concrete, larger beam sizes (600 mm length with a 450 mm span, 150 mm width, and 150 mm depth) under four-point bending tests became preferred in various countries.

In the 1980s, Germany extended the beam length to 700 mm (600 mm span, 150 mm width, and 150 mm depth), continuing with the four-point bending method. For decades, the whole scientific community focused on beam testing to investigate SFRC, aiming to characterize the material's ductility.

The first articles highlighting the errors and biases in these methods appeared in the mid-1980s, authored by C. Johnston and colleagues at the University of Calgary. However, it should be noted that the structural ductility of SFRC applications was never thoroughly analyzed by the scientific community.

DOI: 10.1201/9781003188315-2

During this period, some major steel fiber manufacturers designed fiber products to exhibit superior SFRC behavior in small prismatic specimens. The main goal was to derive SFRC properties for design purposes based on flexural testing of prismatic specimens.

Steel fiber producers were adamant to adopt small prismatic specimens in flexion as the standard for defining steel fiber and SFRC performance. Each steel fiber product was expected to demonstrate its performance under these restricted laboratory conditions, which were then used to derive design rules. This approach effectively created an "identity card" for each steel fiber, detailing its material composition, geometrical parameters, and residual flexion strength factors, which were intended to define material ductility.

The designer can use these standard data and the advantage was to limit the responsibility of steel fiber manufacturers to the supply of a fiber type of a given flexion strength according to a standard test method. Attractive results from prismatic specimen testing often shifted the blame for poor outcomes to on-site concrete quality, contractor practices, or batching plant defects. Ultimately, this lack of accountability inhibited the further development of SFRC structural applications.

Indeed, in the end, nobody is clearly responsible of the end result, so that it becomes a serious hindrance to develop further SFRC applications.

There is no better way to have a satisfied end-customer than through an SFRC solution with a single source of responsibility for the application, from the constituent materials of the SFRC to the design and the installation, including a 5- to 10-year or longer guarantee by the concrete contractor with maintenance for guaranteed serviceability.

Note that, in general, SFRC structures fail in serviceability (SLS) long before the ULS becomes a question.

Testing the steel fiber concrete should take into account the structural ductility of the SFRC as well. Therefore, full-scale testing of applications is needed to understand the structural ductility.

The scientific community desired to show that SFRC is able to display a 25.10^{-3} strain as rebar does, but they wanted it in the small prismatic specimen only and across one single crack!

The structure behavior up to its possible collapse results from the sum of the material ductility and the structural ductility.

The EN 14651 standard introduced a notched prism specimen (500 mm span, 150 mm x 150 mm section, with a 25 mm deep notch at mid-span) for three-point bending tests. While this method aimed to reduce the variability, it failed to address the real-world structural behavior.

In the early 1980s, I organized a full-scale slab-on-grade test in Luxembourg for a major NATO flooring contract (100,000 m² of slabs). The test demonstrated that even with a low dosage of 20 kg/m³ of 1 mm x 60 mm undulated fibers (EUROSTEEL fibers by ARBED), a highly ductile slab could be achieved, exhibiting multiple cracking and a collapse load 500% higher than the first cracking load. This result was attributed to both material and structural ductility. Following this, we decided to adopt small square slab tests (600 mm x 600 mm x 100 mm) supported at the perimeter and subjected to center-point loading. These tests established new stress levels in service for steel fiber dosages of 20, 30, and 40 kg/m³, which proved robust and competitive against wire mesh-reinforced solutions. Jointfree floors on grade were successfully completed in 1983 despite being considered impossible based solely on material properties. SFRC jointfree floors, patented by me, have since achieved worldwide success. Today, the technique and its application are commonplace.

By the mid-1980s, we transitioned to testing small circular slabs (600 mm diameter x 100 mm thickness) and, in the early 1990s, larger circular slabs (1500 mm diameter x 150 mm thickness) under center-point loading. In 1986, the French railroad company SNCF adopted our square slab testing protocol for shotcrete, which later became an EFNARC recommendation and eventually a Euronorm. I must also mention that square slab testing was already being conducted in Western Canada in 1982 for shotcrete control. The shotcrete business was always run with a practical mentality, so that the scientific community did not interfere too much with its development. The result is that today, it is a very widely accepted method of SFRC shotcreting and later also for tunnels built with SFRC prefabricated segments.

From 1994 to 2020, we conducted numerous full-scale tests on SFRC suspended slabs to demonstrate their structural ductility and practical advantages. To address the errors and bias in prismatic testing in flexion, the scientific community invented the notched beam specimen in 1995 in order to decrease the coefficient of variation by forcing the single crack to form at mid-span. The procedure became EN 14651, but is now acknowledged that the coefficient of variations is even lager (35 to 50%) when using un-notched specimens.

The recent FIB Model Code 2010 sticks to EN 14651, using its characteristic flexural strengths. The structural application of slabs is affected by the EN 14651 flexion characteristic strengths, which are plagued by a huge scatter of up to 50% that is never observed in real structures.

A quite impractical size factor K, of up to 1.4 at most, is then introduced. This is considered impractical because it needs to use the real collapse load of the SFRC structure, which is generally unknown unless the SFRC is prefabricated in a large number of pieces in a plant.

There is a recent Swedish reinforced concrete standard for design structures (EN-SS 812 310 - 2015) that includes a structural indeterminacy factor: equal to 1 for statically determinate beams, 1.4 for cases with multiple parallel yield lines as in one-way slabs, and to 2 for cases with two-way yield lines in top and bottom as in two-way slabs.

2.2 UNIAXIAL TENSION SOFTENING

The composite is considered strain-hardening if the apparent tensile stress continues to increase after cracking, and strain-softening if there is a decrease in stress after cracking.

Most, if not all, steel fiber-reinforced concretes used so far are strain-softening in uniaxial tension.

A simple example calculation is provided below to demonstrate this with a 50 kg/m³ dosage rate of a SHE(Single Hook End) deformed steel fiber (1 mm diameter, 1200 MPa steel wire tensile strength). Assuming 50% anchoring efficiency, the steel wire tensile stress reaches 600 N/mm². Under these conditions, the resulting composite exhibits strain-softening behavior:

$$D := 150\,mm$$ — Cylindric sample diameter

$$f_t := 4\,\frac{N}{mm^2}$$ — Tensile strength of concrete

$$V_m := 50$$ — Dosage rate of steel fibers in kg/m³

$$d_f := 1\,mm$$ — Fiber diameter

$$\eta_f := 0.5$$ — Efficiency of steel fiber anchoring in Average of all directions

$$f_s := 1200\,\frac{N}{mm^2}$$ — Consistent steel wire strength

$$F_f := \frac{\pi \cdot D^2 \cdot V_m}{4 \cdot 122^2} \cdot \frac{\pi}{4} \cdot \eta_f \cdot f_s = 2.797 \times 10^4\,N$$ — Total fiber pull-out force of steel fibers

$$F_c := \frac{f_t \cdot \pi \cdot D^2}{4} = 7.069 \times 10^4\,N$$ — Concrete matrix tensile force at rupture

$$\frac{F_f}{F_c} = 0.396$$ — Ratio < 1, it is strain softening

$$s := \frac{122 \cdot d_f}{\sqrt{V_m}} = 0.017\,m$$ — Average distance between two adjacent fibers

$$N_f := \frac{\pi \cdot D^2}{4 \cdot s^2} = 59.364$$ — Number of steel fibers sticking out the Rupture surfaces of the cylindric sample

$$F_f := N_f \cdot \eta_f \cdot \pi \cdot \frac{d_f^2}{4} \cdot f_s = 27.975\,kN$$ — TOTAL fiber pull-out force of steel fibers

To achieve strain-hardening behavior, the ratio

$$\left(V_m \times \eta_f \times f_s\right) / f_t$$

must increase This can be achieved through a higher dosage rate (V_m) of steel fibers, improved anchoring efficiency (η_f), and higher steel wire strength (f_s). The tensile strength (f_t) of the matrix should not increase; preferably, it should decrease or at least remain constant.

At the same fiber dosage rate, reducing the fiber diameter brings fibers closer together, needing a reduction in the aggregate size of the concrete. Consequently, more intergranular mortar is needed, which increases the tensile strength of the concrete. This increase is counterproductive for achieving strain-hardening behavior.

Another example calculation is provided below:

For a 60 kg/m³ dosage rate of a SHE (single hook-end) deformed steel fiber (1 mm diameter, 60 mm length, 1500 MPa steel wire strength), a 75% anchoring efficiency ensures a tensile stress of 1125 N/mm². Under these conditions, the resulting composite exhibits behavior closer to strain-hardening, as shown in the example below:

$D := 150\,mm$	Cylindric sample diameter
$f_t := 4\,\dfrac{N}{mm^2}$	Tensile strength of concrete
$V_m := 60$	Dosage rate of steel fibers in kg/m³
$d_f := 1\,mm$	Fiber diameter
$\eta_f := 0.75$	Efficiency of steel fiber anchoring in average of all directions
$f_s := 1500\,\dfrac{N}{mm^2}$	Constituent steel wire strength
$F_f := \dfrac{\pi \cdot D^2 \cdot V_m}{4 \cdot 122^2} \cdot \dfrac{\pi}{4} \cdot \eta_f \cdot f_s = 6.294 \times 10^4\,N$	Total fiber pull-out force of steel fibers
$F_C := \dfrac{f_t \cdot \pi \cdot D^2}{4} = 7.069 \times 10^4\,N$	Concrete matrix tensile force at rupture
$\dfrac{F_f}{F_c} = 0.89$	Ratio < 1, it is strain softening
$s := \dfrac{122 \cdot d_f}{\sqrt{V_m}} = 0.016\ m$	Average distance between two adjacent fibers
$N_f := \dfrac{\pi \cdot D^2}{4 \cdot s^2} = 71.237$	Number of steel fibers sticking out the Rupture surfaces of the cylindric sample
$F_f := N_f \cdot \eta_f \cdot \pi \cdot \dfrac{d_f^2}{4} \cdot f_s = 62.943 \cdot kN$	Total fiber pull-out force of steel fibers

---

The resulting composite SFRC is still tensile softening.

2.3 FLEXION HARDENING BEHAVIOR

The first example was: $F_t/F_c = 0.396$ thus $f_{tf}/f_{tc} = 0.396$, where the total forces F_t and F_c are translated in section stresses for the fiber concrete and the plain concrete, respectively.

The plain concrete flexion strength f_L is about 1.5 times the f_{tc} tensile strength:

Thus,

$$f_L = 1.5\,f_{tc}, \text{and we have}: f_{tf} = 0.396\,f_L\,/\,1.50 = 0.264\,f_L$$

f_{r3} is the flexion strength (of an equivalent elastic section) at CMOD = 2.5 mm (EN 14651).

As the Model Code 2010 and the "Eurocode 2" draft for 2024 defines $f_{tf} = 0.33\,f_{r3}$, we must have $f_{r3} = 0.264\,f_L/\,0.33$ so that $f_{r3} > 0.396\,f_L/0.33 \times 1.5$ thus, $f_{r3} > 0.80\,f_L$.

To achieve flexural hardening in a standard beam specimen, it is necessary to have $f_{r3} > f_{r1}$, as in the case where $f_{r3} > f_{r1} = 1 = 1.25 \times 0.80$. This results in $f_{tf}/f_{tc} = 0.396 \times 1.25 = 0.50$, meaning the post-cracking tensile strength of SFRC is only half the uniaxial tensile strength of plain concrete.

This demonstrates that an SFRC with tensile-softening behavior can still develop flexural-hardening behavior in a small standard beam specimen. For this to occur, the standard f_{r3} flexural strength of SFRC must be approximately 50% of the standard tensile strength of plain concrete. In such cases, the equivalent elastic flexural stress of a small, statically determinate standard beam increases with deflection after cracking.

Unlike traditional reinforced concrete, where the rebars remain elastic after cracking, fibers embedded in the concrete matrix change more rapidly to a plastic state due to their deformation (fiber anchor deformation plus steel wire body elongation) and the gradual debonding of fibers.

At this stage of composite section plastification, the neutral axis of the member section migrates gradually from half-depth of slab to higher or lower positions within the section, depending on whether the structure is subjected to positive or negative moments.

This behavior is already evident in the standard method for determining the post-cracking flexural strength of steel fiber-reinforced concrete as a material.

2.4 EN 14651 STANDARD FLEXION STRENGTH

2.4.1 Description

The main advantage of the EN 14651 standard is its apparent simplicity, coupled with the small size of the test samples, making it feasible for any testing laboratory with minimal equipment to implement the procedure.

The EN 14651 standard specifies a three-point bending flexion test for small prismatic specimens with dimensions of L=550 mm in length, b=150 mm in width, and t=150 mm in thickness.

Prior to the testing, the small beam specimen is provided with a 25 mm deep notch at mid-span, (created using a saw cut to a depth of 25 mm, resulting in a net thickness of h_{sp} = 125 mm) at the bottom in order to ensure the flexion crack initiates at half span where the section is the weakest.

The load is applied at the mid-span of 500 mm directly above the 25 mm deep notch. The beam is supported by rollers at each end to prevent any arching action and ensure no direct load transfer to the supports.

The load intensity is controlled in a closed-loop system, driven by the notch opening under the beam. This opening is referred to as the crack mouth opening displacement (CMOD).

The EN 14651 beam size is shown in Figure 2.1.

A diagram is recorded where four key acting force intensities (later to become a sort of flexion stresses) F_1 to F_4 are measured at corresponding displacements, CMOD1 (0.5 mm) to CMOD4 (3.5 mm), as shown in Figure 2.2

The rupture pattern of the standard beam consists of two blocks of steel fiber-reinforced concrete connected by a plastic hinge initiated at the bottom

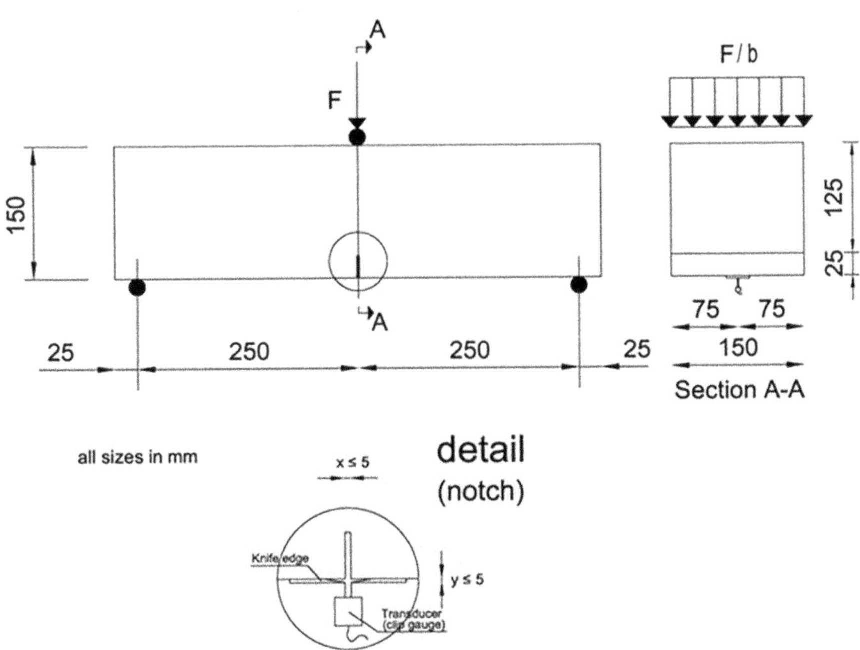

Figure 2.1 EN14651 flexion test specimen and setting-up. (in mm)

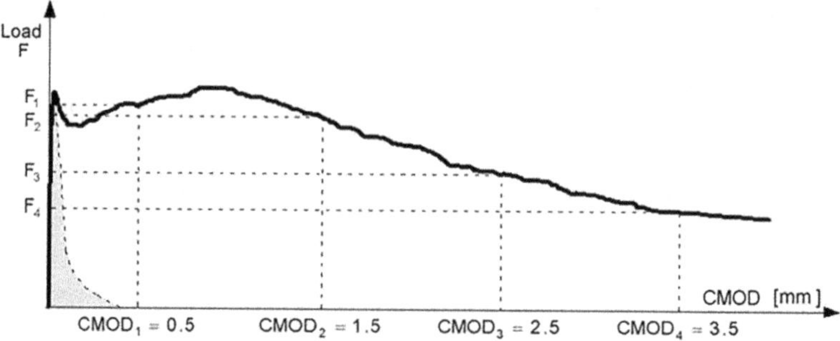

Figure 2.2 A typical EN 14651 flexion diagram.

notch. The entire deformation of the beam occurs within this hinge, without any elastic deformation in the two blocks.

This rupture pattern of the composite concrete is straightforward and consistent across specimens.

It is clear that all, F_1 to F_4, are derived from a no-longer elastic cracked specimen in flexion.

The limit of proportionality (F_L) is defined to be at CMOD = 0.05 mm or a deflection of 0.08 mm

The beam is here below back-calculated by means of elastic formulas to determine the flexion strengths f_{r1} to f_{r4} corresponding to F1 and F4.:

$$Fr_i = 1.5\, F_i\, L\, /\left(b\, h_{sp}\right)^2 \text{ where } L_n = 500\,mm\,length;$$
$$b = 150\,mm\,width \text{ and } h_{sp} = 125\,mm$$

Example: F_i = 10 kN, hence f_{Ri} = 1.5 x 10,000 x 500 /(150 x 125²) = 3.2 N/mm².

Remember that a 10 kN load intensity in the EN 14651 method results in a 3.2 N/mm² elastic flexion stress, even though the specimen's behavior is already plastic due to the plastic hinge at the notch.

A strict minimum of six samples with the same mix design is required to calculate mean and characteristic values.

To design any SFRC structure at the Service Limit State (SLS), the f_{r1} strength is used, derived from CMOD1/F1 with a deflection of 0.47 mm. At the Ultimate Limit State (ULS), the f_{r3} strength is used, derived from CMOD3/F3 with a deflection of 2.17 mm. Note that for slabs, I have always preferred using the average value instead of the characteristic values of f_{r1}.

The FIB Model Code 2012 stipulates that steel fiber reinforcement can substitute, either partially or fully, conventional reinforcement at the ultimate limit state if the following conditions are simultaneously satisfied:

$$f_{R1k}\, /\, f_{Lk} > 0.4 \text{ and } f_{R3k}\, /\, f_{R1k} > 0.5$$

The fiber-reinforced concrete shows deflection-softening behavior when $f_{r3}/f_{r1} < 1$ or deflection-hardening when $f_{r3}/f_{r1} > 1$.

The design of structural slabs and rafts requires $f_{r3}/f_{r1} > 0.7$ to be competitive with the traditional reinforcement in suspended slab and raft applications.

2.4.2 The material ductility of SFRC

Indeed, the area under the graph from CMOD 0 to CMOD 4 represents the energy of rupture of the SFRC material. The statically indeterminate 600 mm x 600 mm x 100 mm square plates (SIA 162; EN 14488), subjected to center-point loading, are used for shotcrete applications to measure the energy of rupture of the composite up to a 25 mm deflection. This test was specifically designed for shotcrete applications, as the form could be shotcrete directly on-site.

The mechanical purpose of this test is to estimate the resistance of the material against falling blocks from the ceiling in a tunnel. The ASTM 1550 test uses round, determinate slabs subjected to center-point loading, supported at three points spaced 120° apart. While this test is more complex than EN 14651, it still does not allow for random cracks, as only three main radial cracks develop. Its back-calculation process is significantly more complicated, though the scatter of results is slightly lower than that of EN 14651, ranging from 15% to 25%.

In my opinion, this test has limited general interest.

2.4.3 EN 14651 imperfections

As outlined in the literature, the EN 14651 standard and similar small-size beam test procedures result in flexural stress measurements plagued by high scatter, typically ranging from 15% to over 50%. Much of this variability is attributable to the test method itself, rather than the properties of the steel fiber-reinforced concrete.

The notched EN 14651/RILEM flexion test, originally designed to reduce the scatter seen in un-notched small prismatic specimens, has failed to achieve this objective.

Clearly, the notch doesn't eliminate the scatter of the un-notched small prismatic specimen.

The high scatter arises from various factors related to the small sample volume of fiber concrete used in these tests, including challenges in sampling, mixing, filling, compacting, and testing. When the test bench lacks sufficient stiffness, it absorbs deformation energy during loading, which is then suddenly released upon the first crack formation, similar to the action of a bow and arrow. It increases the damages in the crack and prevents further cracking.

Several specific difficulties are encountered:

- Filling the form and releasing air bubbles requires external vibration.
- The EN 14651 standard, and similar procedures, fail to specify vibration parameters such as frequency, duration, and amplitude.
- The standard does not provide guidelines for the specifications of the vibrating table necessary to compact the mix and remove excess entrapped air.
- The vibration waves bounce back on the form, causing constituent material of the form to influence the vibration's effectiveness. The EN 4651 is silent about the constituent material of the form regarding its stiffness; every other parameter is kept identical, and the results differ depending on the forms constituent material like wood, synthetic resin, or steel.

When the fiber concrete is vibrated, it shows minima and maxima of amplitude, similar to the behavior of a liquid, leading to visible segregation of fibers and aggregates. The flexion test shows that, across the cracked section after complete rupture, some section areas lack fibers entirely, others contain areas with bundles of fibers and between, the rest has a more randomly distribution of fibers.

The smaller the fiber dosage rate, the greater the likelihood of sections free from fibers.

When the fibers segregate, they tend to do so along with larger aggregates in the mix, which pull the fibers downward during compaction.

In Figure 2.3, typical X-ray photographs of a steel fiber-reinforced concrete standard flexion test beam reveal very poor fiber distribution. Most fibers are pushed toward the extremities, while the mid-span area noticeably lacks steel fibers.

The use of a specific form-release oil is important, as it can minimize the occurrence of large bubbles and reduce the need for excessive vibration of the specimen. However, this is not specified in the EN 14651 standard flexion method.

It is not uncommon for beams with a 20 kg/m^3 dosage rate of steel fibers to perform better than those with a 40 kg/m^3 dosage rate of the same steel fiber.

We recommend that before a test series is conducted by a laboratory inexperienced in SFRC testing, the laboratory's engineers and technicians first develop their own testing procedure. This allows them to minimize scatter and establish a reliable, robust protocol. A sufficient number of tests is needed to understand how to reduce scatter and achieve consistent f_{ri} values with specific concrete types, fiber dosage rates, and fiber types.

It is also worth noting that EN 14651 results are significantly influenced by the mix design, even when the same cement type and quantity are used.

2.5 STANDARD BEAMS VS. REAL SLAB STRUCTURES

A larger and longer beam specimen provides more accurate results, as the concrete is less affected by the friction of the form. It is also easier for the

Figure 2.3 X-Ray photographs of standard specimen. (from Stroeven and Carlsward).

concrete to flow inside such a form, simulating conditions closer to a real installation of steel fiber-reinforced concrete.

In Figure 2.4, a summary of flexion test diagrams is shown, arranged from top to bottom for fiber dosage rates of 20, 25, 30, and 40 kg/m³, using 1 mm diameter, 60 mm length undulated wire fibers with a tensile strength of 1200 N/mm² (EuroSteel 1/60 type, currently known as Tabix 1/60).

The left column refers to five specimens of 900 mm in length, 300 mm in width, and 150 mm in thickness. The right column is about five specimens of 600 mm in length, 150 mm in width, and 150 mm in thickness, arranged from top to bottom with fiber dosage rates of 20, 25, 30 and 40 kg/m³ using 1 mm diameter, 60 mm length undulated wire fibers with a tensile strength of 1200 N/mm² (EuroSteel 1/60 type).

These test diagrams are from FachHochschule Aachen, Germany, in 1984.

The left column with the longer and wider beam specimen shows an improvement in post-cracking stresses as the dosage rate increases from 20 to 30 and 40 kg/m³. In contrast, the right column, with smaller specimens, does not exhibit significant improvement.

From that time, we found it very difficult—if not impossible—to analyze the post-cracking properties of steel fiber-reinforced concrete based on small prismatic specimens in bending. It was clear to us that steel fiber-reinforced concrete would not be economically viable as the principal flexural reinforcement for a statically determinate beam. A too-high dosage rate must be needed, with some additional costs to adapt to maintain workability, which can also increase the shrinkage content of the mix.

Nevertheless, the Model Code 2010 used small statically determinate beam specimens in flexion to define the flexural and shear strength of all types of steel fiber-reinforced concrete structures. The information required

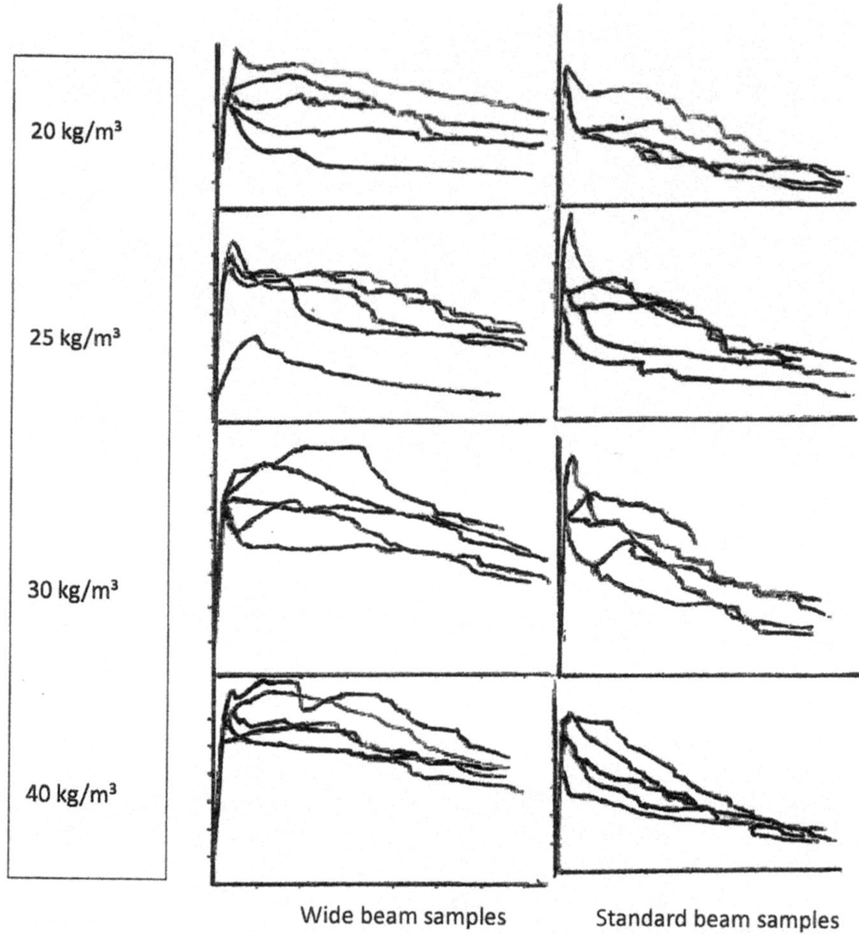

Figure 2.4 300 mm width beams vs. 150 mm width, flexion diagrams.

for SFRC structures is limited to f_L, f_{r1}, and f_{r3}. However, some special types of steel fibers with multiple hooks at each end can show a severe drop in resistance between f_L and f_{r1}, a phenomenon not accounted for in EN 14651-derived design theories. A continuous constitutive model would be more effective than the current three-point dependent model.

I recall a British concrete contractor commenting on a piled slab project. He struggled to understand claims of the complete superiority of a competitor fiber, as the design required nearly twice as many rebars in addition to the fiber dosage to achieve the same results. The imperfections of the EN 14651-based discrete constitutive model significantly underestimate the capability of steel fibers to reinforce many statically indeterminate structures. Sometimes it even becomes a deterrent argument that practically eliminates steel fibers from many otherwise viable applications unless combined with large quantities of traditional reinforcing bars.

In such cases, a small fiber dosage rate (15–25 kg/m³) is often used along-side traditional reinforcement. This practice is more prevalent in countries like Germany, where steel fibers are rather seen primarily as an anti-cracking enhancement of the traditional reinforced concrete. It is a backward point of view that makes it impossible to design durable, economical, and sustainable SFRC structures to fit the needs of the future.

Our message has always been to focus on real indeterminate structures and their structural ductility rather than analyzing the material solely through small beam specimens in flexion.

On the other hand, these standard tests bias was very attractive to the fiber and concrete industries, as these tests rely on simple, inexpensive procedures to demonstrate the performance of a specific SFRC type. Suppliers guarantee these laboratory performances, while consulting engineers design SFRC sections based on standard laboratory material data, thereby limiting their liability.

In real structures like ground-bearing slabs, pile-supported slabs, or elevated suspended slabs, SFRC demonstrates a phenomenal capacity for multiple cracking in both the top and bottom layers, visible well before yield lines develop. This multiple-cracking behavior, essential for structural design, is not considered in the EN 14651 method, despite its critical importance.

It is worth noting that the strain-hardening material shows multiple cracking that is taken into account at the design stage. Flexion hardening materials are not supposed by standards to show multiple cracking.

But in reality, it does.

2.6 IMPORTANT QUESTIONS ARE REGARDING THE TESTING ON SMALL BEAM SPECIMENS

What mix design to use? Think about the fiber type (shape, L/d aspect ratio, and steel strength) and dosage rate as the mix design needs to be adapted to obtain reliable and economical results.

Therefore, we often see results in the public where 50 or 60 kg/m³ dosage rate vs. 35 kg/m² doesn't increase the f_{rI} values.

- How fluid or stiff should be the FRC in the test procedure? How to adapt the vibration intensity and frequency?
- Define the energy of vibration available. How long should it be vibrated?
- How to avoid segregation of fibers or fibers concentrated in bundles of 3 or 5 fibers across the section?
- How is the fiber concrete poured into the form? Is it from the center the right method?
- Which form releasing oil is used?

Once the procedure is defined, resulting from a large experience on a high number of series, precisely continue with the well-defined protocol.

It is also clear that the sampling of the specimen on-site is prone to generate variations because of curing conditions, temperature variations, weather conditions at the time of pouring, and the handling of the specimen when removing them from the site. All of these factors add to further variations.

In general, it is risky to accept the steel fiber concrete f_{ri} results based on on-site sampling sent to a unskilled EN 14651 laboratory. The results are often completely unpredictable and more likely to cause litigation than provide a solution.

The EN 14651 standard was meant to reveal the material properties of SFRC but I would suggest calling it, "EN14651 laboratory SFRC flexion hinge properties."

By the way, the traditional reinforced concrete is never tested from the jobsite concrete. Instead, the concrete matrix, the type of steel and, more rarely, its positioning are checked in the form.

2.6.1 Effect of larger beam sizes in flexion

The four diagrams shown in Figure 2.5 are the results of a flexion test conducted at the Magnel Laboratory at the University of Ghent. These tests were commissioned by the Sidmar/Arbed plant before employing Eurosteel fibers (1 mm x 60 mm, 1200 MPa, with undulations) in a slab application.

The beams tested had spans of 3.60 m, widths of 300 mm, and heights of 600 mm. They were subjected to third-point bending in a C30-37 concrete mix design, with a fiber dosage rate of 40 kg/m³. The purpose of the test was to compare the performance of Eurosteel fibers with other types, including hooked-end fibers (0.8 mm x 60 mm), steel cuttings fibers (approximately 35 mm in length), and Thibo fibers with flattened ends resembling a swallow tail design (0.8 mm x 60 mm).

2.6.2 SFRC standard beam specimens don't represent real scale structures

In small beam testing, hooked-end fibers with a diameter of 0.8 mm have often proven superior in performance compared to Eurosteel fibers of 1 mm diameter, based on our experience.

As shown in Figure 2.5, this is no longer the case in these tests. The 1 mm x 60 mm undulated Eurosteel fibers outperform the hooked-end fibers of the smaller diameter but the same length.

Note that at 143 kN loading intensity at its elastic limit, the beam undergoes a 4.7 N/mm² flexion stress.

Smaller fibers, due to their larger number, tend to exhibit a more regular distribution within the cracked section of a small prism specimen. However, in a larger beam size such as 300 mm x 600 mm, the natural placement of fiber-reinforced concrete results in better natural compaction and a more even fiber distribution. Smaller standard specimen-size beams don't even represent larger sizes of real beams.

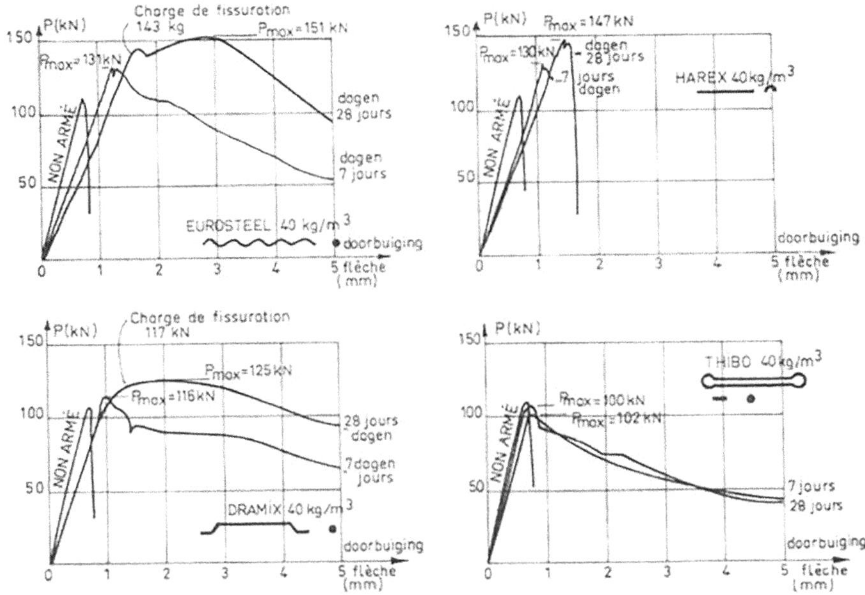

Figure 2.5 Third point loading flexion diagrams of SFRC beams of 3.60 m x 300 mm width x 600 mm height with 40 kg/m³ dosage rate of four different sizes and shapes of steel fibers.

Diagrams made by the Magnel Laboratory of T.U.Ghent for Sidmar-ArcelorMittal client.

2.7 DOSAGE RATE CHECKING

2.7.1 Apparatus

The steel fiber dosage rate control is done easily in the fresh concrete, by pouring 8 liters minimum volume of the fresh SFRC mix in a stainless-steel tube-like apparatus, provided with a magnetic cylindrical core where all fibers stick.

The magnetic core is mounted and welded on a small door so that the sticking fibers can be easily removed at the opening of the small door, as shown in Figure 2.6. It is now widely available on the market today.

Generally, no less than three samples are taken from a truck mixer on-site: one from the first third, one from the midpoint, and one from the second third of the mix. The dosage is determined as the average value.

The overall dosage of the day is the average value of all trucks tested. The number of trucks tested should be defined in the contract or the job specifications.

In general, no single value should be 30% less than the average, and the global average should not be less than 5% of the nominal value in the job specifications.

Moreover, it is very imprecise and difficult to verify the dosage rate from 150 mm or 200 mm diameter cores extracted from hardened concrete.

Figure 2.6 The fiber separator.

These cores are subject to a very large scatter, up to 50% of the fiber dosage rate. Indeed, all concrete constituents (from large aggregates to cement and density) vary significantly in small samples.

In general, I would recommend relying on the concrete contractor used to implement its own tight quality control. This is the best way forward for SFRC.

Don't think SFRC is simply about dumping bags of steel fibers into a C25-30 truck mixer, as too many cases over the past decades have shown issues such as excessive cracking, poor shrinkage, of joint curling, bad fiber appearance, and more deficiencies that have negatively impacted the acceptance of SFRC in the markets.

Even better is to use a SFRC contractor capable of offering a design-built-maintain solution with an insured guarantee. This ensures that the client enjoys a single source of responsibility and guarantee, rather than dealing with different partners playing together where the responsibility will bounce back and forth in case of litigation.

This often happens when standard solutions are offered, where each partner hides behind their business standard to shift problems onto others.

2.7.2 Think about the background

- Cement always conforms to standards everywhere.
- Ready-mixed concrete also consistently meets the specified strength and other property classes outlined in the specifications. Difficulties arise later during delivery to the site and installation.

- The fibers are known to be compliant with the specified fiber standard type and post-cracking strength, based on the manufacturer's laboratory test results.
- Design made according to design standards, where numbers replace letters in standard formula calculations, often leads to a solution that conforms to standards but may not ultimately satisfy the real requirements of the owner or client. In the end, everyone is deemed correct, the standards are considered inadequate, and the customer or end-user is typically dissatisfied.
- In case of litigation, all parties will make their best effort to demonstrate compliance with the standards governing their contributions to the project.

Experts will be needed to analyze each case, distribute responsibilities, and/or propose solutions through remedial work. It is often observed today that even engineers tend to stick blindly to design standards instead of creating innovative solutions that address real problems and take responsibility for their outcomes.

Indeed, innovative minds developed reinforced concrete applications at the end of the 19th century and later introduced post-tensioning concrete applications, all without any established or available standards.

Recently, I came across a 250-page design report for slabs in an automotive industry plant, covering both ground-bearing slabs and elevated slabs. I

Figure 2.7 A 35 kg/m³ SFRC with top and bottom mesh abiding by German design standards.

was surprised to see in the numerous sophisticated finite element diagrams that the 8.75 m span elevated slabs were calculated to bend up to 8.75 m/250 = 35 mm. Are robotics' translation movements still feasible with such ups and downs? I really doubt it. Sure, it complies with EC2 standards. But does the owner know about this? I really doubt it. I hope I'm wrong.

Will the owner enjoy the benefits of a real solution to solve the real requirements for a smooth operation "boring" plant where there is nothing to do but producing and ignoring the concrete slabs?

In Figure 2.7 of the report, I saw photographs of 250 mm hybrid slabs on grade (reinforced with heavy wire meshes and 35 kg/m^3 steel fibers) at three months of age, before final delivery. These slabs, installed in 1,000 m^2 bays separated by engineered construction joints, already showed curling at the joints and visible joint openings. The solution was strictly compliant with all German materials and design standards, as well as the code of good practice!

2.8 SHEAR STRENGTH

The Swedish standard EN-SS 812310 is specific for slabs as shown below in a calculation example of a 220-mm-diameter pile under an SFRC slab. Note that the 0.45 factor is generally associated with f_{r1} flexion strength to determine the f_{t1} tensile strength.

For fiber concrete ground-supported slabs and column bases without conventional bar reinforcement:

$d := 200\text{mm}$	Slab thickness
$R := 110\text{mm}$	A 220 mm diameter pile.
$k := 1 + \sqrt{\dfrac{R}{d}} = 1.742$	Geometric factor of max.2
$f_{cflR3} := 4.0 \dfrac{N}{\text{mm}^2}$	Flexion strength EN14651 EN14651
$C := 0.45$	A standard data.
$V_{Rdf} := \dfrac{k}{2} \cdot C \cdot f_{cflR3} = 1.567 \cdot \dfrac{N}{\text{mm}^2}$	Shear strength of SFRC (Eq. 6.4)
$t_d := V_{Rdf} \cdot d = 313.492 \cdot \dfrac{kN}{m}$	Shear resistance per meter rur
$V_{RD} := 2 \cdot (R + 1.5 \cdot d) \cdot \pi \cdot V_{Rdf} \cdot d = 807.588 \cdot kN$	**Shear resistance of SFRC EN-SS 812310 in case of a 200-mm-thick pile-supported slab**

The ACI 554 6R 15 provides a calculation of the shear strength for fiber-only reinforced concrete slabs.

$t := 200\,\text{mm}$	Slab thickness
$f_{ck} := 30$	Cylinder strength (characteristic) (N/mm²)
$f_{ckc} := 37$	Cube strength (N/mm²) (N/mm²)
$f_{ctm} := 0.3 \cdot \left(f_{ck}\right)^{\frac{2}{3}} = 2.896$	Tensile strength of plain concrete (N/mm²)
$\omega := \dfrac{f_{ck}}{f_{ctm}} = 10.357$	Compressive tensile strength ratio (N/mm²)
$f_{r3} := 4.00$	EN14651 (N/mm²)
$\sigma_{cr} := f_{ctm}$	
$\mu := \dfrac{f_{r3}}{3.104 \cdot f_{ctm}} = 0.445$	ACI544 8R16
$u := 1\dfrac{N}{\text{mm}^2}$	
$D := 180\,\text{mm}$	Pile diameter
$V_{pc} := \left[2 \cdot \pi \cdot \left(\dfrac{3 \cdot t + D}{2}\right) \cdot t \cdot 0.66\right] \cdot \left(\mu \cdot \sigma_{cr}\right) \cdot$ $u = 416.828 \cdot \text{kN}$	SFRC slab shear capacity (Eq. 7.6a)

By SS 812310, the 807 kN design reaction is more than what can offer a 220-mm-diameter pile with a bearing capacity of 350 kN to 500 kN at most.

ACI 544-6R15 also provides a shear strength of SFRC alone without any principal rebars.

It shows that the shear/punching-out of an SFRC slab is not critical in that case; thus, the slab design will be based on critical flexion and serviceability.

It is confirmed in reference (1, 2) where the authors confirmed "The failures observed on the full-scale experiment were clearly associated with bending."

Note also that after 30 years of design cases, I have known over 2 million piles supporting SFRC slabs and not a single case of pile or loads punching-out has been reported to us.

Figure 2.8a and b show a punching-out test of an E-SFRS at 60 kg/m² HE 1/60 1450 MPa. A 60% cancellation of the slab section along the critical punching-out perimeter is needed to try to punch out at 600 kN point loading intensity maximum jack capacity. Such a high600 kN punching-out resistance at the edge with 60% of the section cancelled as well: thus in the middle of the slab, the slab should resist to 600 kN/0.6 x 2 = 2000 kN.

Punching out collapse at 2000 kN. Figure 2.8 shows the Bissen (Luxembourg) test in 2004.

In the case of a 160 mm slab at 30 kg/m³ of HE+1/60 steel fibers with 1500 MPa steel wire tensile strength, a 700 kN Point loading was needed to punch out.

Steel fibers are an excellent reinforcing against punching-out thanks to 3D distribution in all directions of the closely spaced steel fibers.

Steel fibers at a dosage rate of 40 kg/m³ HE+ 1/60 steel fibers and higher cancel the need for stirrups and shear studs in slabs.

Figure 2.8 (a): Punching out test of the slab in its middle: No punching-out no cracks at 600 kN. (b): With 60 % slab section removed, still, no punching out under 600 kN point loading is shown.

2.9 FATIGUE AND DYNAMIC LOADING

2.9.1 Cast in-situ SFRC

Steel fiber reinforcing together with or without traditional reinforcing increases the fatigue and dynamic loading resistance significantly.

One of the first industrial applications was the reinforcing of the temporary sleepers under the railways of the Channel tunnel construction site in 1987 where the reinforcing was only 45 kg/m³ of 1 mm diameter, 60 mm length at 1200 N/mm² wire strength of an undulated shape fiber.

The dynamic resistance of the steel fiber reinforced concrete was used in 1994 for a high-speed train run at 320 kph, a skewed bridge slab subjected to torsion, by the use of 50 kg/m³ of Twincone fibers in 1 mm diameter by 54 mm length made of 1200 N/mm² steel wire tensile strength, provided with conical extremities as anchoring detail.

It is worth noting that there was no SFRC standard available regarding the design of the f_{r1} values, but only about Re,3 the residual factor in flexion. This factor averages post-cracking flexion diagram up to 3 mm deflection of the 150 mm x 150 mm x 450 mm prismatic specimen as an elastic strength.

Two applications are shown in Figure 2.9.

Damad (3) showed that slabs in flexion made of plain concrete resist up to 323 cycles at 70% of the maximum load, while those with 30 kg/m³ steel fibers resist more than 3 million cycles.

When precracked, these slabs, under 50% of the maximum load, still resist more than 5 million cycles.

Fatigue tests in labs are long and costly because at 4Hz (4 cycles/second), it takes almost 15 days to reach 5,000,000 cycles. To achieve 40,000,000, as in the case of the HST bridge, approximately 120 days are required.

Fatigue resistance applications critical examples:

1- Bridge deck: Brussels to Paris High-Speed Train line (a bridge deck subjected to torsion moments.)

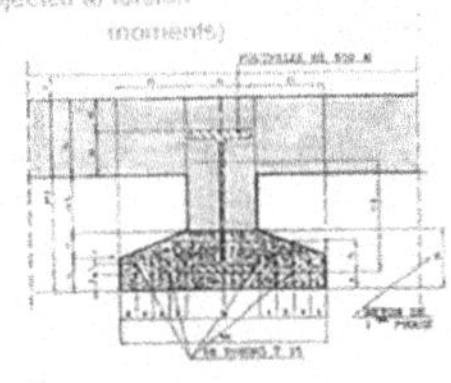

A 500 mm slab spanning 3.48 m between preflex beams with 50 kg/m³ ArcelorMittal Steelfibres: the bridge was completed in 1995 so that it is 28 years in service at 320 kph - 220 kN axle load of an equivalent static loading of 12 Tons/m² and ca. N = 40.000.000 loading cycles till today. The solution proposed by us was investigated and approved by TUCRAIL, the head engineer, even though no SFRC design standard existed at that time

2- Channel tunnel sleepers (UK side): 60.000 SFRC prefabricatd sleepers in C 40 @ 45 kg/m³ of 1 mm x 60 mm AM steel fibres were produced and installed in the tunnel track. Dynamic tests were conducted under TransManche supervision to compare SFRC to traditional: after 3 million cycles of loading up 80% of flexion strength, SFRC didn't show any cracking unlike in traditional reinforcing.

Figure 2.9 Two applications of the high SFRC fatigue resistance.

Testing real structures requires high forces combined with displacements of several millimeters, resulting in high energy demand, including powerful hydraulic oil flow and high electrical currents to pump it.

Real fatigue reference cases, like the one shown in Figure 2.9, are very valuable and should not be ignored, as they provide knowledge when compared to fatigue test on small, simply supported beam samples.

2.9.2 Prefabricated SFRC

A perfect example of the high fatigue resistance of SFRC being exploited is its design and use at the TransMancheLink in the English section of the tunnel jobsite. In the French section, it was excluded, as it was not acceptable due to not being part of the DTU (Documents Techniques Unifiés).

A temporary track was built using SFRC sleepers with 45 kg/m³ of Eurosteel undulating fibers, 1 mm in diameter and 60 mm in length, with a tensile strength of 1200 MPa.

The SFRC sleepers were tested for fatigue resistance at the University of Liège (Belgium) and compared to traditional reinforced concrete sleepers subjected to millions of loading cycles at a frequency of 4 Hz.

The traditional sleepers were significantly cracked, while the Eurosteel fiber sleepers showed no cracking. A total of 100,000 SFRC sleepers, including ca. 300 tons of steel fibers (as shown in Figures 2.10 and 2.11), were manufactured by the Byson prefabrication plant in the UK.

The thin edge of 60 mm thickness was the critical area at the advantage of the steel fibers against cracking. The project has been a success.

After the tunnel project completion, the temporary SFRC sleepers were removed and used somewhere as dolosses, an ideal application of high-strength steel fiber reinforced concrete.

2.10 STRUCTURES SUBJECTED TO EARTHQUAKES

Important earthquakes can ruin reinforced concrete structures even in developed countries where stringent design codes are applied, especially to prevent collapse of columns and slab-to-column connections.

Figure 2.12 shows an important collapsed bridge structure after the 1989 San Francisco earthquake. Note how the concrete between the rebars has been emptied, creating a birdcage effect. Without proper confinement from the rebars and stirrups, the rebars deform and go "bananas." A very sad outcome.

The comment under the original published photos was: "*Concrete piers supporting the Cypress Street Viaduct in Oakland, California, broke like matchsticks during the Loma Prieta earthquake in 1989.*" (H.G. Wilshire/ USGS.)"

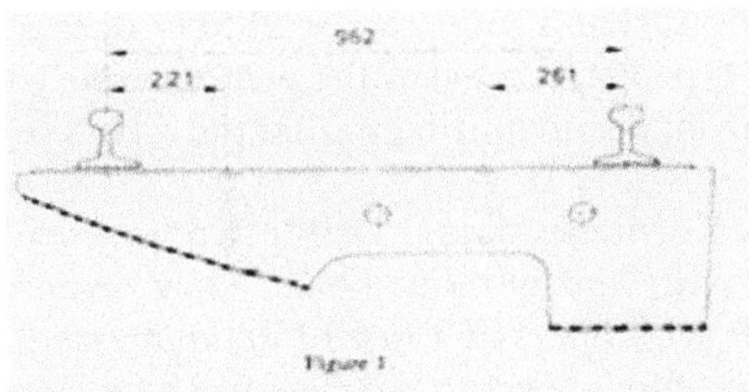

Figure 2.10 Typical section of the SFRC TransManche link sleepers, at 45 kg/m³ dosage arete of steel fibers.

Figure 2.11 SFRC sleepers test bench at the University of Liège (1986).

Figure 2.12 Seismic collapse caused by the Loma Prieta earthquake in 1989.

To me, it is obvious that a moderate steel fiber dosage rate like 30 or 40 kg/m³ of HE1/ 50 type could prevent the pier concrete from peeling off and the core concrete from disintegrating so that the rebars become unsupported and burst out.

In reference (4), Kheni, Scott, and Deb tested beam-columns connections in traditional RC with and without fibers. They demonstrated that traditional RC, together with fibers at a rate of 78 kg/m³ steel fibers and 2 kg/m³ polypropylene microfibers, performed best to show significantly less damage than the control RC specimens.

The Kobe viaduct collapsed like matchsticks in the 1995 earthquake as shown in Figure 2.13 despite the huge diameter of each columns.

Each massive pilaster, approximately 2.50 m in diameter, was reinforced with 40 mm diameter rebars along the perimeter, spaced 100 mm apart (around 2% geometrical reinforcement), but with no reinforcement inside.

The skin concrete peeled off, leaving the rebars unsupported and causing them to "go bananas," as seen in Loma Prieta.

The viaduct collapsed instantly in a brittle manner.

In Figure 2.14, a photo I took after the 1980 El Asnam (Algeria) earthquake shows a typical column in a multistory school building.

We see here the typical rupture of stirrups, resulting in the rebars going "bananas."

I remember that after the El Asnam earthquake, I saw the main hospital completely flattened, with its floors piled up like pancakes, resulting in more than 1000 deaths.

In my opinion, all these three buildings lacked a minimum of 30–40 kg/m³ of suitable steel fibers!

Figure 2.13 The Kobe viaduct collapsed in 1995.

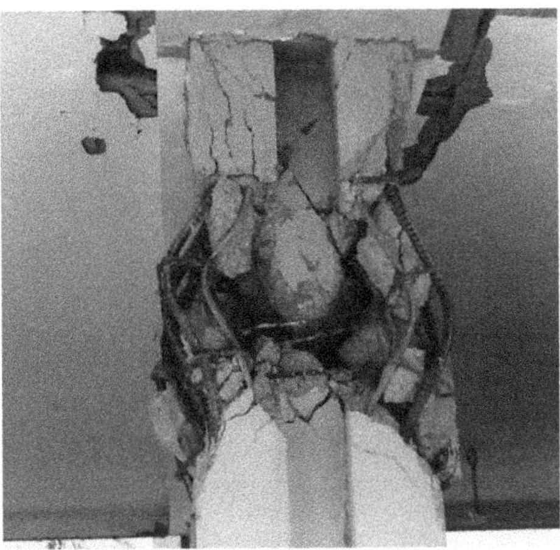

Figure 2.14 A column after the El Asnam earthquake, Algeria.

Each time after severe earthquakes, design codes are upgraded with more steel reinforcing so that it gets more difficult to place the concrete between rebars.

Steel fiber reinforcing adds a lot of ductility by improving crack control to conventional reinforced concrete structures. More ductility means releasing more mechanical energy, more deformations before fatal collapse, and more time to flee.

In Ref (4), the authors concluded that *"The results indicate the significant and positive effect of steel fiber confinement in improving the bond performance and the ductility of the mode of failure of normal strength concrete beam-column connections…"*

I'm convinced, however, that a good move would be to include 30 or 40 kg/m^3 of steel fibers in columns to keep the concrete skeleton intact inside the rebars at the perimeter, unlike stirrups. This could prevent the rebars from going "bananas."

Some references, such as in Reference (4), are even more radical, suggesting the complete elimination of ties! Indeed, 50 years ago, M. Charles Henager, a prominent member of the AC 544 fiber-reinforced concrete committee, concluded (as shown in Figure 2.15) the replacement of stirrups in beam-column joints, and 38 years ago, Lakshmipathy and Santhakumar concluded that (Figure 2.16) a 130% increase in energy dissipation of a frame where steel fibers are additional reinforcing in the column-beam joints.

The literature opens the door to such a positive move:

In reference (5), the authors concluded that "the result of this study confirms that' SFRC, walls can exhibit a performance comparable to that of conventionally reinforced walls in terms of strength and deformation capacities under seismic induced loading" In Reference (6), the authors conclude

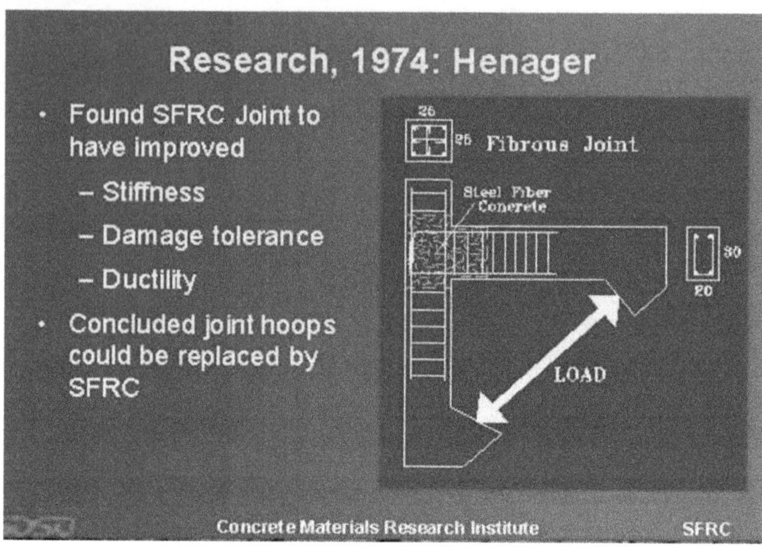

Figure 2.15 Possible elimination of stirrups thanks to SFRC.

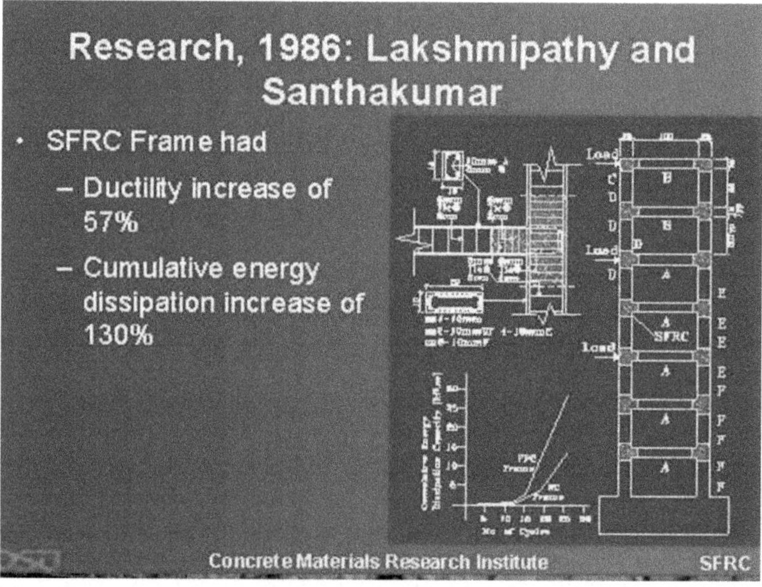

Figure 2.16 SFRC influence on energy dissipation at joints.

that the *"Results of reverse cyclic loading tests on full scale exterior and interior beam-column joints indicate that the fiber reinforced concrete is a possible alternative to conventional confining reinforcement to provide adequate ductility in reinforced concrete framed structures. Steel fibers bridging across cracks in the concrete matrix increase the joint shear strength and can reduce, or even eliminate for exterior joints, the requirements for closely spaced ties."*

I can't understand that why it is yet to be set as a standard to include a ductile SFRC in columns or column-slab joints in California or other regions that are prone to earthquakes.

The code writers believe, maybe, that steel fibers are detrimental to structures or they are not interested although the advantage has been known for 50 years.

Each time the code is revised, they add more steel rebars so that it becomes impossible to compact the concrete around. But it will be like ever that some last code version abiding buildings will still unfortunately collapse.

At last, here I included a photo of a beam reinforcing so congested that the concrete mix couldn't be placed between and around the reinforcing. The reinforcing schema had to be revised to ensure good concrete compaction.

Figure 2.17 shows a shot of Concrete International, in December 2022 (p. 57).

Sometimes, steel fiber reinforcing has been named "rebar decongester." Figure 2.17 also shows that sometimes engineers don't even understand the practical implications of what they design.

I had once had a similar experience with a large raft under a 12-floor high building in Norway. I designed it as usual in SFRC only for the satisfaction of the general contractor and the owner. The difficulty was that the engineer was a prominent officer in Germany. After a lengthy discussion, they agreed

Figure 2.17 Example of extreme rebar congestion.

on the SFRC design. Hopefully, I was on-site at the first pour of 500 m³ of SFRC. Everybody was ready to begin, and the first readily mixed concrete trucks arrived and the pumps were ready to pump.

The G.C. called from the pit for some explanations, where I saw 25 mm rebars spaced 20 mm apart in two layers across the entire width of the raft. Even plain concrete could not flow between rebars placed so closely together.

The G.C. decided to stop the operations and called the German engineer, who arrived urgently on-site the following day. He admitted to an error made by his office and himself and requested that the excess rebars be removed.

Although we had good discussions and reached a satisfactory agreement, the gentleman had added these rebars without even consulting me for an explanation!

2.11 DURABILITY AND CORROSION

Let's be very clear: steel fiber reinforcement significantly increases and improves the durability of a durable matrix. The main reason is that steel fibers effectively control the growth of microcracks, preventing aggressive elements from penetrating the concrete matrix. Microcracks provide an ingress path for all aggressive chemicals.

Moreover, the fibers do not form a continuous network in the concrete since steel fibers are a discrete array. Adjacent fibers do not touch each other, so no current can develop inside the matrix, nor can stray currents form.

Thirdly, any potential fiber corrosion near the surface involves a very small volume that does not break the surrounding mortar, preventing further corrosion. If the matrix is durable, corrosion remains localized to the very surface of the concrete.

However, this could pose an aesthetic issue for decorative applications because fibers are visible at the surface of the structure. Industrial applications, in general, are not concerned with aesthetics.

In contrast, rebar behaves oppositely in all three cases mentioned above. For example, a 20 mm diameter rebar at a depth of 30 mm could, under corrosion, experience a diameter increase of just 0.05 mm—enough to provoke a microcrack along the entire length of the rebar. Over time, the concrete cover will spall and chip away.

To protect rebars from corrosion with time, a traditional solution is to increase the cover thickness to up to 125 mm in marine environments for 50 years or more. It should be considered that such a cover is in plain concrete, subjected to temperature variations and shrinkage so that it will crack. Are the rebars inside protected? A prudent solution should be to use the fiber reinforcing to inhibit bad cracking despite' the heavy traditional reinforcing inside.

Steel fibers should also be used to improve the protection of traditional rebars from corroding in aggressive environments.

In reference (7), the authors mainly concluded that in saturation of 3 g Cl-/kg concrete, (very large indeed as 7.2 kg of chloride per cu.m of concrete: it is similar as 7.2/0.63= 11.5 kg/m³ $CaCl_2$ calcium chloride in concrete, a completely excessive and forbidden quantity in concrete):

"Rust from fibres could be found in the non-cracked concrete only at the immediate surface, in the carbonated zone."

"The crack width up to 0.5 mm did not have an adverse effect on corrosion."

Note that the 0.5 mm crack opening is the crack opening at f_{r1} EN 14651 flexion strength for SLS, service condition.

Also note that the chloride ion saturation at 3 g Cl-/kg concrete is an extremely high level deleterious ion concentration that normally doesn't happen in the usual reality of structures.

Note that 3 g chlorides ion Cl- is about 1/12 mole, thus 1/12 mole of NaCl or 5 g NaCl /kg concrete or 12 kg NaCl par m³ of concrete!

If the concrete mix was prepared with sea water at 35 g/l NaCl, 180 liters of mixing water brings 6.3 kg NaCl per m³ of concrete. But nobody would use seawater instead of potable water for concrete.

In other words, this means that steel fibers embedded in a durable matrix (for example, 350 kg/m³ cement, with W/C = 0.45) could never corrode.

In Reference (8), ACI reported that with "*zinc coated fibers, no staining at the surface*" and "*The crack width of up to 0.5 mm seemed to have no adverse effect on corrosion*"

Hence, zinc-coated fibers are used for aesthetic reasons that don't apply to most structural concretes unless according to accepted specifications.

Zinc coating is also not risk-free as in bad conditions Zn can generate H_2 gas so 65 g of Zn could generate 22.4 liters of H_2 gas. In Figure 2.18, a fractured surface of SFRC shows the microbubbles in the empty trace of one fiber.

To avoid such a case, the Zn coating must bond firmly to the fiber so that it doesn't get loose; however, the concrete mix must be well compacted,

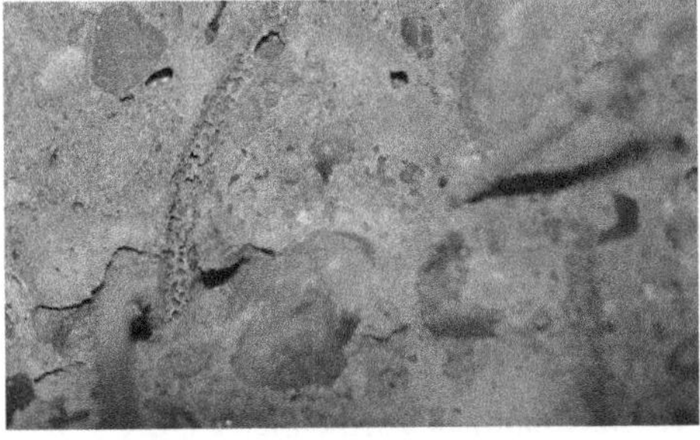

Figure 2.18 Gas bubble in an empty trace of steel fibers.

dense with a rather well-controlled water content together with a low chloride ions concentration.

Almost 40 years ago, we observed a strong case where the finished slab surface was plagued by big bubbles popping out. The investigation confirmed Zn-coated fiber steel induced the generated gas to form pop-outs.

In Reference (9), G. Chanvillard and X. Destrée concluded that *"It seems that in the crack opening range to be expected in structures, there is no fear of mechanical properties loss as the fibres are well protected."*

In Reference (10), A. Erdelyl concluded that *"after 4 years exposition of 25, 50 and 75 kg/m³ dosage rate of cold drawn wire steel fibres demonstrates the superior durability of SFRC to **plain** concrete."*

Thus add rebars to plain concrete and same conditions imposed, you understand that the SFRC could be a lot more durable than traditional R.C. since the rebars and its concrete cover are the culprits in case aggressive environments.

It is about what we can understand from the authors (Raupach and Dauberschmidt) who concluded in Reference (11) that *"In artificial chloride containing pore solutions indicate a distinctly increased resistance against chloride inducing corrosion compared with conventional reinforcing steel for high pH-values. With decreasing diameter of wires, the critical chloride content increases gradually."*

I must say that in 40 years, I have known only one case of litigation regarding the durability of SFRC. It occurred in Belgium in 1984, where a concrete contractor installed an external SFRC slab with sawn cuts every 6 m apart in C25-30 concrete with 30 kg/m³ of steel fibers. C25-30 is a very insufficient specification for external concrete use.

The litigated 7000 m² area was an open-air silo for sugar beets, a livestock food, to be handled by a 35-ton front bucket payloader machine. Over time, the liquid seeping from the sugar beets became acidic, and the C25 mix could not withstand the scraping of the concrete surface by the bucket.

Many steel fibers near the surface were included in the livestock feed, resulting in the death of 144 cows. The matrix was neither durable nor suitable for SFRC in this case.

In any event, steel fibers should not be used if the concrete comes into direct contact with food. Remember, steel fibers cannot turn bad concrete into good concrete.

REFERENCES

1. B. Parmentier, P. Van Itterbeek, "The flexural behavior of SFRC flat slabs: The Limelette full-scale experiments for supporting design model codes" *FRC 2014, Joint FIB-ACI Workshop: from Design to Structural Applications.*
2. M. di Prisco, P. Martinalli, B. Parmentier, "On the reliability of the design approach for FRC structures according to the Model Code 2010: the case of elevated slabs." *Structural Concrete*, 17, 588–602, 2016.

3. I.M.A.A. Damad, "Fatigue Behavior of Steel Fibre Reinforced Concrete Pavements." February 2020, the University of New South Wales, Sydney, Australia.

4. D. Kheni, R.H. Scott, S.K. Deb, A. Dutta, "Ductility enhancement in beam-column connections under cyclic loading using HYFRC." *ACI Structural Journal*, 112 (2), 167–178, 2015. (American Concrete Institute).

5. A. J. Carrillo, S. Alcocer, J. Pinchiera "Shaking Table Tests of Steel Fiber Reinforced Concrete Walls for Housing." In *15 WCEE*, Lisboa 2012.

6. B. Filiatrault, Massicotte, J. Houde, Dpt Civil engineering, Univ. of Montréal Campus, 1995. *Application of Steel Fiber Reinforced Concrete in A seismic Design of Beam-Column Joints*. 2nd.

7. A. Lambrechts, D. Nemegeer, J. Vanbrabant, H. Stang. "Durability of Steel Fibre Reinforced Concrete", ACI SP212-42.

8. "Report on the Physical Properties and Durability of Fiber-Reinforced Concrete." ACI 544.5R10.

9. G. Chanvillard, X. Destrée, "Quelques aspects de la durabilité des bétons renforcés de fibres métalliques à ancrage total." DGCB/ENTPE- ARBED, R. M. Audin F 69518 Vaulx en Velin CEDEX.

10. A. Erdelyl, "Durability Studies on SFRC", Budapest University of Technology, Hungary.

11. M. Raupach, C. Dauberschmidt, "Untersuchungen zum Kritischen Korrosionsauslösended Chloridgehalt von Stahlfasern in künstlicher Betonporen lösung". *Materials and Corrosion 53*, 408–416, 2002.

Chapter 3

The need to test real structures, from ground bearing to elevated suspended slabs

3.1 BACKGROUND

To the material ductility, the structural ductility is added resulting in the overall ductility of the concrete structure being the combination of both material ductility and the structural ductility.

Structural ductility is a characteristic of statically indeterminate structures, where the redundancy in the structures provides multiple load paths from the roof down to the foundations.

The higher is the degree of indeterminacy, the more redundancies will the structure offer and the safer will it be.

A statically determinate beam, in other words, a simply supported beam, doesn't have any redundancy as there is no load redistribution possible after cracking: remove one support and the beam falls down to the ground instantly.

In contrast, a statically indeterminate beam on three supports, thus with two adjacent spans, can redistribute the load from one badly cracked span to the next. This redistribution occurs through controlled rotation over the central support. To enable such controlled rotation, a minimum amount of reinforcement, whether in steel fibers or rebars, is required.

Suspended slabs are two-ways slabs, similar to piled suspended ground slabs or elevated suspended slabs, and have a much higher degree of indeterminacy as they can show multiple cracking due to M_{xx}, M_{yy}, and M_{xy} moments in two main directions; however, it also shows circular cracks around the column supports, including diagonal cracks, often occurring in both bottom and top surfaces.

Suspended slabs also enjoy the arching effect to transfer a portion of the loading pattern directly to the supporting piles or columns, as shown in Figure 3.1.

Note after reading Reference (1), the reader also needs to read Reference (2) to have a complete picture. It is all about academic theory vs. experience.

By calculations today, only some advanced elastoplastic volume finite elements software can attempt to simulate the complexity of an elastoplastic slab in the serviceability stage (SLS) up to the ultimate stage (ULS) and further to collapse.

DOI: 10.1201/9781003188315-3

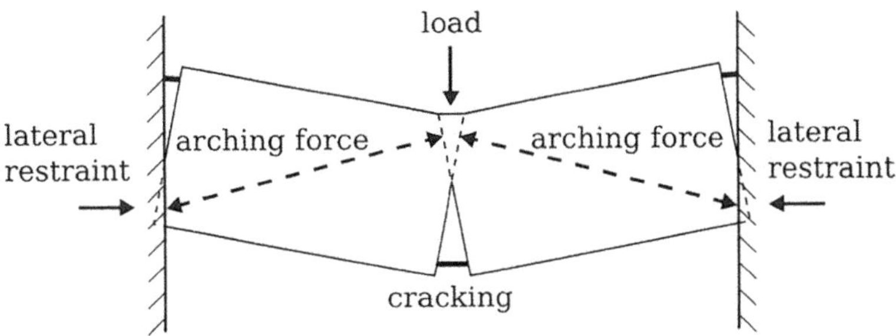

Figure 3.1 Typical arching action of a suspended slab.

Many current finite element software are still based on the use of plane elements with a thickness instead of much more complex volumic elements.

Such software, regarding steel fiber concrete, still needs to be fed with data based on the highly complex analysis of the simplistic small standard notched beam specimen in flexion, providing an overly restrictive view of the slab application.

Most engineers and designers in their day-to-day business can't even afford or don't have the time to run these sophisticated software so a more practical and experienced method, derived from the real observation of full-scale testing of similar real structures, can help completely. It is also a way to calibrate the above software regarding the two-ways suspended application and design.

The aim is to design a reliable and safe structure that meets the code level performance, achieving this at the cheapest cost possible, in order to avoid overdesigning, which causes a waste of resources that otherwise could be productive.

Let's be clear that an optimal concrete structure is safe provided it meets the code requirements in SLS and at ULS so that it doesn't need to be safer. Overdesigning results in heavier solutions rather than necessary and causes a waste of productive resources with financial and environmental losses.

From the material and standard flexion test material properties, how do we design a real structure?

It is often done by the development of models and then by standard design methods, although many real structures tests were still missing. There is a need for more full-scale tests as confirmed in Reference (3) by Orbe, Cuadrado, Losada, and Rolf:

Most researchers focus on the study of molded specimens which are not representative of real casted structures. They do, however, provide an understanding of the different parameters such as wall effects and the orientation factor. At present, there is a huge lack of experience in real-scale structures, with few references on the literature.

3.2 WHY FULL-SCALE TESTS ARE NEEDED?

Robert Maillart, a brilliant Swiss structural engineer, was considered a pioneer who tested full-scale suspended slabs in the early 20th century in order to understand their design process.

Figure 3.2 shows the Maillart full-scale test slab in 1910 and Figure 3.3 shows his first three-story design grain silo elevated slabs warehouse (1912) situated in Altdorf, Switzerland, build for the Swiss Confederation. The photo of the test slab shows a very thin slab of approx. 100 mm thickness with almost 4 m span, subjected to 3.6 m height of sandbags resulting in a UDL of 3.6 m × 16 kN/m³ density, approximately 50 kN/m², and this without visible significant deflection.

Maillart had three basic principles of design: simple analysis with assumptions made based on common sense, to consideration of the structure type and its construction process and finally, testing a real structure while keeping in mind the observation from previously built existing structures. Indeed, it is a "design method assisted by a testing route" as explained in the Technical Report 63 of *The Concrete Society* Reference (4). This is a practical approach based on observations of real slabs under test loadings.

By doing so, Maillart became a pioneer in flat bottom suspended elevated slabs.

Already in Maillart's time, German engineers and scientists had developed elaborate mathematical techniques and were confident that they did not need practical load tests for their designs, which were developed using these techniques.

Why can't we miss SFRC full-scale testings?

Figure 3.2 Maillart full-scale slab test in 1910.

Figure 3.3 Typical Maillart elevated slab of a grain silo in 1912.

The experience we gained teaches the following points:

- The full-scale test process is an experimental process to analyze, under real loads, the behavior of a real structure regarding its SLS as well as its ULS and further to the final collapse of the structure.
- The full-scale test method, together with a careful numerical analysis, leads to the so-called Design Assisted by a Testing Route.
- All real effects and parameters, which can't be included in material standard testings, are indeed taken into account: all material variations, all heterogeneities of the real structure, the conditions of supports, at the edges, the real shrinkage stresses, the creep of concrete including the time effect, and more.
- Unlike for mathematical models, no hypotheses are needed nor do they need to be verified.

3.3 LIST OF FULL-SCALE SLAB TESTS

Before going into a more detailed analysis of some full-scale tests of SFRC slab structures in the following chapters, let's draw their list in (Table 3.1) which includes all full-scale tests organized by third parties according to our specifications of SFRC.

Letter **G** denotes ground-suspended slabs (indeed pile-supported ground slabs), and E denotes elevated suspended slabs. The ACI 544 6R15 denotes them as **G**-SFRC and **E**-SFRC.

The only difference between E-SFRC and G-SFRC is that E-SFRC includes a set of Anti-Progressive Collapse (APC) rebars unlike G-SFRC as it can't support a progressive collapse. G-SFRC is a suspended slab without personnel activities underneath.

Two slabs are on grade, denoted as **sog**, that is, slab on grade, where **w** denotes walls.

Table 3.1 A list of SFRC full-scale slab tests

Location and year of test	Thickness of slab	Ground bearing or Piled. Spans or Elevated Suspended	Fiber type Dosage rate Kg/m³	1st crack load	SLS load Center Point Loading or UDL (contract data)	Collapse load. Or deflection under UDL during n days	No. of test
NATO Luxembourg 1983	130 mm **sog**	Ground Bearing	20 kg/m³ of 1/60 crimped	250 kN	50 kN	1000 kN	1
Ternat Belgium 1994	160 mm **G**	Piled slab/ES 3.1 m x 3.1 m	45 kg/m³ Twincone	100 kN	50 kN	500 kN	2
Townsville Australia 2000	160 mm **G**	Piled slab/ES 3.1 m x 3.1 m	45 kg/m³ Twincone	100 kN	50 kN	500 kN	3
Bissen Luxembourg 2004*	200 mm **E**	ES 6 m x 6m	100 kg/m³ 1.3 mmx 50 mm crimped	125 kN	10 kN	500 kN	4
Niew Vennep Holland 2006	180 mm **G**	PS/ES 3 m x 3.6 m	40 kg/m³ +1/60 crimped	29 kN/m²	15 kN/m²	Unknown 150 hours at 29 kN/m², δ=1.5 mm	5
Tallinn Estonia 2007*	180 mm **E**	ES 5 m x 5m	100 kg/m³ 1.3mm 50 mm crimped	100 kN	10 kN	Unknown > 650kN	6
Vaasa Finland 2009	100 mm **sog**	Ground bearing on insulation	35 kg/m² HE 0.75/50 mm hooked ends	150 kN	50 kN	➢ 270kN	7
Eindhoven Holland 2011	180 mm **E**	ES 1-way 3 x 5.40m	50 kg/m³ HE 0.9/60	4 kN/m²	3 kN/m²	16 kN/m²	8

Klaipeda, Lithuania, 2012	210 mm **G**	PS 4 m x 4 m	45 kg/m³ Twincone	Not observed	30 kN/m²	None 90 d @ 30kN/ m² δ=1.0 mm	9
Tingstat, Sweden, 2014	220 mm **G**	PS 4 m x 4.70 m	45 kg/m³ HE +1/60	None	40 kN/m²	None 14 days @ 40 kN/m² δ=1.0 mm	10
Goteborg, Sweden, 2019	250 mm	PS/ES 3,8 m x 3.8 m	45 kg/m³ HE + 1/60	30 kN/m² @ 40 kN/m² 7 d crack opening = 0.4 mm max.	40 kN/m²	78 kN/m² collapse 7 d @ 40 kN/ m² δ = 5 mm	11
Stockholm, Sweden, 2011 Friends Arena	300 mm **G**	PS 4 m x 4 m 5 m x 5 m	45 kg/m³ HE+1/60	None	None	Test loaded by several 700 tons cranes – 200 kN/m²	12
Poprad, Slovakia, slab 2019 Fire test *	180 mm **E**	ES 1-way Statically determinate	50 kg/m² HE +1/60	EN ISO	3.5 N/mm² flexion stress	3 hours fire resistance	13
Poprad, Slovakia 2019 Fire test wall	180 mm **w**	Wall of 3 m height Statically determinate	50 kg/m² HE +1/60	EN ISO	330 kN/m	> 4 hours	14

* denotes when full-scale slabs are provided with a set of APC rebars according to the ACI 544-6R 15 report and the Canadian Standard Reference (4)

We must summarize some very important observations about each test of the list, regardless of the span length and the dosage rate, or the type of steel fibers:

- All tests are organized by a third-party organization such as Technical University or Research Institute.
- All tested slabs have been loaded with the most onerous SLS loading case and beyond, up to ULS and further beyond, up to collapse. When it was impossible to load the slab up to ULS or collapse, a load smaller than the ULS load was applied over several days or weeks or months to only observe stable and limited deformations.
- Each slab was able to sustain loads up to the ULS loading and beyond, up to a multiple of the SLS loading intensity.
- Each slab showed beyond the SLS loadings, a multiple cracking process in the bottom, in the top, and in all directions, in a manner typical of a very high ductility, which is the key to building safe structures. Al test slabs exhibited a deflection hardening behavior.
- In all tested slabs, the standard crack opening limit of 0.3 mm, typical of traditional reinforced concrete of the Eurocode, was observed under load intensities beyond the SLS loading intensity that was applied.
- In all tested slabs, no sign of punching-out was observed even under extreme point loading intensity.

A normal man of art, an engineer having attended these testings would conclude that the steel fiber reinforced concrete (SFRC) is fit to be used as a reinforced concrete for suspended slabs and suspended elevated slabs. SFRC Foundations slabs are accepted, as being typical of the structural ductility that a simply supported, statically determinate beam doesn't enjoy.

All competent engineers understand that safe and reliable concrete slab structure depends on the combination of the ductility of its constituent materials and its structural ductility.

A structure is considered ductile when there are multiple possible load paths in the event of a local failure, when moments can redistribute, and when a local failure does not lead to a global failure.

The first crack of a slab doesn't lead to final collapse beyond the ULS stage.

3.4 QUASI-FULL-SCALE TESTS: ROUND INDETERMINATE SLAB TESTS

I believe that to efficiently design SFRC slabs, we can't overlook the observations and analysis of real full-scale slabs tested at SLS, ULS, and up to the final collapse.

I had quite often an objection to the cost of the organization of full-scale slab tests. Hence, a quasi-full-scale test has been defined, where both the material and structural ductilities can be observed, at a smaller scale, thus at reduced experimental cost and time.

Testing round indeterminate slabs is a quasi-full-scale test as outlined in the ACI 544-6R15 report on the design and construction of SFRC elevated slabs.

Its size should be such that the fabrication of the slab sample shall not impair the nature of the fiber concrete by:

- A size that is too small for the form resulting in erratic fiber distribution, as shown in the standard beam in flexion, where the fiber concrete is refrained from free-flowing in the form.
- The need for mechanical vibration to compact the fiber-reinforced concrete.
- A slab section that is too thin not representative of real-world applications.

Figure 3.4 shows a possible set-up of the round indeterminate panel test. In general, we kept mostly a span-to-depth ratio of 10. A slab of 200 mm thick should be no less than 2 m net diameter.

The test rig is shown in Figure 3.5.

Such an ideal test slab should be round, without corners, with a span of ten times its thickness, supported along its round perimeter, as shown in Figure 3.4.

We should not be confused by the ASTM C 1550 shown in Reference (5) round slab test in flexion, as it is a three-point supported round slab, statically determinate, with an 800 mm small diameter and of an unrealistic 80 mm thickness, in which only three radial cracks can develop at mid-angle between the supports.

It downplays the multiple cracking behavior of the SFRC observed in full-scale or quasi-full-scale slab test. With three as well, as up to eight, support points under the round slab, the cracks develop at a predetermined location in a lesser number. Under a continuous perimeter support, as shown in Figure 3.8, a high number of micro-cracks (in red) between the main yield lines.

3.4.1 Manufacturing of round slabs

- External dimensions are given as: diameter = 1650 mm; thickness = 150 mm; clear span = 1500 mm.
- The form bottom consists of smooth marine plywood installed on the ground at constant level so that the thickness of the hardened slab doesn't vary.

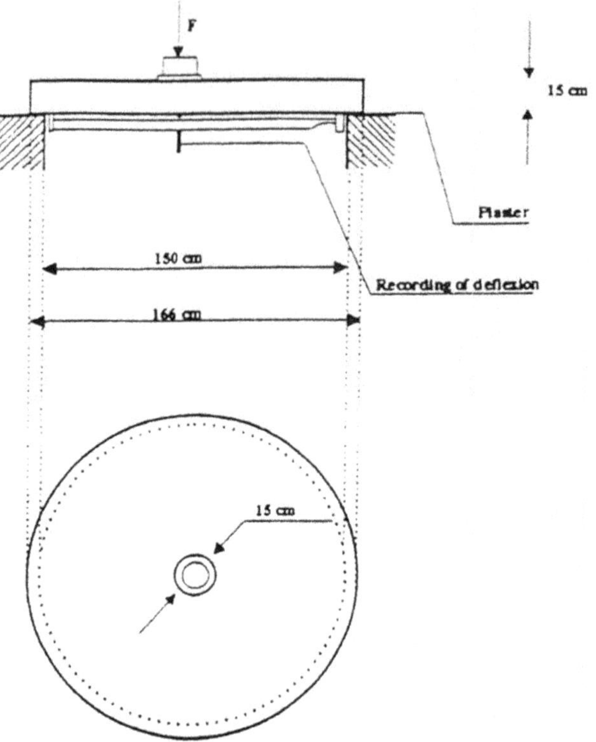

Figure 3.4 **Typical round indeterminate slab test set up.**

- Mix design: the final mix shall be a F5 flow concrete, dumped from the mixer into the form. It flows from the center of slab to the edges to ensure full contact with the edge form. A vibrating needle shall be used externally and briefly on the lateral form of the round slab.
- Three stirrups of 8 mm diameter are plunged into the fresh concrete along the edge each at one-third of the perimeter, in order to lift up the hardened slab.
- The surface is straight-edged and bull floated, and finally cured by covering it with a wet burlap or a polythene sheet for 28 days at 20°C.

3.4.2 Round indeterminate slab test set up

- The slab is simply supported along the circular rim of a concrete pipe of 1500 mm internal diameter. To ensure flexion without torsion, a bed of fresh plaster is laid on top of the pipe rim before placing the round slab.

Figure 3.5 Test rig at the University of Sheffield.

- The load is applied with a 250 kN actuator working under displacement control through a closed loop. The applied load is measured by a calibrated load cell as shown in figure 3.5.
- The rate of displacement of the stroke is 1 mm/min.
- The system of actuator and load cell is hinged at the top and bottom.
- Beneath the lower hinge, the load is applied on the slab through a solid steel cylinder of 150 mm diameter, resting in a bed of fresh plaster.
- Several LVDT transducers are used to deduce the net deflection of the slab. All measurements of the upper face of the plate are referred to the floor of the laboratory.
- The vertical displacement of the 150 mm diameter steel cylinder is the average of three 50 mm stroke transducers radially disposed at 120° apart.
- The settlement of the slab on its circular support is measured by three transducers of 5mm stroke situated 120° apart along a circular rim with 1500 mm diameter.
- The deflection recorded is the difference between the vertical displacement of the steel cylinder and the average settlement of the outer rim.

A finite element analysis, as shown in Figures 3.6 and 3.7, shows that at 5 N/mm² flexion stress (at the limit of proportionality typical of a C30-37 concrete with a dosage rate of 45 kg/m³ HE+1/60 steel fibers) the slab should carry 61.5 kN with a deflection of 0.28 mm under the center point loading.

A deflection of 0.28 mm calculated by the FEM software at 62 kN center point loading imposed on the round indeterminate slab is completely verified experimentally, as shown in Figure 3.8.

A Timoshenko formula calculation leads to a similar loading intensity at limit of proportionality:

$$f_L := 5.00 \cdot \frac{N}{mm^2}$$

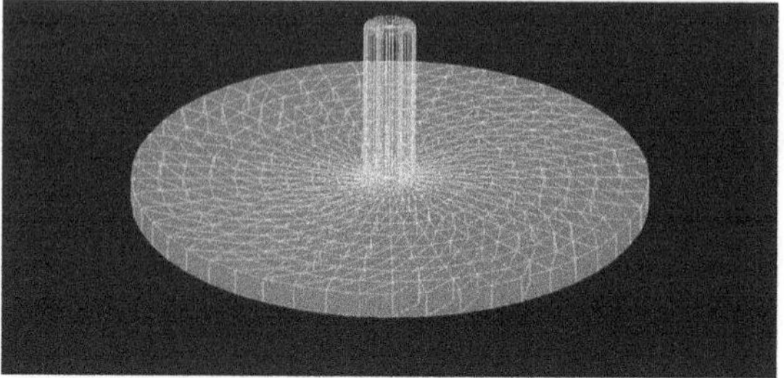

Figure 3.6 Thick plane finite elements and center point loading.

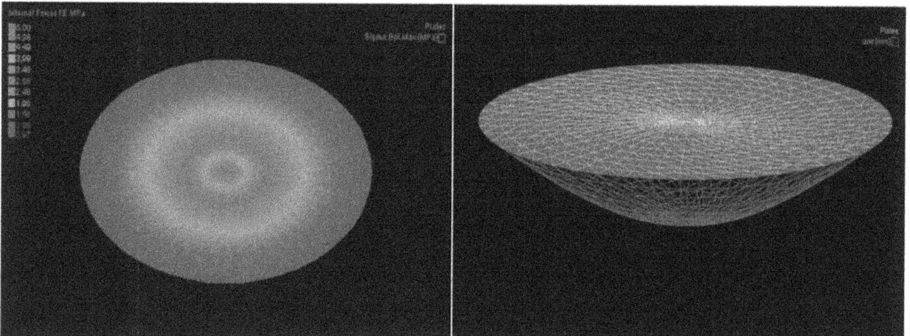

Figure 3.7 At 5 N/mm² flexion stress, the slab carries a center-point loading
intensity of 61.5 kN with a deflection of 0.28 mm.

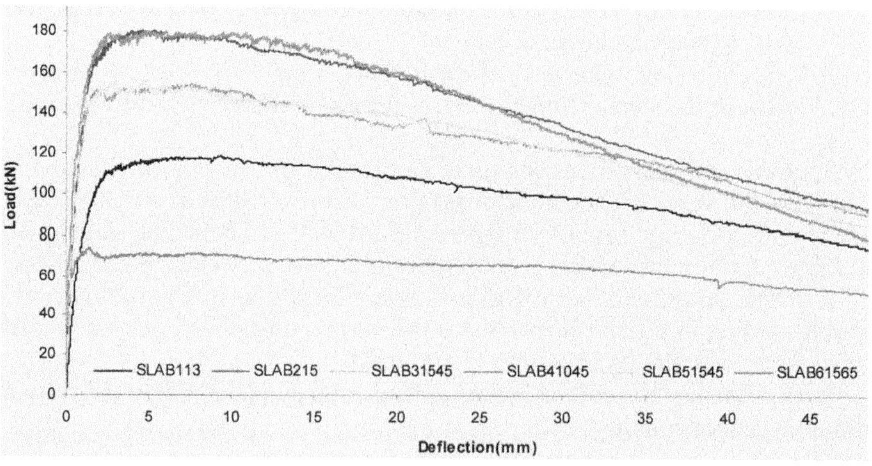

Figure 3.8 Load deflection diagrams of the round indeterminate panel test.

$$a := 1500 \cdot mm$$

$$h := 150 \cdot mm$$

$$\nu := 0.20$$

$$Q := f_L \cdot h^2 \cdot \left[(1+\nu) \cdot \left(0.485 \cdot \log\left(\frac{a}{h}\right) + 0.52 \right) + 0.48 \right]^{-1}$$

$$Q = 66.726 kN$$

where a is the diameter of the plate, h is the thickness, ν is the Poisson coefficient of concrete, and f_L is the limit of proportionality flexion stress.

Q is, as calculated at 5 N/mm², the loading intensity reaches the limit of proportionality and causes the first very small crack of the plate under the center point loading.

3.4.3 Experimental findings

A testing program, detailed in Table 3.2, of round indeterminate slabs was carried out at the University of Sheffield, UK (DTI-KTP report) on slabs made of C30-37 concrete with a 1500 mm clear span and thickness of 100 mm, 130 mm, and 150 mm. The slabs were reinforced with dosage rates of 45 and 65 kg/m³ of ArcelorMittal HE+1/60 steel fibers, featuring hooked ends, a 60 mm length, 1 mm diameter, and 1500 N/mm² wire tensile strength. These fibers have been systematically used for suspended slabs, suspended elevated slabs, and foundation rafts since 1997.

A 45 kg/m³ dosage rate of HE+1/60 steel fibers is frequently used to reinforce piled slab (ground-suspended slabs), and raft foundations, while 50 kg/m³ dosage rate is the minimum to reinforce suspended elevated slabs.

The load vs. deflection diagrams are shown in Figure 3.8.

The slabs identification is: Type of Slab X number; XX thickness in cm; XX dosage rate in kg/m³.

Table 3.3 shows a summary of the results. The limit of proportionality is observed in Figure 3.9 by scaling up the abscissa.

Table 3.2 Testing program

Number of slabs type	Thickness	Dosage rate (kg/m³)
2	150 mm	45
2	150 mm	65
1	130 mm	65
1	100 mm	45

Table 3.3 A summary of the results of the testing program

Type of slab	Limit of proportionality	Maximum loading intensity
150 mm - 65 kg/m³	62,5 kN	180,15 kN
150 mm - 45 kg/m³	62,5 kN	155,10 kN
130 mm - 65 kg/m³	47,5 kN	118,60 kN
100 mm - 45 kg/m³	28,1 kN	79,70 kN

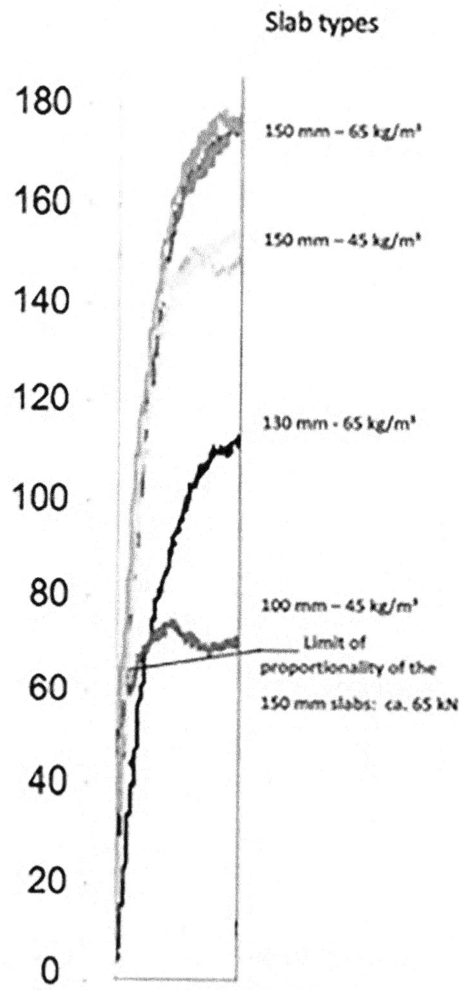

Figure 3.9 Scaling up of abscissa to observe the limit of proportionality (kN vs. deflection).

We observe that the limit of proportionality loading intensity doesn't increase when the steel fiber dosage rate increases from 45 kg/m³ to 65 kg/m³.

Assume that the slab doesn't reach its flexion strength in service, that is, that the most onerous point loading intensity in service is 62.5 kN under a 5N/mm² flexion stress. Then, the safety to rupture is 155.1/62.5 = 2.48, which is an acceptable ratio under static loading for a reinforced concrete slab.

We can say that a manhole cover of the same size and concrete is able to carry a 62 kN static load at SLS, and the ULS load is 62 kN x 1.5 = 93 kN and that the collapse load is 155 kN. This is a very safe design, assisted by a testing route, where the cover shouldn't crack under a 62 kN center-point loading intensity. It is designed to be flexion and punching-out crack-free under static SLS conditions.

Typical slabs after the test are shown in Figure 3.10.

At 45 kg/m³ dosage rate, 19 small cracks are visible in red, together with 6 radial yield lines while at 65 kg/m² dosage rate, 22 small cracks and 7 radial yield lines are visible.

At 45 kg/m³ of HE+1/60 steel fiber type (Hooked ends of 1 mm, diameter of 60 mm, and a wire strength of 1500 MPa), the plastic tensile strength is calculated as f_{tu} = 2.39 N/mm²

At dosage rate of 65 kg/m³ and plastic tensile strength of f_{tu} = 2.54 N/mm², the fiber dosage rate increase by 40% which only increase the plastic tensile strength by 6.5%.

The mix design at the Sheffield University Laboratory was the same, and therefore very likely not the optimal one for the higher dosage rate of 65 kg/m³.

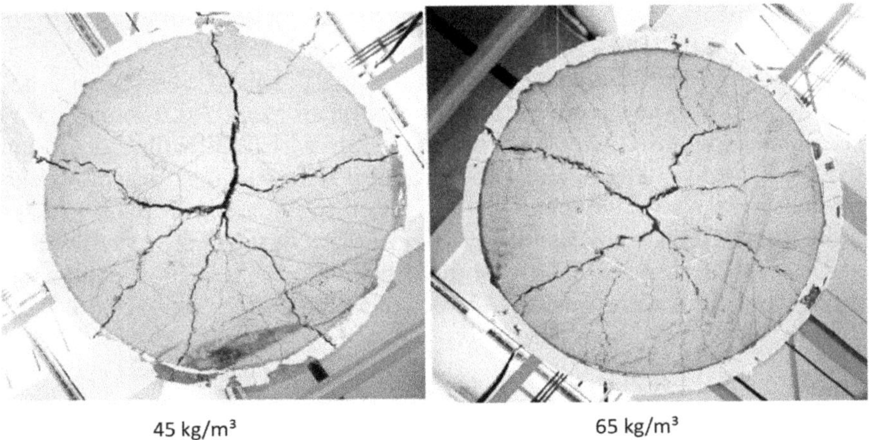

Figure 3.10 Cracking and yield line pattern of the round inderminated pannel test with 45 kg/m³ (LEFT) and 65 kg/m³ (RIGHT) dosage rate.

Figure 3.11 Flexion cracking of 4 round indeterminate pannels in the side face.

Note that the average spacing between two adjacent fibers is of 18 mm in case of the 45 kg/m³ dosage rate and 15 mm in case of 65 kg/m³.

It is very likely that when using a smaller aggregate size, the f_{tu} of 65 kg/m³ could have been about 7% higher to attain 2.72 N/mm², which is 14% higher than the 45 kg/m³ dosage.

The f_{tu} increase in function of the dosage rate of the same fiber type is smaller than the linear increase with the dosage rate. There is a kind of fiber saturation.

The reduction of the aggregate size would also require a larger mortar content in the mix, and would cause an increase in shrinkage, a negative trend in real applications.

Very similar results and observations have been made on more round indeterminate slabs with the same span–to-depth ratios at IUT-Bethunes-France, ULB-TU Brussels, Polytechnic of Montréal, and TU Bilbao, Spain.

The conclusion drawn by TU Sheffield is very clear: "*Statistical analysis was carried out on the post crack tensile strength. The results show that the standard error and the confidence level (95%) of the tensile strength are 0.079 and 0.202 which fall into the highly reliable range.*"

Figure 3.11 shows the round indeterminate panels (1.50 m span x 150 mm depth) after the flexion tests at Polytechnic of Montréal. The undulating steel fibers were used at 100 kg/m³ dosage rate with TABIX fiber of 1.3 mm diameter, 50 mm length, and 850 MPa steel wire tensile strength. A dense microcracking is visible on all slabs, with up to 36 cracks at the very edge. Note that the cracks are limited to at most 90% of the depth after testing.

The plastic moment along the yield lines is obtained by the Johansen model for round plates subjected to a typical fan pattern of cracks:

In the 65 kg/m³ dosage rate case:

$$P_{max} = 180.15 kN$$
$$c = 150 mm$$

$$M_p = \frac{\left(1 - \dfrac{2 \cdot c}{3 \cdot R}\right)}{2 \cdot \pi} \cdot P_{max} = 26.76 kN \cdot \frac{m}{m}$$

R is the radius (= span/2) of the round plate and c, radius of the imposed load

As observed during the testings, the post-cracking tensile strength calculations assume, that 90% of the section thickness is cracked while the 10% remains in compression. Hence, we can write Reference (8):

$$h = 150 \text{mm}$$

$$f_{tu} = \frac{M_p}{0.45 \cdot h^2} = 2.643 \frac{N}{mm^2}$$

In the case of 45 kg/m³ dosage rate, slab thickness of 150 mm, and P_{ult} = 155.1 kN,

M_p = 23.039 kNm/m with f_{tu} = 2.28 N/mm²

where f_{tu} is the plastic tensile strength of the fiber-reinforced concrete of the example. It has been derived from a statically indeterminate slab model that you can't compare to the notched beam EN 14651 specimen in flexion.

Let's assume that the round indeterminate test slab of 150 mm thickness and 1500 mm span is a cover of a round pit in a factory. What static loading intensity can it carry in service conditions?

With 65 kg/m³ dosage rate, P_{Rd} = 180.15 kN/1.5 = 120 kN. With a 1.5 variable loading factor and a 0.85 longtime factor, the maximum allowable SLS load is 0.85 x 120 kN /1.5 = 68 kN

In the case of 45 kg/m³ dosage rate, P_{Rd} = 155.1 kN/1.5 = 103.4 kN, and the SLS loading limit is 0.85 x 103.4/1.5 = 59 kN /1.10 = 54 kN.

Under such a loading intensity, the pit cover has not yet reached its limit of proportionality in flexion, and is therefore not going to crack under the most onerous service loading.

It doesn't, however, exclude cracks for other reasons (such as poor curing conditions).

Table 3.4 shows that at a 45 kg/m³ HE+1/60 steel fibers dosage rate, the design resisting moment (under factored loads) M_{Rd} is a function of h.

Table 3.5 shows the same at 65 kg/m³HE+1/60 steel fibers dosage rate.

We can conclude that the round indeterminate panel test is a robust method with little scatter to determine an SFRC slab's yield moment and define the upper limit of the service domain.

At the University of Pretoria in South Africa, more typical round indeterminate panel tests (100 mm thick × 700 mm span) were tested under the 100 mm diameter center-point loading. The disk had an 800 mm external size, and the SFRC contained 40 kg/m³ of HE 75/50 steel fibers (Hooked ends, 0.75 mm diameter × 50 mm length).

The back calculations here below are derived from six slabs:

$$P_{ult} := 79.6 \ kN$$

$$h := 100 \ mm$$

$$a := 50 \ mm$$

$$R := 350 \ mm$$

Table 3.4 Design resisting moment of a C30 mix with 45 kg/m³ dosage rate of HE+1/60 (simple hook ends, 1 mm diameter by 60 mm length) of 1500 MPa tensile strength

thickness (mm)	130	150	170	200	220	240	250	270	300
M_{Rd} (kNm/m)	11.58	15.42	19.88	27.41	33.18	38.89	46.42	54.14	66,84

Table 3.5 Design resisting moment of a C30 mix with 65 kg/m³ dosage rate of HE+1/60 (simple hook ends, 1 mm diameter by 60 mm length) of 1500 MPa tensile strength

Thickness (mm)	130	150	170	200	220	240	250	270	300
M_{Rd} (kNm/m)	13,44	17,91	23,1	31,82	38,51	45,82	49.72	60,00	71,60

$$M_L := \frac{P_{ult}}{2\,\pi} \cdot \left(1 - \frac{2 \cdot a}{3 \cdot R}\right) = 11.462 \ kN$$

$$f_{tu} := \frac{M_L}{0.45 \cdot h^2} = 2.547 \ MPa$$

The resulting flexion diagrams are shown in Figure 3.12.

The complete theory behind the round indeterminate panel test has been developed by Mobasher in Reference (9), in (chapter 16, pp. 295–214).

PRIMEKSS ROUND PANEL TESTING

Project:	Trial run 1
Testing facility:	Engineering 4, Concrete laboratory
	University of Pretoria
Casting date:	30 January 2024
Testing date:	29 April 2024
Age:	90 days
Fibre content:	40 kg/m³
Panel size:	800 mm x 100 mm thick
Testing machine:	500 kN MTS

- *Primekss takes meticulous steps to ensure quality throughout the entire process. This includes sourcing high-quality raw materials, optimizing the local mix in Primekss laboratories, approving a mix design, and conducting thorough quality control onsite. Onsite, there is careful integration of steel fiber and additives, with constant testing of fresh concrete to maintain quality standards.*

FRESH PROPERTIES OF CONCRETE IN TESTED PANELS

	Batch 3	Batch 4
	31/01/2023	
Air Content (%)	1.8	1.9
Density (kg/m³)	2512	2520

HARDENED PROPERTIES OF CONCRETE IN TESTED PANELS

90-day cube strengths

Batch	Weight Air (g)	Weight Water (g)	Density (kg/m3)	Strength (MPa)
	8485	5099	2505.9	65.5
3	8540	5186	2546.2	65.3
	8565	5204	2548.3	66.3
	8530	5154	2526.7	65.8
4	8550	5200	2552.2	68.0
	8740	5342	2572.1	56.1
	Average:		2541.9	64.5

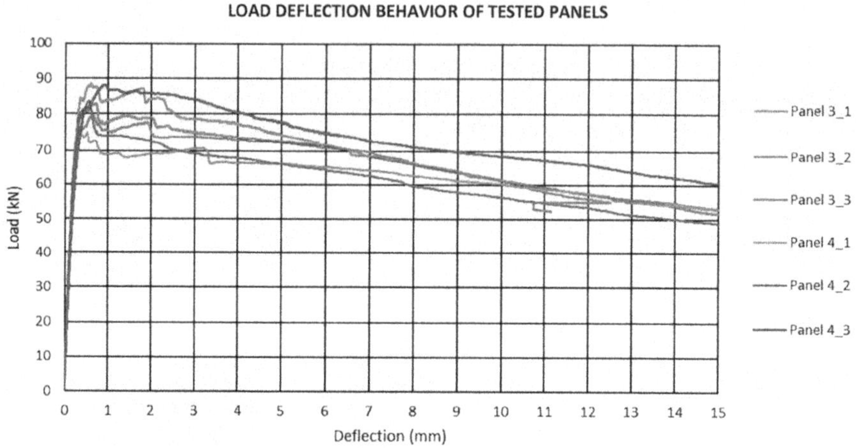

Figure 3.12 Diagrams of 700 mm span x 100 mm thickness of 40 kg/m³ HE 75/50 steel fiber flexion.

3.5 LESSONS LEARNED FROM THE FULL-SCALE TESTS

3.5.1 The NATO slab on grade test, Sanem, Luxembourg 1983

It is the first in our list as it is the first of my own experience and the first fiber-reinforced concrete slab-on-grade full-scale test.

The NATO defense organization, at the time of the Cold War, had a program of building weaponry warehouses across Belgium, Holland, and Luxembourg. The first project consisted of 100.000 m² spread across 20 warehouses of 5000 m² each.

The original specification was to build a 200 mm slab with one layer of 8 mm x 8 mm wire mesh, spaced 150 mm distance apart in a C 25 concrete with shrinkage joints sawn every 5 m apart.

The base was the natural clay soil, lime-treated to dry it out and prevent expansion, topped by 500 mm compacted granulated slag of 0 to 50 mm aggregate grading, supplied by the Arbed steel plant almost next to the site. The mix design was also required to be "all slag", meeting no natural sand or natural aggregates, with an all-slag cement.

In order to be "attractive," I proposed to the SILIDUR slab contractor a solution of 130 mm thickness slab reinforced by a 20 kg/m³ dosage rate of the patented EUROSTEEL 1/60 undulated steel wire fiber (1 mm diameter × 60 mm length). This fiber features an 8 mm pitch of undulation, 1 mm depth at the central axis, and a 1200 N/mm² steel wire strength. It was manufactured under license by the ARBED steel wire plant.

It proved very difficult to make it acceptable by the administrations overseeing the project, but I remember the top NATO generals were quite attracted by the huge savings generated by such a solution: savings of 7.000 m³ of concrete and 500 tons of wire mesh replaced by 250 tons of steel fibers.

Under the pressure of the top NATO Generals and with the support of office for **Security** and **Control** of projects (SECO), the technical controller from Belgium, a full-scale test process was organized to check the safety factor against the most onerous point loading: a truck wheel of 50 kN inflated at 0.625 N/mm², creating a footprint of 200 mm x 400 mm size.

During the loading test, a single crack, even a minute one, was not acceptable under twice the most onerous load, that is, 100 kN imposed anywhere on the slab—at the center, the edge, or a corner.

A trial ground area was then prepared exactly according to the specification, ensuring that the k-Westergaard value was a minimum of 0.1N/mm³, quite a high value, so that no rutting was observed under concrete truck mixer wheels.

On top of it, a 7 m x 7 m size slab of 130 mm in C25 concrete of 0–25 mm slag aggregates grading, with 20 kg/m³ dosage rate of EUROSTEEL fibers, was built in reality, including complete finishing and curing.

The 320 kg/m³ of cement was also of slag-type cement consisting of 70% of slag and 30% of OPC. The water demand of the granulated slag is very high so 210 kg water was needed together with an addition of superplasticizer to reach a 150 mm Slump.

The test slab was instrumented to measure the deflections both upward and downward, as well as the invisible cracking at the bottom face, thanks to ultrasonic electrodes to measure the travel time of the sound wave, which increases noticeably after the crack formation.

The Westergaard-type calculations was performed, as described in Technical Report 550, and based on the k-value = 0.1 N/mm³, where the base is considered as a dense liquid, including the hypothesis that pressure p= k x w, with w being the elastic settlement, and that the base, as a liquid, is devoid of any shear strength.

The calculation results obtained were very pessimistic, as the test slab was expected to pass the center point loading but fail in the corner and at the edge, as shown by the calculations and Table 3.6.

a) **Load case "Internal load"—Westergaard's equation**

$$\sigma_{Q^M} = \frac{0.27 \cdot Q}{h^2} \cdot (1 + \mu) \cdot \left[4 \cdot \lg\left(\frac{l}{b}\right) + 1.069 \right]$$

Table 3.6 Flexion stresses under unfactored wheel loading intensity

Load type		Fork-lift	Truck	Others 1	Others 2
Concentrated load Q	(N)	0	50000	0	0
Contact pressure p	(N/mm²)		0.63	0.00	0.00
Subgrade modulus k	(N/mm³)	0.094	0.094	0.094	0.094
Loading radius a	(mm)	0	290	0	0
Equivalent radius b	(mm)		290	42	42
Internal stress σ_i	(N/mm²)	0.00	1.79	0.00	0.00
Edge stress σ_e	(N/mm²)	0.00	3.22	0.00	0.00
Corner stress σ_c	(N/mm²)	0.00	3.25	0.00	0.00

b) **Load case "Edge load"—Kelley's equation**

$$\sigma_{Q^R} = \frac{0.519 \cdot Q}{h^2} \cdot (1 + 0.54 \cdot \mu) \cdot \left[4 \cdot \lg\left(\frac{l}{b}\right) + \lg\left(\frac{b}{25.4}\right) \right]$$

c) **Load case "Corner load"—Pickett's equation**

$$\sigma_{Q^E} = \frac{0.412 \cdot Q}{h^2} \cdot \left[1 - \frac{\sqrt{\frac{a}{l}}}{0.925 + 0.22 \cdot \frac{a}{l}} \right]$$

where

$$a = \sqrt{\frac{Q}{\pi \cdot p}} = \text{radius of are of loading}$$

$$b = \sqrt{1.6 \cdot a^2 + h^2} - 0.675 \cdot h \quad \text{for } a < 1.724h \qquad \text{or}$$
$$b = a \qquad \text{for } a > 1.724h$$

The internal stress doesn't exceed the limit, as 2 x 1.79 = 3.58 N/mm² < 4.5 N/mm², but the edge and the corner stress should not pass 2 x 3.22 = 6.24 N/mm² and 2 x 3.25 = 6.50 N/mm², both values are greater than 4.5 N/mm², which is the limit of proportionality at the formation of the first crack.

3.5.1.1 Experimental observations

In Figure 3.13, we can see the loading vs. deflection diagram of the NATO test slab at the center-point loading of the 7.70 m x 7.70 m slab.

The travel time of the sound waves jumps up at 320 kN loading intensity and 1.3 mm deflection, visible on the diagram in negative abscissa on the Y-axis, confirming the first bottom crack.

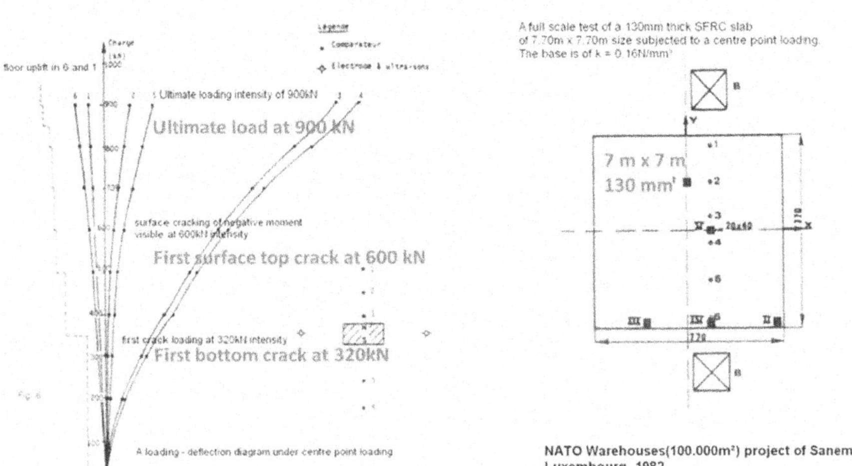

Figure 3.13 NATO test slab (Sanem, G-D. Luxembourg, 1983) load vs. deflection diagram.

According to the Westergaard method, the first bottom crack should have happened at 50 kN x 4.5/1.79 = 126 kN, showing a 320 kN/126 kN = 2.54 factor of underestimation by the Westergaard theory.

The loading intensity could be increased much further to 650 kN, at which point two circular negative moments cracks developed at 1 m and 1.50 m distance around the point loading. Upon unloading, these two cracks completely disappeared as they closed.

The slab could be reloaded up to 900 kN, at which point major radial cracks developed, along with the reopening of the 650 kN cracks. At 950 kN, the slab collapsed into six radial, significantly open cracks as shown in Table 3.7.

Considering the traditional 1.7 safety factor for slabs on grade, the Westergaard theory predicts that collapse should have happened at 1.7 x 126 kN = 214 kN. This results in an underestimation factor of 950 kN/214 kN = 4.44!

Table 3.7 Summary of all point loadings applied

	Experimental		Westergaard theory	
	First crack	Ultimate	First crack	Ultimate
Center Point Loading	320 kN	950 kN	126 kN	224 kN
Edge Point Loading	250 kN	300 kN	69 kN	127 kN
Corner Point loading	200 kN	275 kN	69 kN	127 kN

In the corner case, the first crack was observed at a 200 kN load, while the slab collapsed at 275 kN. At collapse, the slab began to lift up at a 2 m distance around the point of loading.

For the edge case, at 200 kN, the slab showed a 2.38 mm deflection, and minute cracking started to become visible at 250 kN loading intensity to form an open circular crack around the point loading at 300 kN intensity, and this crack remained open after unloading the slab.

Table 3.8 provides a comparison of the "Westergaard based theory" calculated values with the experimental full-scale test values.

The underestimation of slab resistance by the Westergaard theory ranges between a factor of 3 and 5, depending on the loading case.

Table 3.6 provides a summary of the Point loadings.

We can now also compare this to the recent Technical Report 34 (TR 34), 4th edition calculated ultimate loading: P_{ULT} = 232 kN; underestimation factor by the TR34 4th edition is = 950/232 = 4,1!

$h := 130 \, mm$ — Slab thickness

$E := 25000 \, \dfrac{N}{mm^2}$ — Young's modulus

$f_{ck} := 25$ — C 25

$f_{r4} := 1.6 \, \dfrac{N}{mm^2} \qquad f_{rl} := 2.4 \, \dfrac{N}{mm^2}$ — EN 14651 20 kg/m³ Eur 1/60

$a := 160 \, mm$ — Radius of contact of load of 200 mm × 400 mm contact area

$k := 0.1 \, \dfrac{N}{mm^3} \qquad \upsilon := 0.2$ — Westergaard coefficient / Poisson coefficient of concrete

$L := \left[\dfrac{E \cdot h^3}{12 \cdot (1 - \upsilon) \cdot k} \right]^{0.25} = 0.489 \, m$ — Radius of rigidity of the slab

$\dfrac{a}{L} = 0.327 \qquad u := 1000 \, mm \qquad \gamma_m := 1.5$ — Unit of length, material factor

$M_{up} := \dfrac{h^2}{\gamma_m} \cdot (0.29 \cdot f_{r4} + 0.16 \cdot f_{rl}) = 9.554 \, kN \cdot \dfrac{m}{m}$ — Residual positive resisting moment

$f_{cm} := 2.5 \, \dfrac{N}{mm^2}$ — Mean uniaxial tensile strength

$f_{ctdfl} := f_{ctm} \cdot \dfrac{\left(1.6 - \dfrac{h}{u}\right)}{\gamma_m} = 2.45 \, \dfrac{N}{mm^2}$ — Design compressive flexion strength

$$M_{un} := \frac{f_{ctdfl} \cdot h^2}{6} = 6.901 \; kN \cdot \frac{m}{m}$$

Residual negative resisting moment

$$P_u := 4 \cdot \frac{\pi \cdot \left(M_{up} + M_{un}\right)}{\left[1 - \left(\dfrac{a}{3 \cdot L}\right)\right]} = 232.088 \; kN$$

Ultimate Loading, Technical Report 34 (TR 34) 4th edition

Despite its more complicated and cumbersome calculations, the TR 34 4th edition doesn't improve the picture, although it was published many years after the Technical Report 550 cited in Reference (6). This is even more disappointing that the TR34 methodology of calculations is supposed to take into account the slab ductility.

The TR 34 4th edition Reference (7) underestimates the ultimate load of the full-scale NATO slab point loading by a factor of 4. (indeed 900 kN/232 kN = 3.88)

Why so?

This is because similar to the TR 34, most slabs on grade design guides use the k-Westergaard theory to represent the soil underneath.

The Westergaard theory solves the plate's differential Laplace equations of the 4th order, with a simplifying second member hypothesis that the vertical displacement is proportional to the pressure so that p = k.w, where k is the Westergaard coefficient.

Hence, the ground support is modeled as a dense elastic liquid, represented by the k-value, which assumes no shear strength or shear stresses in the slab and the base. This results into a circular flexion with identical vertical displacements on top and bottom of slab. In addition, the theory assumes no friction at the base/slab interface, which significantly reduces the stresses transferred to the base. Consequently, the system is modeled as much weaker than it is in reality.

The effect of the Westergaard theory is to cancel the structural ductility attributable to the base and to the ductile fiber reinforced concrete slab. The Westergaard theory is indeed entirely elastic.

Is the slab-on-grade elastic?

How long does a slab remain elastic? Likely only for a few days at most after the installation.

Keep in mind that most slabs today are still designed by the Westergaard theory, even when finite elements software is used that includes the p = k x w hypothesis condition. Should we say then that all slabs on ground are overdesigned by a factor of 3 to 4 so that the thickness is 50% to 100% thicker than what it should be?

Not really as despite that high calculated safety factor value, we know that there are many litigations in the concrete slab business due mostly to

Table 3.8 Test results of the full-scale test cases 1 and 2 in Ternat and Townsville

Year/ Location	fields nr./ columns nr.	Span span/depth = 19	Thickness of C30-HE+1/60 45–50 kg/m³	Column footprint	First crack load/ deflection id and corner spans	ULS load/deflection
1994 **Ternat Belgium**	**9/16**	**3100 mm**	**160 mm**	**210 mm x 210 mm**	**mid span: 110 kN/ 2 mm** **corner span: 80 kN/ 2 mm**	**mid span: 450 kN/19 mm** **corner span: 180 kN/ > 20 mm**
2000 **Townsville Australia**	**9/16**	**3100 mm**	**160 mm**	**210 mm x 210 mm**	**mid span: 110 kN/ 2 mm** **corner span: 80 kN/ 2 mm**	**mid span: 400 kN/19 mm** **corner span: 180 kN/ >20 mm**

cracks and joints. This is true even with the existence of numerous standards, design guides and recommended practice in many countries about slabs-on-grade design and construction.

The culprit is the overall shrinkage caused by both hygral and thermal variations. The detrimental effect of shrinkage is the cause of cracking, opening, curling, and rocking of joints. Its origins lie in the nature of the standard-abiding constituent materials of the concrete, as well as in the design of the slab, including joint types and numbers, mix specifications, installation processes, and the concrete supply chain—from the plant, through transit, to final completion and curing process.

If all shrinkage stresses were eliminated, significant strength reserves would become available to resist flexion moments, shear, and punching-out, as calculated following the codes. This could allow up to half the traditional thickness to be eliminated, resulting in a longer-lasting, more durable slab.

Hence, a much better environmental footprint is achieved, together with significant cost savings, when the volume of concrete used is drastically reduced. The reduction of shrinkage to the point that it becomes no longer detrimental to the completed slab should be a major goal in any slab design methods or codes. Sadly, all existing slab codes, such as TR 34, 4th edition, are designed merely to ensure that past practices continue, without promoting any potential concept revolution.

The advantages of these codes are essential to exonerate the designer of any responsibility for the final result and the concrete contractor of any design responsibility. Consequently, there are multiple sources of responsibilities from material constituents and materials supplied, to standard design, and installation. This results in suboptimal or poor quality slabs and frequent, long, and costly litigations when the level of quality and serviceability becomes questionable to the end user.

In our opinion, all slab design codes should promote the concept of a single source of responsibility. This approach is only possible when a full design-built process is applicable. It allows for more design freedom with non-standard abiding materials and design methods with the advantage of advanced solutions being more economical, and more reliable with a smaller environmental footprint.

Such processes are commonplace in most industries but are mostly ignored in the construction businesses, where the responsibilities are mainly distributed among owners, the material suppliers, general contractors, and design firms.

In Chapter 9, you'll read about Primekss fiber concrete, which is chemically self-stressed so that the overall shrinkage becomes no longer detrimental. It results in a possible significant reduction of thickness for both slabs-on-grade and piled slabs.

Today, the most decisive inventions in the construction industry are developed by material suppliers whose responsibility often ends when the new materials arrive at the job site.

The overall fragmentation of responsibilities in the construction industry, particularly in countries such as Germany and France, significantly hinders progress.

We must observe as well that, in Europe, at highest level, most building design standards are the product of the academic world; however, their vast knowledge practically serves more as a limitation than an incentive to explore beyond the comfort zone.

We would have suffered similar disadvantages in other industries, for example stage coaches were still in use today instead of cars, trucks, planes, and trains; candles instead of LEDs; and witch doctors instead of modern healthcare.

REFERENCES

Design of floors on Ground, Technical Report 550, J.W.E. Chandler Cement and Concrete Association, 1988.

Concrete Industrial Ground floors, Technical Report 34, 4th Edition, The Concrete Society, 2013.

3.5.2 The Ternat (Belgium, 1994) and Townsville (Australia) Tests (E-SFRC)

Both full-scale tests at Ternat (1994) and Townsville (1999), as shown in Figures 3.15 and 3.16 are identical in size, materials used, and loading procedures. Both slabs simulated the typical piled slab application with respect to span-to-depth ratio, thicknesses, and service loading intensity for commercial/industrial business purposes.

The slabs were square, comprising three consecutive slab spans of 3.10 m, (ca. 9 m² each), with a 160 mm thickness, as shown in Figure 3.15. The slabs were made out of C30-37 concrete with 45 kg/m³ dosage rate of 1 mm diameter by 54 mm length steel fibers provided with conical heads, as shown in Figure 3.14, Twincone fibers. the constituent steel wire is of 1200 N/mm² tensile strength. (T in the diagrams is for metric tons of load 40 T= 400 kN)

Both slabs were tested under center point loading in mid-span and corner span. Table 3.8 is a summary of the observed results.

The loading vs. deflection diagram in the center span and the corners span is shown in Figure 3.16.

The collapse cracking patterns are shown in Figure 3.17a (positive moments) and Figure 3.17b. (positive and negative moments depicted by cracking and yield lines, respectively)

We also have to observe here that a 450 kN point load can lead to the final collapse of the slab, without causing any punching-out or hairline cracks in the top surface!

The round indeterminate panel test revealed a 23 kNm/m ultimate yield moment in case of 150 mm thickness with 45 kg/m³ dosage rate., In case of 160 mm thickness, the yield moment becomes: 23 kNm/m x

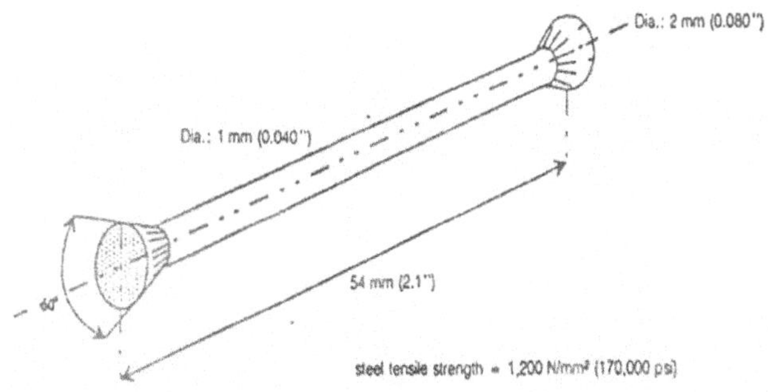

Figure 3.14 The total anchorage Twincone fiber.

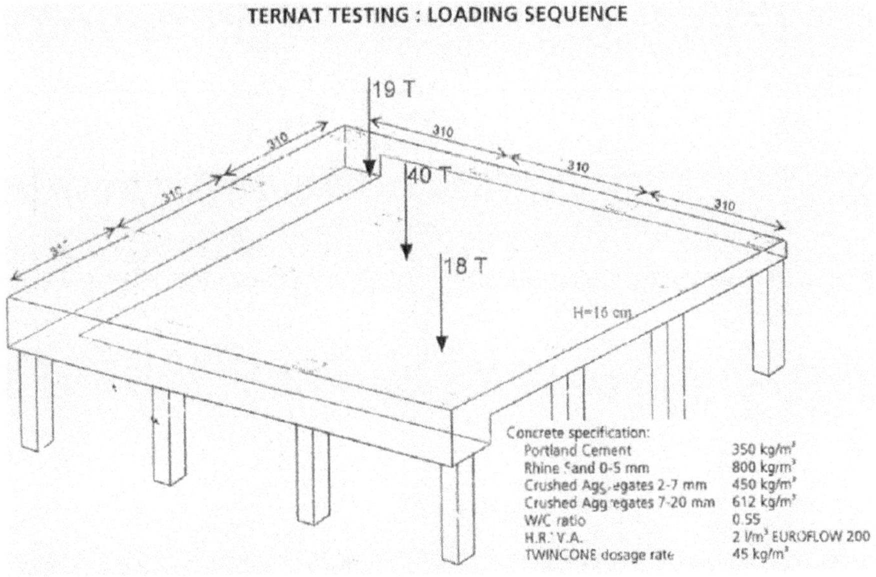

Figure 3.15 Full scale-test sizes and collapse loads in Ternat (in metric tons denoted by T).

$160^2/150^2$ = 26 kNm/m. Hence the yielding center point loading of the center field should be 16 x 26 kN m/m = 416 kN. This was experimentally verified while the corner span showed 156 kN = 6 x 26 kNm/m = < 180 kN experimental. result

A plastic moment calculation is now required to verify if the round indeterminate panel test derived yield moment capacity is correct to be used as limit in the Ternat Test back calculation.

The back-calculation of the plastic tensile strength f_{tu} at the ultimate stage of 50 kg/m³ steel fibers concrete gives even higher values than those derived from the round indeterminate panel test. This can be calculated from the

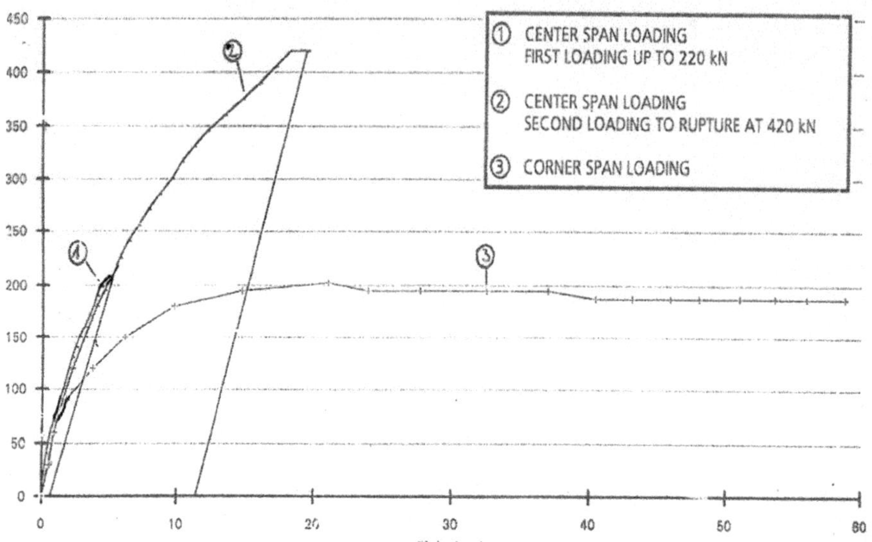

Figure 3.16 The loading-deflection diagrams of Ternat and Townsville are almost identical.

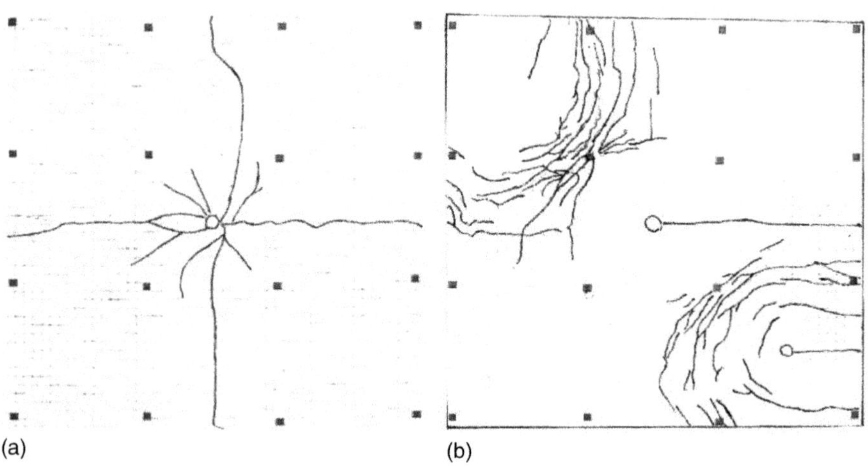

(a)

(b)

Figure 3.17 (a) Center field collapse cracking pattern. (b): Corner fields collapse cracking pattern.

Townsville test, where a clear fan pattern of rupture developed, as shown below:

$t := 160 \ mm$ — Thickness of the Ternat slab

$w_G := 3.84 \ \dfrac{kN}{m^3}$ — Weight of the 160 mm slab

$D := 200 \ mm$ — The pile diameter

$L_x := 3.10 \ m \quad L_y := 3.10 \ m$ — The spans

$\lambda_{DL} := 1.35 \quad \lambda_{LL} := 1.50$ — Dead and variable load factor

$P := 104 \ kN$ — Center Point loading intensity

$L_{rx} := L_x - D - t = 2.74 \ m$ — Net span

$d_P := 3.10 \ m$ — Distance between 2 point loads

$\phi := 0.85$

$M_{Px} := \dfrac{\lambda_{LL} \cdot P \cdot L_{rx}}{8 \cdot d_P} = 17.235 \ kN$ — Design moment of point loads at center point

$M_L := 26 \ kN \cdot \dfrac{m}{m}$ — Round slab-derived yield moment material factor

$\gamma_c := 1.50$

$\dfrac{\frac{M_L}{\gamma_c}}{M_{Px}} = 1.006 \quad > 1$ — The Ternat slab is able to carry a 104 kN center point loading in service.

We can conclude that the plastic design method with three yield lines, one positive in the bottom under the center point load and two negative in the top above the piles, shows a maximum allowable static loading of 104 kN, which is very close to the first crack loading intensity (110 kN) at the Ternat test.

A simple plastic design method is accurate when the resisting moment is derived from the round indeterminate panel test in flexion.

We can calculate according to the ACI 544-6R15 to obtain the allowable UDL in SLS.

$t := 160 \ mm$ — Thickness of the Ternat slab

$w_G := 3.84 \ \dfrac{kN}{m^3}$ — Weight of the 160 mm slab

$D := 200 \ mm$ — The pile diameter

$L_x := 3.10 \ m \quad L_y := 3.10 \ m$ — The spans

$\lambda_{DL} := 1.35 \qquad \lambda_{LL} := 1.50$ — Dead and variable load factor

$q := 17 \ \dfrac{kN}{m^2}$ — UDL intensity

$\lambda_Q := \left(\dfrac{\lambda_{DL} \cdot w_G \cdot t + \lambda_{LL} \cdot q}{w_G \cdot t + q} \right) = 1.495$ — Average loading factor

$L_{rx} := L_x - D - t = 2.74 \ m$

$L_{ry} := L_y - D - t = 2.74 \ m$ — Net spans

$\phi := 0.85$ — Reduction factor of resisting moments

$M_{qx} := \dfrac{(\lambda_{DL} \cdot w_G \cdot t + \lambda_{LL} \cdot q) \cdot L_{rx}^2}{12} = 16.473 \ kN \cdot \dfrac{m}{m}$ — Free edge and corner design moment

$t := 160 \cdot mm$ — Slab thickness

$f_{ck} := 30$ — Cylinder strength (characteristics)

$f_{ctm} := 0.3 \cdot (f_{ck})^{\frac{2}{3}} = 2.896$ — Tensile strength

$\omega := \dfrac{f_{ck}}{f_{ctm}} = 10.357$ — Ratio of strength

$f_{r3} := 5.10 \qquad u := 1 \cdot \dfrac{N}{mm^2}$ — Unit of stress

45 kg/m³ HE+1/60 EN 14651 - average value

$\sigma_{cr} := f_{ctm}$

$\mu := \dfrac{f_{r3}}{3.104 \cdot f_{ctm}} = 0.567$ — ACI 544 8R16

$m_{Rdaci} := \dfrac{3 \cdot \omega \cdot \mu}{\omega + \mu} \cdot f_{ctm} \cdot \dfrac{t^2 \cdot u}{6} = 19.939 \ kN \cdot \dfrac{m}{m}$ — (ACI 544 6R15)

$\phi := 0.85$

$M_{qx} := 16.47 \ kN \cdot \dfrac{m}{m} \qquad \dfrac{M_{qx}}{\phi \cdot m_{Rdaci}} = 0.972$

$$V_{P_c} := \left(2 \cdot \pi \cdot \left(\frac{3 \cdot t + D}{2} \right) \cdot t \cdot 0.66 \right) \cdot \mu \cdot \sigma_{cr} \cdot$$

$u = 370.656\ kN$

Shear resistance at critical perimeter

$$R := \left(\lambda_{DL} \cdot W_G + \lambda_{LL} \cdot q \right) \cdot L_x \cdot L_y = \left(2.805 \cdot 10^5 \right) N$$

Total factored pile reaction

$$R_{SLS} := \frac{R}{1.5} = \left(1.87 \cdot 10^5 \right) N$$

Pile Reaction at SLS

$$\frac{R}{V_{P_c}} = 0.757$$

$$E := 20000\ \frac{N}{mm^2}$$

Long-term modulus of elasticity

$$\delta := \frac{0.185 \cdot q \cdot L_x \cdot \left(\frac{L_{rx}}{t} \right)^3}{E} = 2.304\ mm$$

Deflection ACI 544-6R15
(Eq.J.2)

$$\frac{L_x}{\delta} = 1.345 \cdot 10^3$$

Span-to-deflection ratio > 500

From this, we can conclude that the Ternat test slab of 160 mm thickness and of 45 kg/m³ Twincone fiber reinforcing is fit for a 16 kN/m² UDL at SLS.

3.5.3 The Townsville test (Australia, 2000) (G-SFRC)

As shown in Figure 3.18, extracted from a poster session held at the James Cook University in Townsville, the test slab has been installed by RINOL contractor.

We can calculate the following yield moment at 400 kN ultimate point loading intensity:

$M_L = P_{ULT}/4\,\pi = 400\,/12.57 = 31.82$ kN m/m, and thus:

$f_{tu} = 31{,}820/(0.45 \times 160^2) = 2.76$ N/mm² instead of 2.64 N/mm², which is likely attributable to the a better structural ductility in the multiple spans compared to the round indeterminate panel test.

It was impressive, 25 years ago, to read the conclusions drawn from the Townsville test by Prof. Dr. J. Ginger and W. Karunasena, as shown in the slide in Figure 3.18.

A conventionally reinforced slab would require approx. three times the amount of steel as the SFRC slab to hold identical loads - A conventionally reinforced slab would cost approx. twice as much as a SFRC slab. - Steel fiber reinforced concrete offers a safe and economical alternative to conventionally reinforced concrete.

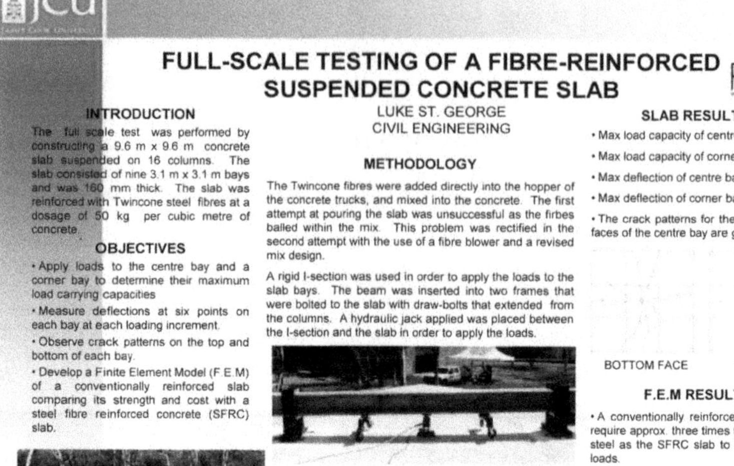

Figure 3.18 Townsville test poster.

These conclusions were at that time already known by us years earlier once the Ternat test was completed in 1994.

3.5.4 The Bissen Test (2004) (E-SFRC)

As shown in Figure 3.19 the 200 mm thick two-way SFRC tested of a flat bottom (thus without drops) slab at the ArcelorMittal plant of Bissen in 2004 was a free suspended elevated slab of 6 m span with three consecutive

Figure 3.19 Bissen (Luxembourg-2004), full-scale slab test.

spans resting on columns of 300 mm x 300 mm footprint (9 fields on 16 columns of ca. 335 m² area) so that it had a span to depth ratio of 30.

The slab test is to simulate the residential or light commercial applications of elevated suspended cast in situ slabs with point loadings up to 10 kN and uniformly distributed loadings up to 5 kN/m².

The mix design is of a C 30-37 with a 0.50 value of the W/C ratio and an F5 fluidity so that the mix was pumpable and self-compacting.

The fiber reinforcing consisted of 100 kg/m³ of TABIX fibers, with a 1.3 mm diameter, 50 mm length, and an undulating shape, made from steel wire with a tensile strength of 850 N/mm².

Initially, the TABIX 1.3/50 type was selected for use in suspended elevated slab applications because it is a very user-friendly fiber that could be easily introduced into ready-mixed concrete without requiring specific equipment. However, the disadvantage of TABIX 1.3/50 is of relatively lower performance, compared to 50 kg/m³ of HE +1/60 -1500N/mm² wire strength steel fibers, indeed a quite cheaper option. The HE+1/60 (simple hook-end, 1 mm diameter, 60 mm length, and 1500 MPa steel wire tensile strength) replaced the TABIX 1.3/50 at half the dosage rate.

A set of three APC rebars, each of 16 mm in diameter was placed continuously from column to column at the bottom of the slab, as shown in Figure 3.20. This was recommended at the time, referenced in reference (10), by the Canadian reinforced concrete code (CSA A23.3–94, page 99), to prevent collapse in the case of a column failure. These APC rebars, acting as continuous bottom steel through the columns, are designed to function as suspension cables, carrying the dead load along with the maximum unfactored variable total load of the field spanning four columns.

Figure 3.21 shows a typical slump flow of the SFRC used.

Despite the fact that APC rebars, as shown in Figures 3.20 and 3.22, were not yet mandatory in many codes, unlike in Canada, we decided to include them in every project involving elevated suspended SFRC slabs.

Figure 3.20 Installation of the Bissen Test slab. SFRC pumping and APC bottom rebars.

Figure 3.21 F 5 flow of the SFR concrete ready for pumping.

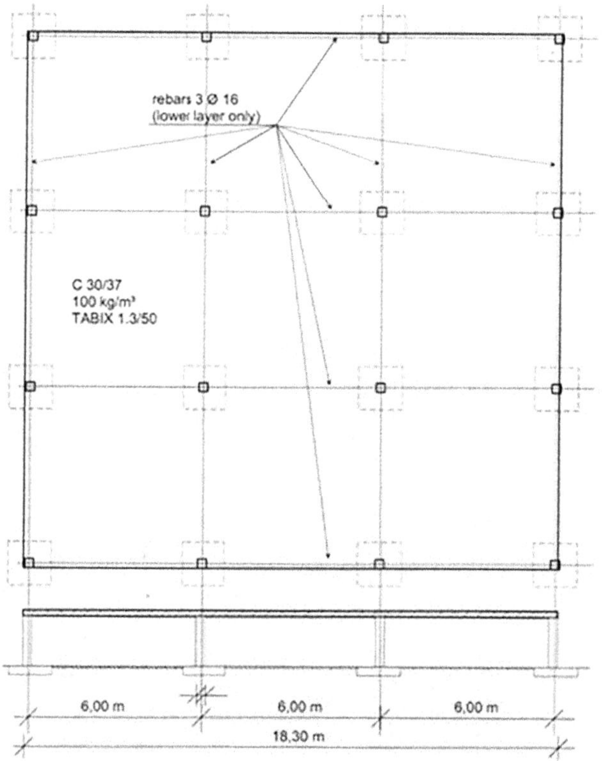

Figure 3.22 A drawing to show the test slab and the set of APC rebars (3 x 16 mm diameter rebars in the bottom).

We often encounter objections from academics, who argued that these rebars might "hide" the real contribution of the steel fiber reinforcement during full-scale loading tests.

Nevertheless, we never included the APC rebars' contribution in the moment design of the slabs of the SFRC, except for what they are meant for,

the APC prevention to prevent a local collapse from extending to a dispro-
portionate collapse.

Right now, the reader needs to know and admit that an SFRC suspended
elevated slab completion is a whole process that follows a reliable design
procedure with a reliable installation, from mixing steel fiber concrete to
placing it to curing, with suitable materials and forms used in accordance
with precise specifications checked by an experiences professional.

It is in no way about a simple supply of steel fibers pallets to a customer
who is provided with the dosage rate, to the ready-mixed concrete plant, or
to the concrete contractor who lets the worker sprinkle the bags or boxes of
fibers, glued or not, into the truck mixer for a strength-specified concrete,
assuming everything else follows traditional practices.

Such a lax practice opens the door to potential disasters, similar to those
sometimes experienced in standard-compliant traditional reinforced concrete where nobody is really responsible as each of them claims to have followed the standards.

All diagrams presented here are taken from the final report to ArcelorMittal,
prepared by Dr. Prof. U. Gossla (TU. Braunscweigh and FH. Aachen) who
organized and attended the testings.

Unfortunately, Dr. U. Gossla passed away some years later after this
report, much too early in his brilliant and high-potential academic life.
Today, we all deeply regret this loss.

As shown in Figure 3.23 taken from Reference (11), the deflections under
3.5 kN/m² and 6 kN/m² uniformly distributed loadings on the central row
and central field were limited to less than 2 mm and 4 mm, respectively,
including some top cracking opening that is limited to 0.2 mm.

The load vs. deflection diagrams in Figure 3.23 show the loading case of
6.00 kN/m² imposed on the 6 m x 6 m central span for the deflectometers
from no. 1–13.

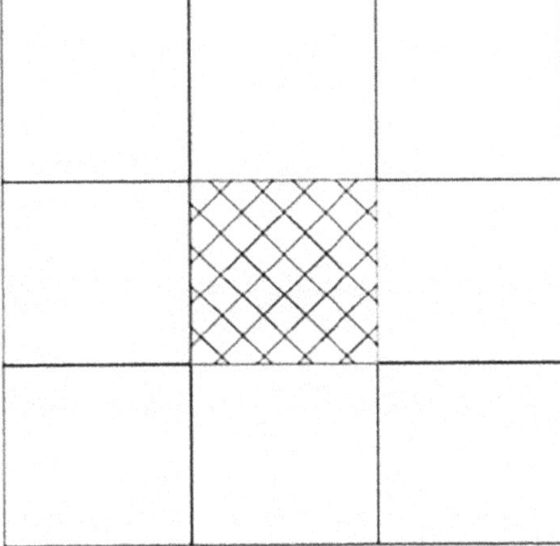

Figure 3.23 Center span load case.

The experimental deflections have been somewhat smaller than those calculated by the use of a Civil Design (Montréal) Finite Element software, as shown in Figures 3.25 and 3.26: 3 KN/m² on the central row, δ = 3.5/3 x 2.6 mm = 3 mm > 2 mm.

The corner field showed a 2.2 mm deflection under 3.5 kN/m² UDL, thus at 6 kN/m², the deflection would be expected to be 6/3.5 x 2. 2 mm = 3.8 mm (span/1578, far better than the allowed limit of Span/500)

Figure 3.25 shows the FEM-calculated elastic deflection of 2 mm under 3 kN/m² loading. At 6 kN/m², the deflection was expected to be 4 mm. However, the experimental deflection was limited to 2.6 mm, as shown in Figure 3.24, thus indicating it was stiffer than predicted by calculations. The span-to-deflection ratio was thus limited to 6000 mm /2.6 mm = 2307, which is 4.61 times more than the L/500 standard limit!

The center point loading test showed a 450 kN ultimate loading intensity, while the limit of proportionality was 210 kN with a 7 mm deflection (span/857) under a 200 kN center point load. This is indeed far better than the Eurocode 2 limit of a span/500 for a slab under the most onerous service load, which in reality, in residential applications, doesn't practically exceed · 10 kN at most!

We also observe a 3.0 mm deflection of the central field under a 100 kN center point loading intensity, which is less than 3.6 mm calculated by the Civil Design elastic finite element software.

The rupture pattern exhibits a pure fan pattern, and the yield moment was back-calculated as follows, taking into account the self-weight of the field itself: M_L = (450 + 49) kN/12.57 = 39,70 kNm/m so that f_{tu} = 39700/(0.45 x 200²) = 2.21 N/mm². Note that the 12.57 factor is indeed equal to 4 π.

In Figure 3.26, the flexion test diagrams follow the German standard procedures: a 600 mm unnotched span x 150 mm x 150 mm slab under a third-points loading with a 200 mm distance apart.

Such a full-scale test derived 2.21 N/mm² strength is to be compared with the same concrete from the test site to the average values obtained from flexion tests on L= 600 mm x h = 150 mm x w = 150 mm prismatic specimens according to the German Standard as shown in Figure 3.26: f_{r1m} = 5.2 N/mm² and f_{r3m} = 3.6 N/mm² both calculated from the elastic formula f = 6 M/ w h² while f_{tu} = M /(0.45 x h²) so that f_{tu} = f_{r3m}/ (6 x 0.45) = 3.6/2.7 = 1.33 N/mm².

In Figure 3.26, we see a very wide 3 N/mm² stress gap between the lowest and the highest post-cracking graph. The gap is as large as the f_{r3m} value! It results from the standard test method only. Such a test should not be used to understand the material behavior of the SFRC.

In the "Concrete" UK, magazine of April 2003, in p.15"Fibre distribution", we can read *"The Table 2 clearly indicates a significant movement of fibers away from the critical test area."*

On the other hand, B. Massicotte at the Polytechnic of Montréal, in his report, obtained f_{tu} = 3N/mm² with round indeterminate slabs of 1500 mm diameter x 150 mm thickness and 2000 mm diameter x 200 mm thickness.

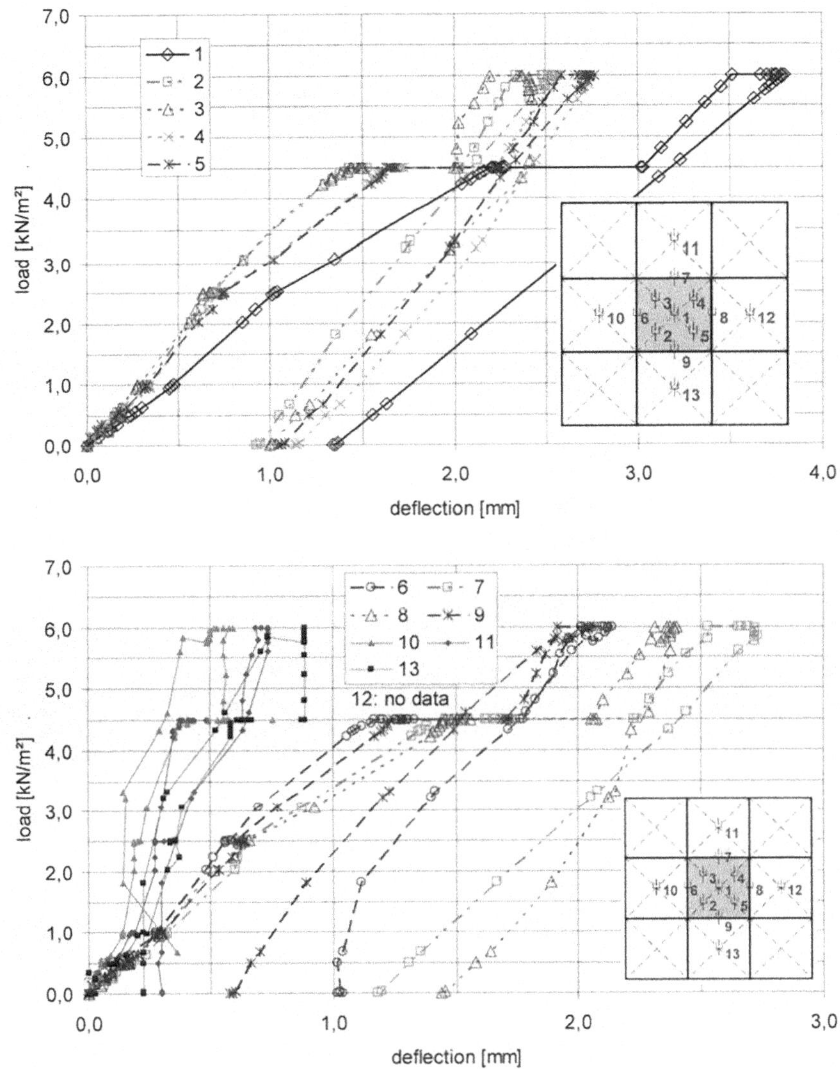

Figure 3.24 Edge and center field load vs. deflection diagrams.

The explanation that "Montreal values" are larger than "Bissen values" although they are of the same mix design, is that the Montreal concrete had a higher density or volumetric weight since it had been defined by using the "Maniabilimetre L.CP.C" to obtain the optimal Gravel/Sand ratio in the mix. Moreover, the local constituent materials of concrete in the Bissen area are suboptimal in size, shape, and grading.

According to the LCPC Baron-Lesage method, as shown in Figure 3.27, the minimum flow time from B to D, the optimum compaction and function

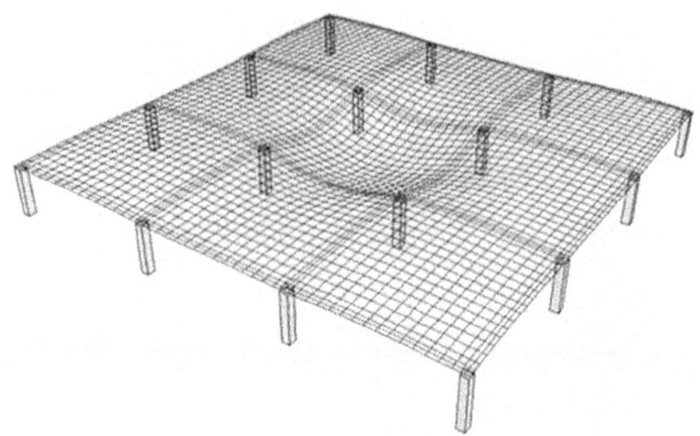

Figure 3.25 Deformed shape under a center field UDL of 3 kN/m².

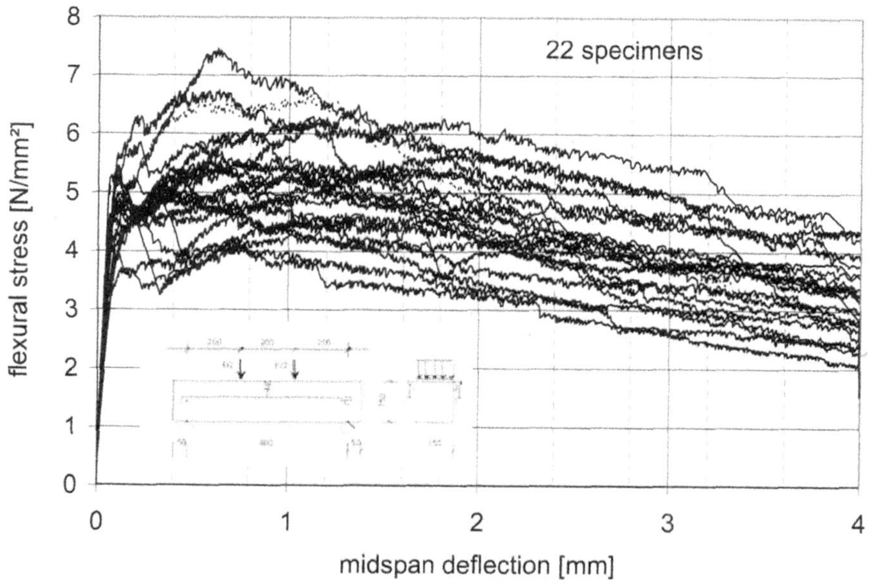

results of four-point bending tests on 150 mm wide specimens

Figure 3.26 German standard flexion tests diagram.

of sand-to-gravel ratio together produce the best workable concrete. Once the optimum mix has been defined, the final workability required is obtained by precisely increasing and adjusting the concentration of the superplasticizing admixtures.

It is worth emphasizing here that the best-performing SFRC with the lowest shrinkage cracking potential should always be of the highest density and/or highest volumic weight. The LCPC-Baron/Lesage apparatus is an excellent tool to define the optimum aggregate grading.

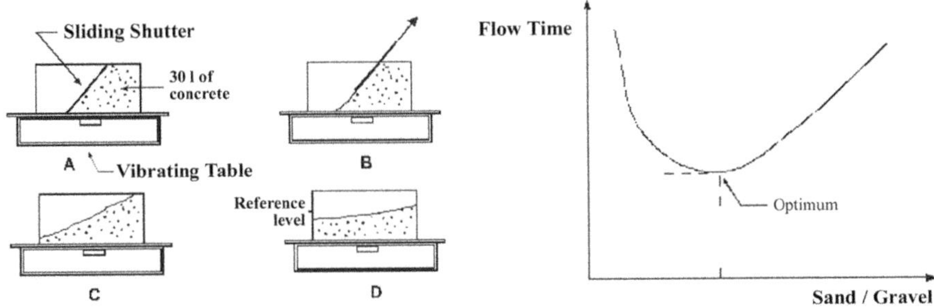

Figure 3.27 EN 14651 flexion test diagram of 100 kg/m² dosage rate of TABIX 1.3 mm diameter by 50 mm length made out of a 850 MPa tensile strength undulating steel wire.

Never forget that, according to standards, the mix design and the ready-mixed concrete are prepared to obtain a given standard compressive strength. However, it is more economical to obtain such a performance with less consumption of valuable and expensive constituent materials and achieve it more cheaply with more air, more water, and finer sand.

We recommend using the optimum mix design of the highest density for SFRC structural applications.

The same cement content per cubic meter in an optimum mix design gives much higher strength.

The Bissen slab test load-deflection diagram and ultimate cracking pattern of the center field under a center point loading are shown in Figures 3.28 and 3.29.

The ductility observed is impressive and so is the result of the addition of both material and structural ductilities.

The corner field, where the redundancy factor is quite smaller than in the center span, demonstrates considerable ductility, as shown in Figure 3.30. The ultimate load tops at 210 kN with 60 mm deflection. It can be unloaded and reloaded up to over 200 mm deflection, at which point the slab looks like a dish while the residual load intensity remains at 140 kN—a great number of times the maximum conceivable load over a residential or commercial slab. At this stage, the residual loading intensity is higher than the first flexion crack load intensity, as 140 kN > 100 kN.

It also shows the exceptional ductility potential of the SFRC slab together with the APC rebars in the bottom from column to column.

There were three rebars of 16 mm diameter resulting in 1.6 kg/m x 3 x 6 x 2 = 58 kg of rebars in a 6 m x 6 m x 0.20 m slab, equating to 7.2 kg/m³. The fiber dosage rate and type in 2004 were 100 kg/m² of TABIX 1.3 mm diameter x 50 mm length at 850 MPa wire tensile strength. Today, this has been replaced with a minimum of 50 kg/m³ of hooked ends fibers, 0.9 mm diameter x 60 mm length, and 1500 MPa wire tensile strength.

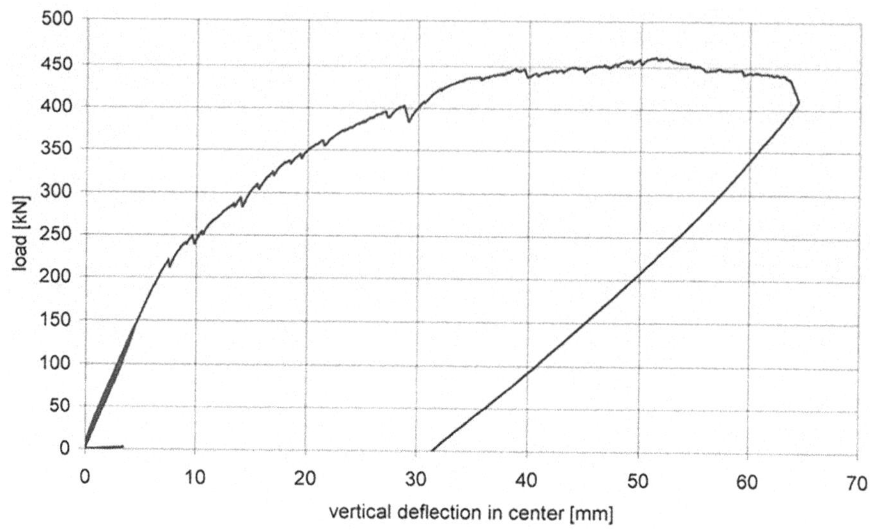

Figure 3.28 Center Point loading vs. deflection diagram.

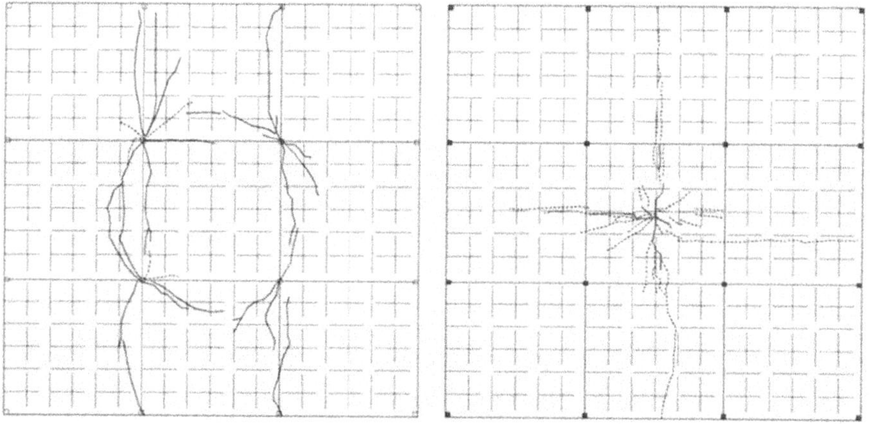

Figure 3.29 Flexion diagram and yield line pattern at the Bissen 2004, slab test.

In total, this amounts to 58 kg/m³ of steel reinforcing, indeed quite an economical and very sustainable number. Moreover, ArcelorMittal now supplies steel fiber with a very small carbon footprint, reduced to 300 kg CO_2 per ton of steel and further down to 100 kg CO_2 depending on the type of green fiber selected.

With traditional steel reinforcing requiring a minimum of 100 kg of rebars or wire meshes per m³ of concrete and a carbon foot print of 1000 kg CO_2 per ton of steel, the possible CO_2 savings in this case are significant. The reduction is from 20 kg/m² slab CO_2 emissions with traditional reinforcement down to 5 kg/m² slab, resulting in a 15 kg/m² CO_2 saving by using ArcelorMittal steel fiber reinforced green steel fibers (@300 kg CO_2 per ton of steel).

Figure 3.30 shows that at 210 kN maximum corner span loading, this is almost half of the central field's ultimate loading. The difference is understandable as only half of the fan pattern in the central field is available, due to the two free edges being unable to offer a resisting moment.

Figure 3.31 shows the cracking and yield lines pattern of the central field under a center point loading. It shows a combined fan and folded plate pattern.

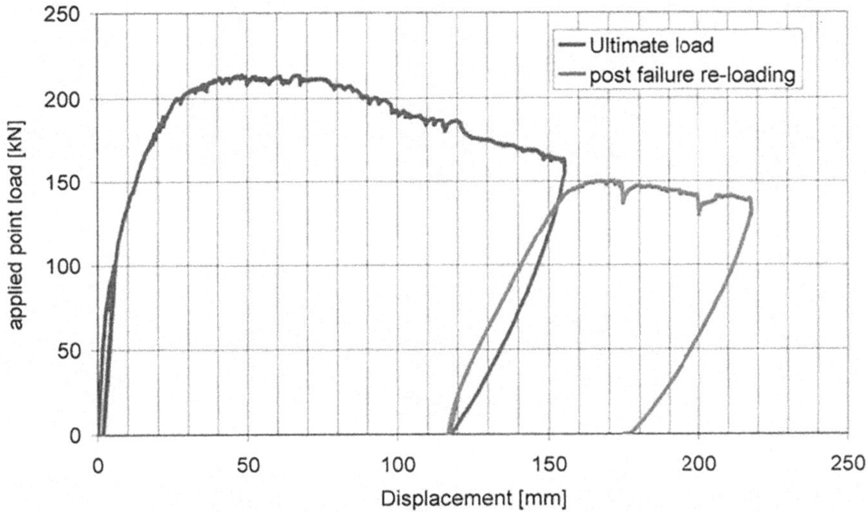

Figure 3.30 Corner field of center point loading vs. deflection diagram.

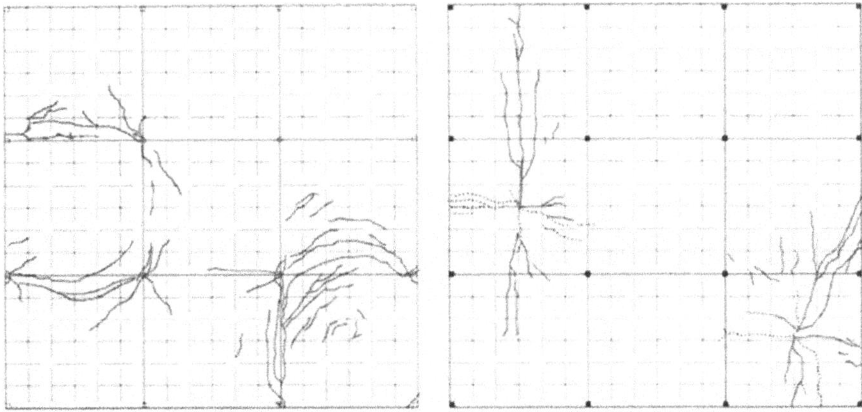

Figure 3.31 The free corner field center point loading cracking and yield lines pattern.

Ten round indeterminate panel tests with a 1.50 m span diameter and 150 mm thickness were fabricated at the Bissen test site and tested by the Brussels Polytechnic School Civil Engineering Laboratory, Test report no. 33.410, to calculate f_{tu} = 2.35 N/mm² as the average value, thus close to the 2.21 N/mm² derived from the full-scale test itself, and indeed, a characteristic value representing 94% of the average value based on 10 samples from the same batch and the same concrete as used at the test site.

This confirms once more that the high scatter observed with the standard small prismatic specimen (EN 14651, ASTM C 1609, etc.) is incorrect and caused by the standard process itself.

Here below, a calculation of the Bissen slab was made following the provisions of the ACI 544 6R15. It shows that a 3.5 kN/m² uniformly distributed load or a point load of 63 kN is the maximum allowable intensity that can be quickly estimated: 3.5 kN/m² x 6 m x 6 m / 2 = 63 kN. This limit is determined by the edge span condition as the resisting moment is smaller there.

$t := 200 \cdot mm$ Slab thickness

$f_{ck} := 30$ C30 concrete

$f_{ctm} := 0.3 \cdot \left(f_{ck}\right)^{\frac{2}{3}} = 2.896$ Tensile strength of C30

$\omega := \dfrac{f_{ck}}{f_{ctm}} = 10.357$ Strenght ratio

$u := 1 \cdot \dfrac{N}{mm^2}$ Unit of stress

$f_{r3} := 5.500$ EN 14651 residual flexion strength

$\sigma_{cr} := f_{ctm}$ Tensile Limit of proportionality of SFRC

$\mu := \dfrac{f_{r3}}{3.104 \cdot f_{ctm}} = 0.612$ Equivalent to $f_{r3} = 1.9\ \sigma_{cr}$

$m_{Rdaci} := \dfrac{3 \cdot \omega \cdot \mu}{\omega + \mu} \cdot f_{ctm} \cdot \dfrac{t^2 \cdot u}{6} = 33.462\ kN \cdot \dfrac{m}{m}$ Resisting moment by ACI 544-6R15

$\gamma_f := 1.5$ SFRC material factor

$\gamma_c := 24 \cdot \dfrac{kN}{m^3}$ Volumic weight of concrete

$q := 3.00 \cdot \dfrac{kN}{m^2}$ U.D.L. intensity

$P := 0 \cdot kN$

$$W_G := \gamma_c \cdot t = 4.8 \; \frac{kN}{m^2}$$

Weight of slab per square meter

$$D := 200 \cdot mm$$

Diameter of column supports

$$L_x := 6.00 \cdot m$$

$$L_y := 6.00 \cdot m$$

Spans center to center

$$b := 6 \cdot m$$

$$\lambda_{DL} := 1.35$$

Loading factors

$$\lambda_{LL} := 1.5$$

Average loading factor

$$\lambda_Q := \frac{(\lambda_{DL} \bullet W_G + \lambda_{LL} \bullet q)}{W_G + q} = 1.408$$

$$L_{rx} := L_x - D - t = 5.6 \; m$$

$$L_{ry} := L_y - D - t = 5.6 \; m$$

Net spans

$$\Phi_p := 0.90$$

Design moment at edge span

$$M_{Px} := \left(\frac{(\lambda_{DL} \bullet W_G + \lambda_{LL} \bullet q) \bullet (L_{rx})^2}{12} + \frac{\lambda_{LL} \bullet P \bullet L_{rx}}{6 \bullet b} \right)$$

ACI 544-6R 15

$$= 28.694 \; \frac{1}{m} \bullet kN \bullet m$$

$$\frac{M_{P_x}}{\Phi_p \bullet m_{Rdaci}} = 0.953$$

< 1

$$V_{pc} := \left(2 \cdot \pi \cdot \left(\frac{3 \cdot t + D}{2} \right) \cdot t \cdot 0.66 \right) \cdot \mu \cdot \sigma_{cr} \cdot$$

Shear/ Punching-out resistance

$$u = 587.834 \; kN$$

$$R := (\lambda_{DL} \cdot W_G) \cdot L_x \cdot L_y + \frac{\lambda_{LL} \cdot P}{4} = 233.28 \; kN$$

Column reaction at ULS

Column reaction at SLS

$$R_{SLS} := \frac{R}{\lambda_Q} = 165.718 \; kN$$

$$\frac{R}{V_{pc}} = 0.397$$

< 1

Long term elasticity modulus

$$E := 20000 \cdot \frac{N}{mm^2}$$

$$\delta := \frac{0.185 \cdot q \cdot L_x \cdot \left(\frac{L_{rx}}{t}\right)^3}{E} = 3.655 \ mm \qquad \text{Deflection (ACI 544-6R 15)}$$

$$\frac{L_x}{\delta} = 1.642 \cdot 10^3 \qquad\qquad << 500$$

It was the first time that such an important test of an SFRC-suspended elevated slab with a span-to-depth ratio of 30 was carried out.

We demonstrated the feasibility and safety of the system from the perspective of both design and installation.

The allowable loading intensity imposed obtained by calculations is significantly smaller than the limit of elasticity of the slab which is about 1/3 to 1/6 of the collapse loading of the slab.

Under the allowable loading intensities, the deflection is much smaller than the typical SLS span/500 slab deflection Eurocode limit.

Arbed/ArcelorMittal invited about 300 international delegates, including engineers, contractors, technical controllers, professors and researchers, owners, and competitors, to the Bissen full-scale test.

Regarding the center-point loading test conducted up to the final rupture, the last of the series of tests to complete, the delegates were invited to guess the ultimate loading intensity. The winners were awarded famous bottles of Champagne and Pomerol wine.

The winner was a highly reputable Canadian structural engineer who accurately guessed 460 kN.

Frustratingly, the lowest guess at 30 kN (Yes, you read correctly—30 kN) was given by an engineer from the main Belgian wire and steel fiber producer, a company with over 50 years of experience and worldwide reputation.

3.5.5 The Limelette test, Belgium, an attempt to replicate the Bissen test slab

In 2011, the CSTC, the CRIC, and Polytechnico de Milano (Italy) replicated the Bissen slab test in Limelette, Belgium, as shown in Figures 3.18 and 3.19. The only difference was the absence of the set of APC bottom rebars spanning from column-to-column support.

As an R&D. consultant to ArcelorMittal, I decided to step out of the whole testing process, ensuring that neither myself nor ArcelorMittal remained involved in anything whatsoever.

The reinforcing used a 70 kg/m³ dosage rate of single-hook-end fibers HE, 1 mm x 60 mm, with 1500 MPa wire strength.

The authors reported that "based on the analysis of nine EN 14651 small beams, they found an average residual strength f_{r3m} = 3.5 MPa (COV =

21%) meaning a reduction of almost 50 % compared to the initial lab characterization."

In my opinion, based on what I know, the high discrepancy is attributable to the installation itself where a CEM V-type cement was used following the concrete supplier's commercial recommendation instead of CEM I. This resulted in an overly flowing concrete with a slump test of up to 260 mm, likely causing segregation of aggregates and fibers.

Indeed, the Limelette full-scale test slab was already cracked on top and bottom with up to 0.3 mm crack opening at the release of the form underneath before any loading was applied.

My own experience with CEM V is that it is very prone to early shrinkage cracking in slabs, as the shrinkage due to water loss, builds up much quicker than strength. CEM V typically contains up to 50% SCM.

The slab was placed in late November in an open-air environment under windy and cold, humid conditions so that the strength developed very slowly, unlike the hydraulic shrinkage which reached its maximum under the windy conditions.

We see here a typical job-site-related event that almost halved the performance of SFRC.

The CEM V standard cement offers the needed specified laboratory compressive strength, as this remains the most important goal in the concrete and cement industry.

Note that the ultimate load recorded at the Bissen slab in 2004, of the same size and thickness, was 50% higher with a similar mix design than that of the CEM V slab tested at the Limelette Full-Scale test in 2011.

In Reference (12), the Limelette (Belgium) tested slab did not include the mandatory APC rebars.

The edge span was loaded in flexion after the earlier loading of the center span.

Under the last loading sequence at the middle of the edge span, which was already cracked from shrinkage before any applied loading, the edge span of the slab collapsed suddenly in a brittle manner. An operating technician sitting on top of the slab was injured and hospitalized with a broken pelvis.

We can understand that the slab had been pre-cracked at form release, particularly on top above the column supports along almost its total length. Without APC rebars, nothing could stop the collapse of an edge slab that had effectively become a statically determinate structure.

It is, once more, a demonstration that APC rebars should be clearly mandatory in all SFRC elevated suspended slab applications. Nevertheless, the EC2 and ACI reinforced concrete codes currently include the use of continuity rebars in all types of suspended elevated slabs but fail to address the use of SFRC.

I have been personally, in my role as an R&D consultant to ArcelorMittal, the supplier of the steel fibers, against the omission of the APC rebars, as confirmed in an official letter prior to the Limelette testing. I decided not to attend the test nor involve ArcelorMittal any further.

The authors report, however, in Reference (11), that *"In real situations, anti-progressive collapse rebars would be used from columns to columns to provide a robust structure."* However, they didn't report the collapse, mention the omission of APC rebars, as well as not even mention the elevated slab, that needs to be included to prevent the collapse happened.

This implies that SFRC is not robust—a completely wrong statement. The present chapter shows how robust SFRC slabs are, provided that SFRC is not only a standard fiber-reinforced material but also forms an adequate part of the structure as a whole, based upon a detailed design and drawings executed by experienced professionals who do not cut corners to save some dollars.

We have to reiterate here, once more, that we decided back in the year 2001, from the initiation of the R&D process, on the mandatory inclusion of the APC rebars in any statically indeterminate SFRC slab, as required by the Canadian reinforced concrete standard. This decision followed discussions with Prof. Massicotte, as referenced in (9) of the Polytechnique de Montréal, regarding any type of suspended elevated slab, whether rebar-reinforced or post-tensioned, prefabricated, or in SFRC.

3.5.6 The Tallinn Test (Estonia) in 2007 (E-SFRC)

3.5.6.1 *Test set-up and slab installation*

Figure 3.32 shows a view of the Tallinn Full-Scale SFRC slab test.

In order to complement and confirm the Bissen test conclusions, a similar full-scale test was organized by the RUDUS ready mixed concrete company in Tallinn (Estonia) regarding a 180 mm thick slab of 5.00 m span resting on 16 columns, at a span-to-depth ratio of 28, in 3 consecutive spans thus of ca. 230 m² surface area. One corner span was supported by two concrete edge walls.

The mix design was similar to the Bissen case with a 100 kg/m³ dosage rate of TABIX 1.3 mm/50 mm steel fiber type and including the set of 3–16 mm diameter APC rebars in the bottom from column to column as shown in the Figure 3.34. The test has been supervised by Prof. Pello of the Tallinn Technical University.

Figure 3.32 Aerial view of the Tallinn SFRC FST.

Figure 3.33 The anti-progressive collapse rebars and the flowing SFRC mix.

Figure 3.34 3 diameter 16 mm APC rebars in the bottom from column to column.

The concrete is at F5 fluidity so it was easily pumped and didn't require any mechanical vibrating to be compacted, as shown in Figure 3.33, left.

The Tallinn slab design is comparable to the Bissen slab and is suitable for any residential or light commercial applications with a maximum variable uniformly distributed loadings of 5 kN/m² and a maximum point loading of 10 kN.

The loading was applied in 20 kN steps, with a minimum waiting time of 5 minutes at each step.

3.5.6.2 Experimental findings

The maximum loads at testing of different spans are as follows:

- load on middle span (E): 595 kN in Figure 3.36—indeed the maximum jack capacity!
- load on corner span with perimeter walls (1): 410 kN;

Figure 3.35 Uniformly distributed loading at the Tallinn test.

- load on edge span (B): 310 kN;
- load on corner span (G): 275 kN.

Under a 5 kN/m² UDL on the corner span with free edges, a 2 mm deflection was observed, equivalent to span/2500, as shown in
Figure 3.35

Back-calculating the yield moment $M_L = P_{ULT}/4 \times \pi = 595/12.57 = 47.34$ kNm/m and $M_{Rd} = M_L/1.5 = 31.56$ kNm/m is such that the plastic average tensile strength $f_{td} = 31560/0.45 \times 180^2 = 2.17$ N/mm².

Like with the EN 14651 standard flexion test, we can calculate the elastic strength although the section is plastic: $f_{rd} = 6 \times 31560/180^2 = 5.84$ N/mm².

Note that $f_{rd} = 2.7 \times f_{tu}$ as $0.45 \times 6 = 2.7$ and that $2.7^{-1} = 0.37$ as in the FIB Modelcode 2012 where the definition is: $f_{t3} = 0.37 f_{r3}$.

By dividing the maximum stress by the 1.5 load factor, the SLS flexion stress is reduced to $5.84/1.5 = 3.89$ N/mm². Under the most onerous allowable service load, the slab should not crack as the modulus of rupture (MOR) is approximately 5 N/mm².

According to the Model Code, we should obtain, assuming $f_L = f_{r3} = 5.84$ N/mm²

$$f_{r3k} = 0.7 \times 5.84 = 4.09 \, \text{N} / \text{mm}^2$$
$$f_{rd} = 4.09 / 1.5 = 2.72 \, \text{N} / \text{mm}^2$$

Shape factor $\tau = 1.4$, the maximum possible value, so that $f_{rd} = 1.4 \times 2.72 = 3.80$ N/mm².

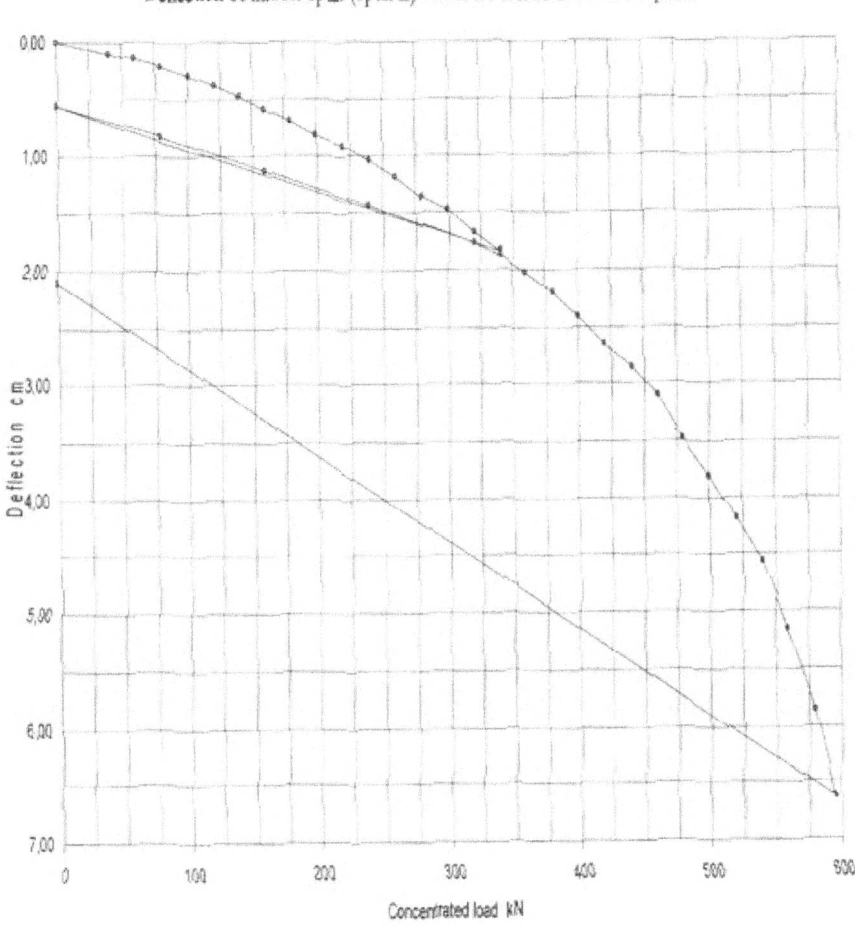

Figure 3.36 Deflection vs. loading diagram regarding the center field.

The maximum flexion stress at SLS = 3.80 N/mm²/1.5 = 2.53 N/mm², which is 35% smaller than 3.89 N/mm², derived from the full-scale test.

The shape factor τ is underestimated in the Modelcode. A more realistic value would be 1.4 x 3.89/2.23 = 1.4 x1.53 = 2.14.

The Swedish standard (SS 812 310) defines η_{det}, the statically indeterminate factor of 2 for two-way suspended slabs.

Note that a full-scale test is characteristic by nature, with no need to reduce the strength to account for material scatter due to the very small sizes of the standard material sample.

- First crack occurred in the center field under P = 120 kN, deflection δ = 3,55 mm, corresponding to a span/1400 ratio that is better than the L/500 standard limit of a 10 mm deflection limit. The crack had a

0.05 mm – 0.1 mm opening, which is less than the 0.3 mm allowed for standard reinforced concrete or 0.5 mm at f_{r1} for SFRC.

- The slab was unloaded at P = 340 kN, with a 5.44 mm permanent deflection.
- At the maximum possible loading intensity, P max = 595 kN indeed (limited by the hydraulic jack capacity, we observed:

  – A permanent deflection after unloading the 595 kN: δ = 22 mm
  – Maximum bottom crack opening under point loading: 3.5 mm under 480 kN, which is approximately the EN 14651 f_{r4} crack opening. In a real scale statically indeterminate slab, both f_{r3} and f_{r4} don't exist at ULS, as a much higher loading intensity than ULS is needed to reach stages 3 and 4.
  – Observation of an extremely dense minute fan pattern cracking under the point loading with on top of slab, a circular multiple cracking passing by the 4 footprints of columns.
  – Along the 4 edges: Under 595 kN minute cracking is visible only at all four edges of the slab together with a 70 mm maximum deflection under the point loading.

This confirms that the 595 kN intensity is still not the collapse loading intensity due to the absence of required wide open yield lines from edge to edge, didn't show-up.

Figure 3.37 shows the history of loading, unloading, and reloading of the test slab.

The crack opening of 0.1 mm, 0.3 mm, and 3 mm at 200 kN, 250 kN, and 595 kN, respectively, are shown in Figures 3.38a and b, and Figure 3.39 for

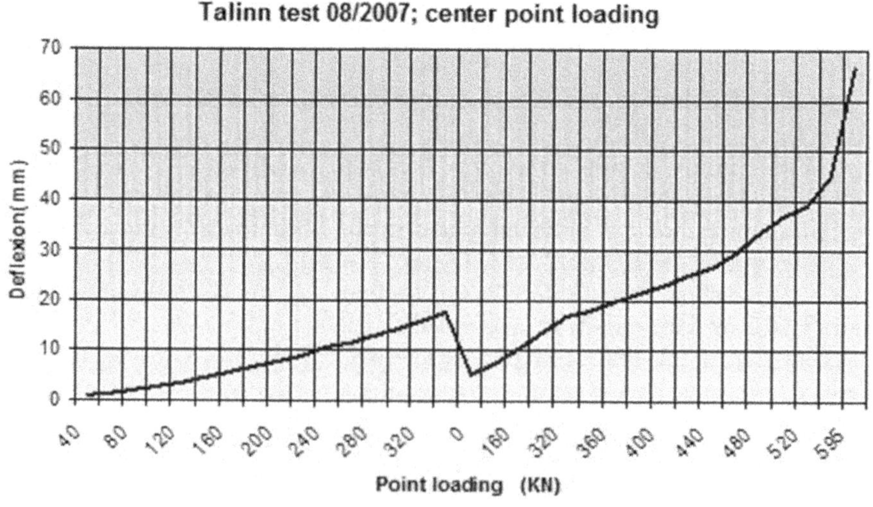

Figure 3.37 History of slab loading.

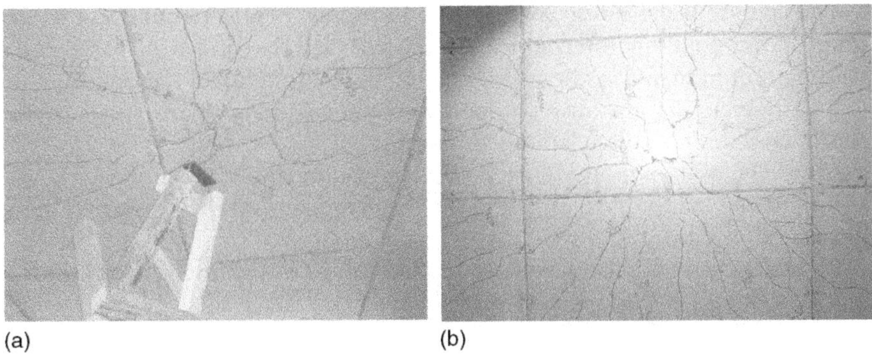

(a)   (b)

Figure 3.38 (a) 0.1 mm crack opening at 200 kN. (b) 0.3 mm crack opening under 250 kN.

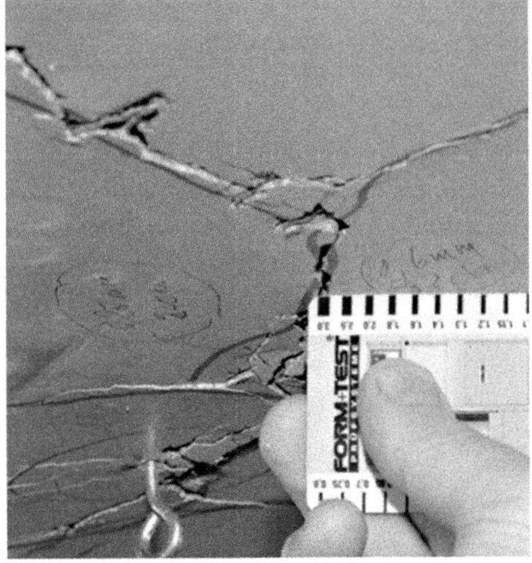

Figure 3.39 3 mm crack opening under 595 kN point loading.

the 3 mm crack opening. The 3 mm crack opening is typical of the ULS stage at CMOD 4 and f $_{r4}$ (EN 14651)

A wrong SLS load could be derived as 595 kN/1.5 = 400 kN. However, the SLS loading should correspond to a 0.3 mm crack opening or 0.5 mm of CMOD1/f$_{r1}$, thus it is likely no more than 300 kN under center-point loading.

The SLS load cannot be that high as determined by a FEM elastic calculation, as demonstrated below.

The span/500 is reached approx. 240 kN under both loading and reloading stages.

We must observe here that the 0.3 mm crack opening limit in Eurocode 2 is attained at 300 kN center-point loading of the middle span and at half of this value, or 150 kN, in the corner span with free edges.

The span/500 deflection limit of the Eurocode is reached under 240 kN center-point loading intensity, an intensity well beyond any foreseeable loading requirements in the residential or light commercial sectors.

All these results are now compared to the calculations provided by the elastic civil design finite element software.

U0 to U4, as shown in Figure 3.40, denotes the application of a UDL = 3 kN/m^2 at the locations depicted in the same figure, and the point loading is illustrated in Figure 3.41.

P1 to P4 denotes a center point loading of 50 kN at the following locations, as shown in Figures 3.41 and 3.42:

U0: in free corner field only	P1: in free corner field
U1: in central field	P2: in central field (also noted P center)
U2: in central row	P3: in edge field
U3: checker board	P4: in wall supported corner field
U4: rows crossing	D: Dead load

A summary of the results is shown in Tables 3.8 and 3.10

We see in all cases that the flexion stress is at most 4 N/mm^2 (D + P1), thus less than the 5 N/mm^2 first crack flexion strength. The loading intensities could thus be increased by 25% to 3.75 kN/m^2 UDL and 62.5 kN

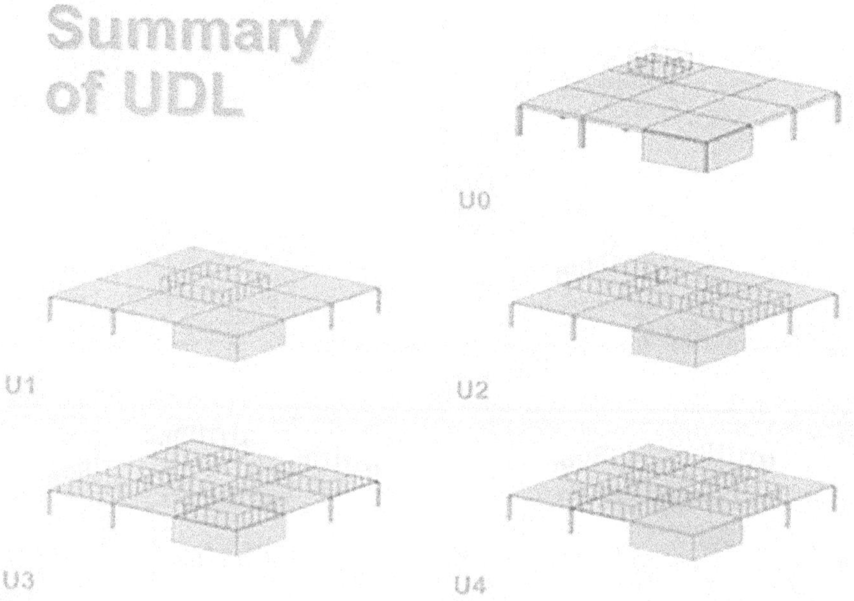

Figure 3.40 View of five different loading patterns calculated.

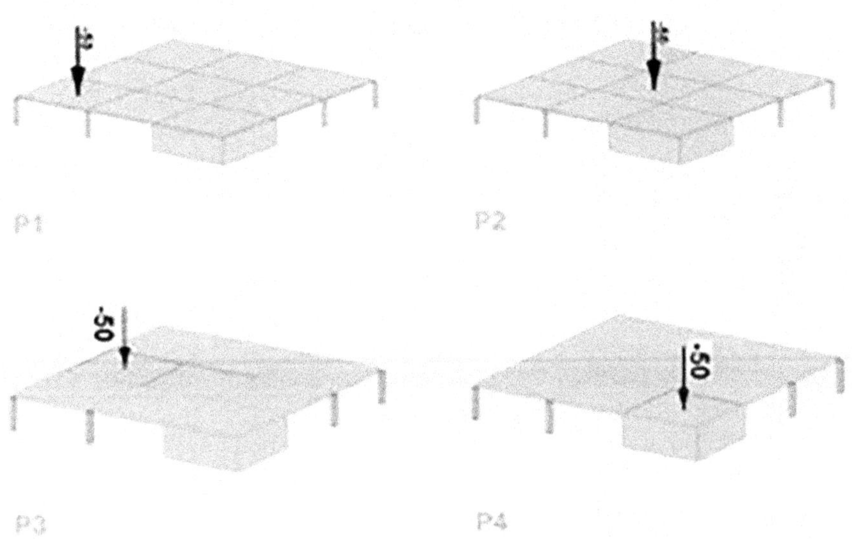

Figure 3.41 Location of the point loadings.

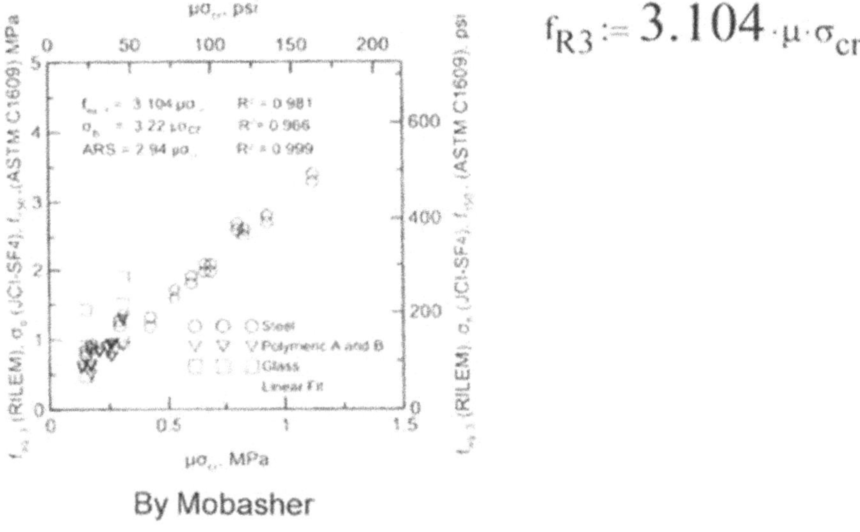

$$f_{R3} := 3.104 \cdot \mu \cdot \sigma_{cr}$$

Figure 3.42 Diagram showing f_{re3} vs. μ.

point loading. The deflection then becomes at most 3.61 mm x 1.25 = 4.51 mm, equivalent to span/1108, which is far stiffer than L/500 in the short term.

At the long term, the maximum deflection is calculated as 8.32 mm x 1.25 = 10.4 mm, equivalent to span/481, taking into account an E modulus of 15,000 N/mm². This is a pessimistic value since SFRC at high dosage rates inhibits and controls cracking and creep. A more appropriate long-term E modulus of 20,000 N/mm² would produce a long-term deflection of 7.8 mm, equivalent to span/641.

Such a design means a global safety factor of 595 / 62.5 = 9.52 on point loadings as shown in Tables 3.9 and 3.10. Regarding UDL, we observe that (D + U0) gives 3.8 N/mm² flexion stress, so the UDL could be increased to reach 5 N/mm² flexion stress instead of 4 N/mm², a 32% increase. This corresponds to: 24 kN/m³ x 0.18 m + 3 kN/m² = 7.32 kN/m², thus up to 7.32 x 1.32 = 9.66 kN/m², resulting in a maximum allowable variable UDL of: 9.66 – 4.32 = 5.34 kN/m².

The deflection at short term then becomes 3.99 mm x 1.32 = 5.6 mm (E = 35,000 N/mm²), and at long term: 9.19 mm x 35,000 / 20,000 = 9.8 mm (E = 20,000 N/mm²), which is less than 10 mm = L/500.

The test data indicate that, at the short term, with a 62.5 kN point loading, the maximum deflection is 2.2 mm, which is significantly smaller than the deflection predicted by the Civil Design FEM calculations.

We will check this slab below according to the ACI 544 6R15 method. There is a relation given by Prof. B. Mobasher between post-cracking flexion strengths of the EN 14651-f_{r3}, RILEM TC162, and the μ factor (ACI 544 8R16, Eq. 6.4.1b and H.8) used in the ACI 544 6R15 design report.

The relation, shown in Figure 3.42, is found in the ACI 544 8R16 report on "Indirect method to obtain stress–strain response of FRC" in Figure 5.4 left of p.13:

$t := 180mm$	Slab thickness
$f_{ck} := 30$	Cylinder strength(characteristic)
$f_{ckc} := 37$	Cube strength
$f_{ctm} := 0.3 \cdot \left(f_{ck}\right)^{\frac{2}{3}} = 2.896$	Tensile strength of plain concrete
$\omega := \dfrac{f_{ck}}{f_{ctm}} = 10.357$	Ratio
$f_{r3} := 7.00$	50 kg/m³ HE+1/60 EN 14651 - average value

$$\sigma_{cr} := f_{ctm}$$

$$\mu := \frac{f_{r3}}{3.104 \cdot f_{ctm}} = 0.779$$

ACI 544 8R16

$$u := 1\frac{N}{mm^2}$$

$$m_{Rdaci} := \frac{3 \cdot \omega \cdot \mu}{\omega + \mu} \cdot f_{ctm} \cdot \frac{t^2 \cdot u}{6} = 33.979 \cdot kN \cdot \frac{m}{m}$$

(Eq.6.4. 1b and H.8)

$$\gamma_c := 1.5$$

Material factor

$$\gamma_c := 24\frac{kN}{m^3}$$

Concrete volumic mass

$$q := 5.5\frac{kN}{m^2}$$

Unfactored UDL

$$W_G := \gamma_c \cdot t = 4.32 \cdot \frac{kN}{m^2}$$

Slab weight

$$D := 250mm$$

Column diameter

$$L_x := 5.40m$$

Span lengths

$$L_y := 5.40m$$

$$\lambda_{DL} := 1.4$$

Load factors on the own weight

$$\lambda_{LL} := 1.6$$

Variable load factor

$$\lambda_Q := \left(\frac{\lambda_{DL} \cdot w_G + \lambda_{LL} \cdot q}{w_G + q}\right) = 1.512$$

Average load factor

$$L_{rx} := L_x - D - t = 4.97m$$

Net span lengths

$$L_{ry} := L_y - D - t = 4.97m$$

$$\Phi_P := 0.90$$

Reduction factor of resisting moment
Design moment of the edge span
eq.7.5.3.c and 6.3.1.1m

$$M_{Px} := \left[\frac{\left(\lambda_{DL} \cdot w_G + \lambda_{LL} \cdot q\right) \cdot \left(L_{rx}\right)^2}{12}\right]$$

$$= 30.563\frac{1}{m} \cdot kN \cdot m$$

$$\frac{M_{Px}}{\Phi_P \cdot m_{Rdaci}} = 0.999$$

$< 1, OK$

punching shear verification:

$$v_{pc} := \left[2 \cdot \pi \cdot \left(\frac{3 \cdot t + D}{2} \right) \cdot t \cdot 0.66 \right] \cdot (\mu \cdot \sigma_{cr}) \cdot$$

Shear resistance at critical perimeter (eq.7.6a)

$$u = 664.921 \cdot kN$$

$$R := (\lambda_{DL} \cdot W_G + \lambda_{LL} \cdot q) \cdot L_x \cdot L_y = 432.968 \cdot kN$$

Total factored reaction

$$R_{SLS} := \frac{R}{\lambda_Q} = 286.351 \cdot kN$$

Total unfactored reaction

$$\frac{R}{V_{pc}} = 0.651$$

$<1, OK$

Deflection

$$E := 35000 \frac{N}{mm^2}$$

Short term modulus of elasticity

$$\delta := \frac{0.185 \cdot q \cdot L_x \cdot \left(\frac{L_{rx}}{t} \right)^3}{E} = 3.305 \cdot mm$$

Eq.J.2 (ACI 544-6R15) deflection

$$\frac{L_x}{\delta} = 1.634 \times 10^3$$

A stiffer span-to-deflection ratio > 500, OK! (better than the Eurocode 2 limit)

Equivalent center point loading intensity

$$P_{eq} := \frac{q \cdot L_x \cdot L_y}{2} = 80.19 \cdot kN$$

Maximum SLS Point loading intensity per field

3.5.6.3 Conclusions: checking the corner span

The Tallinn slab, with a thickness of 180 mm and a 5.0 m column grid constructed in C30-37 concrete reinforced with either 100 kg/m³ of TABIX 1.3/50 or 60 kg/m³ of HE+1/60, under a 5.5 kN/m² UDL, passes the ACI 544 6R15 calculations methodology. The verification is made for the corner span, which is the most critical case.

Table 3.9 With a 15,000 N/mm² modulus of elasticity at short term

Combin.	Forces y kN	€zKnot number of the FEM software	Déplacement (displacement) y mm	maximum flexion stress above column MPa	maximum flexion stress between columns MPa
D+U0	−1271.56	397	−9.19	−3.3	3.8
D+U1	−1256.56	397	−4.34	−1.7	1.9
D+U2	−1406.56	1925	−5.33	−2.0	2.4
D+U3	−1556.56	397	−7.84	−3.0	3.4
D+U4	−1481.56	1153	−4.77	−2.1	2.4
D+P1	−1231.56	428	−8.32	−2.6	4.0
D+P2	−1231.56	397	−4.35	−1.7	2.7
D+P3	−1231.56	412	−5.81	−2.0	3.9
D+P4	−1231.56	1889	−5.09	−1.8	3.7

Table 3.10 A summary of forces vs. displacement with a 35,000 N/mm² modulus of elasticity

Combin.	Forces y kN	Numéro Noeud (Dépl. max)	Dépl. y mm	smax Col MPa	smax Span MPa
D+U0	−1271.56	397	−3.99	−3.3	3.8
D+U1	−1256.56	397	−1.89	−1.7	1.9
D+U2	−1406.56	1925	−2.36	−2.0	2.4
D+U3	−1556.56	397	−3.40	−3.0	3.4
D+U4	−1481.56	1153	−2.12	−2.1	2.4
D+P1	−1231.56	428	−3.61	−2.6	4.0
D+P2	−1231.56	397	−1.90	−1.7	2.7
D+P3	−1231.56	412	−2.54	−2.0	3.9
D+P4	−1231.56	1889	−2.21	−1.9	3.7

The equivalent point loading is 80 kN per field, thus 80 / 275 = 29% of the experimental 275 kN, the ultimate load in the edge span, and 13% of 595 kN, the experimental ultimate load in the central span.

The short-term deflection here is 50% larger than the experimental. The set of APC rebars does not contribute to the static verification of the slab.

The ACI 544 6R15 method, which is quite easy and straightforward, proves to be conservative in all aspects of moments, shear, and deflections when compared to the full-scale test results.

The theory behind the section calculation for M_{rd}, as described in reference (9) by B. Mobasher, takes into account the static indeterminacy of the slabs.

3.5.7 The Niew Vennep test, Holland (G-SFRC)

The SFRC piled slab had been completed 12 months earlier and was fully loaded under service conditions. The floor surface showed crazing, similar to what occurs when the very top layer of a slab develops excessive shrinkage. Everybody knows there are many potential causes, such as an excessive amount of cement paste on the surface, a lack of bleeding, air currents during finishing, or insufficient curing.

The owner was convinced that the cause was structural, so he didn't pay the bill in full. A flooring consultant was appointed to investigate and organized a full-scale loading test. A decision was made to test the slab's loading capacity without ground support underneath. An edge row of fields was selected, and the ground was removed manually with shovels under the building.

The slab was load-tested by TNO, the official Netherlands State Laboratory. The 180 mm thick slab, reinforced with a 45 kg/m³ dosage rate of TABIX + 1/60 undulating shape steel fibers of 1500 MPa wire tensile strength, was located in Nieuw-Vennep (Holland). This piled slab was tested under suspended elevated conditions above a 3.0 m x 3.60 m pile grid, where the piles were provided with 0.8 m x 0.8 m pile head support.

The slab was subjected to q = 29 kN/m² UDL, instead of the 15 kN/m² specified in the contract, thus loaded by 93% more than the SLS loading intensity, for seven days, instead of the 15 kN/m² maximum service loading intensity as per specifications.

In Figure 3.43, the 29 kN/m² loading is shown on the edge span against the wall of the Nieuw-Vennep slab. The load test was organized on the edge span, as shown in Figures 3.43 and 3.44, the first span from the edge of the building, where the flexion demand is at least 33% higher (16/12 = 1.33) compared to any inside span farther from the edge, all other conditions being equal.

Under the 29 kN/m² uniformly distributed loading applied for 150 hours, only a 1.5 mm deflection was recorded, as shown in Figure 3.43.

The resulting span-to-deflection ratio was 3000 mm / 1.5 mm = 2000, which is four times better than the "500" standard minimum limit in the Eurocode 2 design standard, as shown in Figure 3.45.

Indeed, the slab remained uncracked on both its top and bottom faces, as shown in Figures 3.44 and 3.45, after the full load test of 29 kN/m² intensity. The initial crazing cracks did not widen further during the entire test.

The Nieuw-Vennep test demonstrated the high robustness and stiffness of SFRC piled slabs with a span-to-depth ratio of 3000/180 = 17, under almost double the loading intensity specified in its design.

The test load was even 29% higher than the Eurocode factored ULS loading intensity (22.5 kN/m²).

Figure 3.43 At the Niew Vennep test (Holland), application of the 29 kN/m² sand bags uniformly distributed loading.

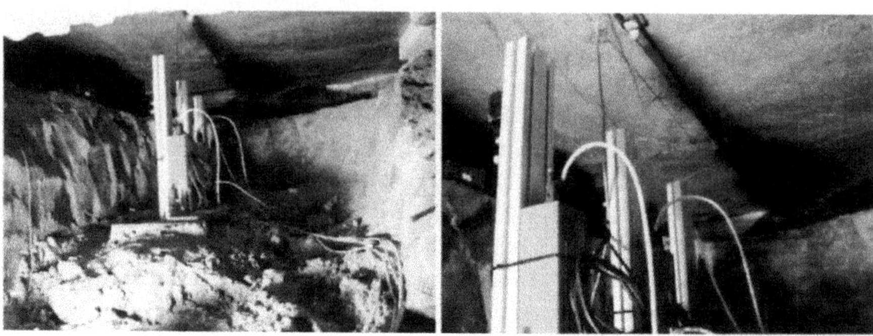

Figure 3.44 Recording of deflections underneath the slab.

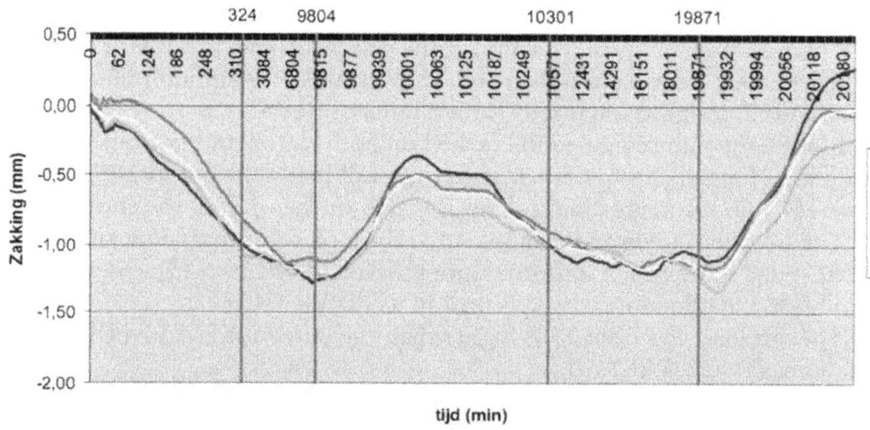

Figure 3.45 1 mm deflection after 150 hours under 29 kN/m² uniformly distributed loading.

Below is an ACI 544-6R15 statical calculation for the edge span.

$t := 180 \cdot mm$ slab thickness

$f_{ck} := 30$ C 30 strength class

$f_{ctm} := 0.3 \cdot (f_{ck})^{\frac{2}{3}} = 2.896$ tensile strength of C30 class

$\omega := \dfrac{f_{ck}}{f_{ctm}} = 10.357$ ratio of strength

$u := 1 \cdot \dfrac{N}{mm^2}$

$f_{r3} := 4.60$ EN 14651 residual flexion strength

$$\sigma_{cr} := f_{ctm}$$

$$\mu := \frac{f_{r3}}{3.104 \cdot f_{ctm}} = 0.512$$

$$m_{Rdaci} := \frac{3 \cdot \omega \cdot \mu}{\omega + \mu} \cdot f_{ctm} \cdot \frac{t^2 \cdot u}{6} = 22.878 \ kN \cdot \frac{m}{m}$$

$$\gamma_f := 1.5$$

$$\gamma_c := 24 \cdot \frac{kN}{m^3}$$

$$q := 19 \cdot \frac{kN}{m^2}$$

$$P := 0 \cdot kN$$

$$W_G := \gamma_c \cdot t = 4.32 \ \frac{kN}{m^2}$$

$$D := 800 \cdot mm$$

$$L_x := 3.60 \cdot m$$

$$L_y := 3.00 \cdot m$$

$$b := 3.6 \cdot m$$

$$\lambda_{DL} := 1.35$$

$$\lambda_{LL} := 1.5$$

$$\lambda_Q := \frac{\left(\lambda_{DL} \cdot W_G + \lambda_{LL} \cdot q\right)}{W_G + q} = 1.472$$

$$L_{rx} := L_x - D - t = 2.62 \ m$$

$$L_{ry} := L_y - D - t = 2.02 \ m$$

$$\Phi_P := 0.90$$

$$M_{Px} := \left(\frac{\left(\lambda_{DL} \cdot W_G + \lambda_{LL} \cdot q\right) \cdot \left(L_{rx}\right)^2}{12} + \frac{\lambda_{LL} \cdot P \cdot L_{rx}}{6 \cdot b} \right)$$

$$= 19.639 \ \frac{1}{m} \cdot kN \cdot m$$

$$\frac{M_{Px}}{\Phi_P \cdot m_{Rdaci}} = 0.954$$

ACI 544-8R16

Resisting moment ACI544-6R15

U.D.L.

pile diameter

spans

load factors

average load factor

net spans

reduction factor of resisting moment

design moment at edge span

< 1

$$V_{pc} := \left(2 \cdot \pi \cdot \left(\frac{3 \cdot t + D}{2} \right) \cdot t \cdot 0.66 \right) \cdot \mu \cdot \sigma_{cr} \cdot$$

shear resistance at critical perimeter

$$u = 741.152 \ kN$$

$$R := \left(\lambda_{DL} \bullet W_G + q \right) \bullet L_x \bullet L_y + \frac{\lambda_{LL} \bullet P}{4} = 268.186 \ kN$$

total factored reaction

$$R_{SLS} := \frac{R}{\lambda_Q} = 182.165 \ kN$$

SLS reaction

$$R := \left(\lambda_{DL} \cdot W_G + q \right) \cdot L_x \cdot L_y + \frac{\lambda_{LL} \cdot P}{4} = 268.186 \ kN$$

$$R_{SLS} := \frac{R}{\lambda_Q} = 182.165 \ kN$$

$$\frac{R}{V_{pc}} = 0.362$$

< 1

$$E := 20000 \cdot \frac{N}{mm^2}$$

Long-term E modulus

$$\delta := \frac{0.185 \bullet q \bullet L_x \bullet \left(\dfrac{L_{rx}}{t} \right)^3}{E} = 1.951 \ mm$$

deflection 3000/1.9 = L/1578

The ACI 544 6R15 method allows, when considering the edge span, a 19 kN/m² UDL intensity with a 1.9 mm long-term deflection in the edge span, taking into account f_{r3} = 4.6 N/mm² (EN 14651), C30-37, ω = 10.36.

A middle field from the second row of piles is suitable to carry (19 kN/m² / 12) x 16 = 25 kN/m².

Clearly, the Nieuw-Vennep slab can support much more load than 15 kN/m² UDL in regular service conditions, as under such a UDL, it undergoes a span/deflection ratio of more than 1500.

Under 25 kN/m², the slab is still stiffer than the EC limit:

1.9 mm x 25 / 19 = 2.5 mm, or 3000 / 2.5 = 1200, equivalent to L/1200, which is much better than the L/500 limit.

It is very likely that the collapse UDL of such a slab falls in the loading range of 50 to 60 kN/m².

Indeed, the ACI 544 6R15 method is both quite conservative and easy to use.

3.5.8 The Vaasa (Finland, 2009) test of a Primekss ground bearing slab over an insulation layer

The test slab had the following characteristics:

The Vaasa test slab was a square slab measuring 4 m x 4 m with a thickness of 100 mm, resting on a 100 mm thick base of EPS 200 polystyrene

Figure 3.46 Vaasa slab test portal frame and hydraulic loading jack.

with E = 10 N/mm², coefficient of Poisson ν = 0, and 0.09 N/mm² long-term compressive strength at 1% deformation onto a rigid 1 m thick concrete raft.

A portal frame, as shown in Figure 3.46, was used to impose, by means of a hydraulic jack, a center point loading intensity using a contact disk plate of 125 mm diameter. The portal frame and test setup are shown in Figure 3.46.

The concrete compressive strength was 32 MPa, and the slab was reinforced with 35 kg/m³ of randomly distributed hooked-end steel fibers, each with a diameter of 0.75 mm, length of 50 mm, and tensile strength of 1100 MPa.

The resulting K_w bearing coefficient of Westergaard was also measured with the 760 mm round plate diameter so that K_w = 30 MPa/m. Hence the radius of rigidity of the slab is calculated:

$$L_r := \left[\frac{E \cdot h^3}{12 \cdot \left(1 - \nu^2\right) \cdot K_W} \right]^{\frac{1}{4}} = 537.285 \cdot mm$$

The point loading intensity was increased in steps of 10 kN up to 270 kN, at which point the hydraulic jack collapsed. Deflection recording was feasible up to an intensity of 213 kN. At 270 kN point loading, the slab did not collapse. The ultimate loading intensity at ULS for the slab was recorded as 270 kN, as it could have resisted an even higher loading intensity.

The slab's collapse load is thus unknown but is likely far greater than 270 kN, as the slab did not show any sign of punching out.

A similar TR 34 4th Edition calculation, as shown in the NATO Test section earlier, estimates an ultimate loading intensity of 180 kN.

As shown in Figure 3.47, when the center point loading intensity increased from 10 kN to 213 kN, the zero-deflection abscissa was located between 500 mm and 600 mm from the load footprint center. Such a distance is the radius of rigidity (Lr), meaning a disk of Lr radius of concrete applies the load onto the insulation with an average contact pressure of 0.11 N/mm² (110 kPa) at 100 kN.

The zero-deflection point, shown in Figure 3.47, is located at a distance equal to the sum of radius of load application and the radius of rigidity (L_r), which is 537 mm here.

At 110 kPa, the EPS 200 (typical 2% deformation under 200 kPa) insulation should experience a total deformation of: 110/200 x 0.02 x 100 mm = 1.1 mm, which matches the experimental data exactly.

We see here that the 110 kN point loading, with a pressure of 110,000/Π x 62.5² = 8.96 N/mm², represents quite a high contact pressure onto the

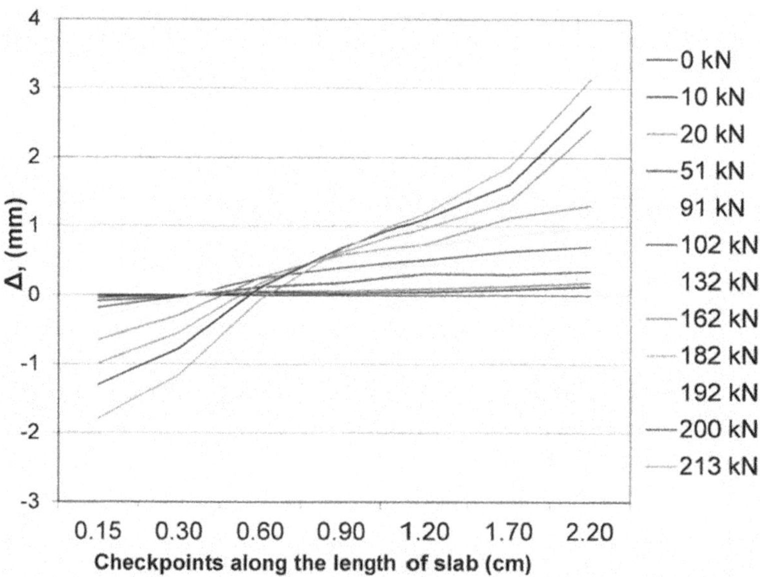

Figure 3.47 Deformed slab diagram of the vertical displacement versus the distance to the center point in meters.

slab. This pressure is distributed onto the insulation within a 12° angle, derived from arctan (537–62.5 = 474), which is far greater than the usual 45° or 60° angles typically assumed.

The lab report also shows that the uplift of the slab under 213 kN point loading intensity starts at a 675 mm distance from the center point. The possible reaction force is calculated as:

0.675^2 x 3.14 x 90 kPa = 129 kN (28 kip), far less than the 213 kN (46 kip). This suggests that the slab is even stiffer than indicated by the elastic back-calculation.

We also observed that no punching-out occurred during the test.

Once again, we understand and confirm our earlier conclusions (see NATO test) that ground-bearing slabs are grossly overdesigned when following documents like TR 34, 4th edition.

However, in real-world conditions, slabs are often under-designed for shrinkage, which is the primary cause of most slab deficiencies. These issues arise due to the materials used, the installation process, and the joints.

Uncontrolled shrinkage cracks are very frequent in slabs on grade, as are joint movements and the induced cracks caused by joint curling and rocking.

The Primekss slab, known as PrimeComposite or PrimX , offers a zero-shrinkage concrete solution through the addition of proprietary binders and steel fibers at high dosage rates, which offset all shrinkage.

A separate chapter is devoted to the Primekss zero-shrinkage mix later in this book.

3.5.9 The TU Eindhoven (Holland, 2011) one-way suspended elevated slab (E-SFRC)

The Technical University of Eindhoven-Holland organized a series of full-scale tests on one-way SFRC slabs installed together with supporting walls.

In September 2008, a technical delegation of the Dutch Concrete Society visited Tallinn (Estonia) to study the two-way E-SFRS system used in the Rocca Tower, a 17-floor building where every floor was constructed with SFRC along with a set of Anti-Progressive Collapse (APC) rebars.

Based on the conclusions of the technical delegation, the Dutch Concrete Society decided to install and test a real full-scale slab in Eindhoven, Holland.

It was, however, decided to test a one-way slab in this case, consisting of three consecutive spans reinforced only with steel fibers and without any APC rebars (also known as structural integrity reinforcing).

The slab was part of a tunnel-form installation, where the slab and the supporting walls were cast together, as shown in Figure 3.48.

The tunnel-forming technique is commonly used in Holland for condos and other residential applications.

Three tunnel-formed slabs with three consecutive spans were installed: slabs 180 mm thick, 2.30 m wide, and with a 5.40 m span, while the supporting walls were 250 mm thick.

Figure 3.48 View of the tunnel-formed test slabs of Eindhoven.

Each form was installed with an 18 mm camber at mid-span, as in the traditional method. The 2.3 m wide tunnel-formed slabs were separated from each other by polystyrene foam, allowing for the testing of nine different spans.

The mix design consisted of 350 kg of CEM I and CEM III per cubic meter, with W/C < 0.50, and included 50 kg/m^3 of Type I hooked-end steel fibers with a 0.9 mm diameter, 60 mm length, and 1100 MPa tensile strength.

The mix had F4 fluidity, requiring only light mechanical vibration during installation. Walls and slabs were poured together as part of the tunnel-formed structure.

As required for this construction method in Holland, form release began 16 hours after concrete installation. This requirement is mandatory to enable the general contractor to reuse the tunnel form as soon as possible for the next floor.

The mix design was formulated to ensure a compressive strength of at least 14 N/mm^2 at 16 hours and 1.4 N/mm^2 in flexion. The 28-day compressive strength at maturity was 65 N/mm^2, recorded on 150 mm cubes. At 28 days, f_{r1} and f_{r4} average flexion strengths, according to EN14651, were 6.1 N/mm^2 and 5.3 N/mm^2, respectively.

The Dutch housing specification requires a long-term variable loading intensity not exceeding 2.5 kN/m^2 in total. The factored loading intensity is calculated as: 1.35 x 0.18 x 24 kN/m^3 + 1.5 x 2.5 kN/m^2 = 5.83 + 3.75 = 9.58 kN/m^2.

The experimental imposed test load was far beyond both the SLS of 2.5 kN/m^2 and the ULS of 3.75 kN/m^2.

Following Eurocode provisions, at SLS, the upper limit of crack opening for traditional RC is 0.3 mm, and the deflection upper limit is 5400 mm/500 = 10.8 mm.

The test slab remained stable at ULS under a (1.5 x 2.5) = 3.75 kN/m² loading intensity, although with deflection and cracking exceeding the SLS limits, reaching values larger than 10.8 mm and 0.3 mm, respectively.

The ULS loading stage serves as a true warning stage, indicating that the building remains stable but exhibits excessive deflections and cracking, necessitating evacuation of the building.

The test loading consisted of a distributed loading intensity applied in incremental steps using a number of concrete blocks of varying sizes to achieve a final surface load of up to 12 kN/m², as shown in Figure 3.49.

Under a 2.5 kN/m² UDL, representing the SLS loading, the total loading was calculated as: 2.5 kN/m² x 5.4 m x 2.3 m = 30 kN total loading intensity. The slab showed only a minute deflection of 0.4 mm with no visible cracking.

Under a 3.75 kN/m² UDL, representing the ULS loading intensity, the total loading was calculated as: 3.75 kN/m² x 5.4 m x 2.3 m = 47 kN total loading intensity. The slab showed only a minute deflection of 0.9 mm, still without any visible cracking in both positive and negative moment areas.

Under an 8 kN/m² uniformly distributed loading intensity, as described above, the deflection of the central span was recorded as 8 mm initially and increased to 10 mm after 24 hours. Five parallel minute bottom cracks, each up to 0.4 mm in width, were visible.

At 8 kN/m² loading intensity, the slabs remained stable.

The slabs collapsed at a loading intensity exceeding 11 kN/m² UDL, with an average imposed collapse loading of 12.18 kN/m². This corresponds to:

12.18 x 5.4 m x 2.3 m = 151 kN total loading intensity, as shown in Figure 3.49.

Figure 3.49 The blocks were deposited by the crane on lumbers at 500 mm spacing apart.

A long-term loading test conducted over six months with a 7 kN per m² intensity (87 kN) on an edge span confirmed no significant visible creep.

We see that the collapse loading intensity is five times greater than the Dutch standard housing limit and that the collapse load is 3.19 times higher than the Dutch standard housing limit at ULS.

The positive moment cracking in the slab bottom began to develop at a loading intensity of 9.67 kN/m² (120 kN), as shown in Figures 3.50 and 3.51.

The negative moment cracking occurred first at 6.5 kN/m² (81 kN) loading intensity.

It is very reassuring to observe the multiple cracking in the bottom of this one-way slab, unlike the behavior observed in all standard-size beam flexion tests, where a second crack never develops, regardless of the fiber type, dosage rate, or concrete mix design.

The maximum allowable UDL on this 180 mm slab, spanning three consecutive spans of 5.40 m and supported by 250 mm thick walls, is 6.0 kN/m² (111 kN), ensuring the slab does not crack under flexion. This shows a safety factor of more than 1.5 up to the ULS load of 9.66 kN/m² (120 kN).

The average loading factor is

$$\Upsilon_{Qm} = (1.35 \times 4.32 + 1.50 \times 9.66) / (4.32 + 9.66) = 1,45.$$

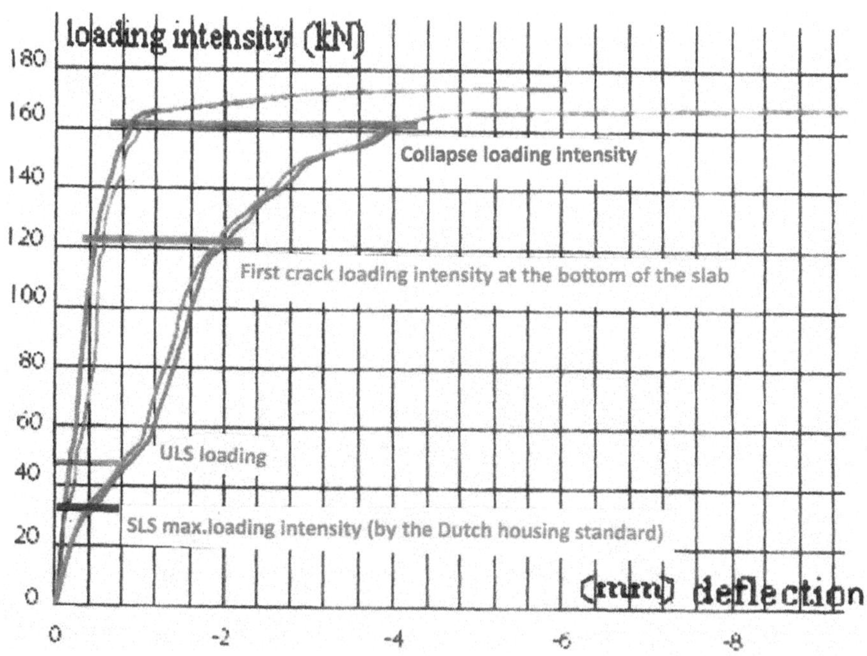

Figure 3.50 Load vs. deflection diagram at SLS and ULS.

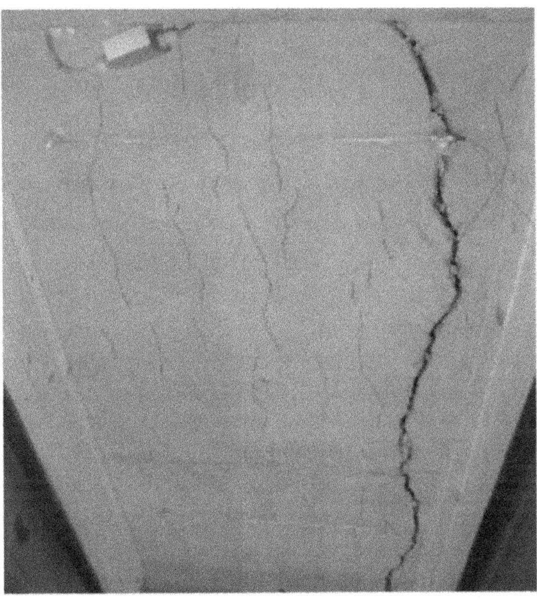

Figure 3.51 View of the SFRC positive moment slab bottom multiple cracking observed.

The ULS moment

$$M_d = 1.45 \times 13.96 \times (5.40 - 0.25)^2 / 16 = 33.55 \, kNm / m$$

We can assume f_{tu} as a constant plastic tensile strength at cracking over 90% of the section depth (Massicottte 2003, Espion 2004):

$M_{Rd} = 33,550$ N mm/mm / (0.45 x f_{tu} x 180 ^{2}) or f_{tu} = 33550/(0.45 x 180^2) = 2.30 N/mm^2

As derived from full-scale tests, such as 2.30 N/mm^2 strength is characteristic

Equating both elastic and plastic moment:

$$0.45 \times f_{tu} \times h^2 = f_r \times h^2 / 6 \, gives \, f_r = 2.7 \, f_{tu} \, thus \, a \, design \, strength \, f_r$$
$$= 2.7 \times 2.3 = 6.21 \, N / mm^2$$

As the TU Eindhoven slab test concrete showed (from the EN 14651 standard flexion test) $f_{r3\,k}$ = 1.53 N/mm^2 at 28 days and 3.73 N/mm^2 at 129 days, it is easy to understand that a multiplying factor for f_{r3k} is needed to maintain the credibility of the EN 146551-derived flexion strength.

Such a factor has already been defined in the EN-Swedish standard for structural SFRC applications (SS 812310). It is called η_{det} (a statical indeterminacy factor), and is equal to 1.4 and 2, respectively, for one- and two-way slabs.

It should be noted that $f_r/1.4 = 6.21/1.4 = 4.43$ N/mm² a realistic value of f_{r3k}

3.5.9.1 *Conclusions of the full-scale test carried out at cold temperature*

- All nine spans tested exhibited quite ductile behavior resulting from significant moment redistribution and a multiple-cracking pattern under sagging moments.
- The deviations between the nine spans tested are significantly smaller than those recorded during the EN14651 beam test.
- The EN14651 test method shows deviations that are inherent to the test method itself and not reflective of the real behavior of the structure as tested.
- The first-crack experimental loading intensity in all nine spans occurred at more than 300% of the standard housing service loading intensity in Holland (2.5 kN/m²).
- The ultimate loading experimental intensity in all nine spans occurred at more than 500% of the standard housing service loading intensity in Holland (2.5 kN/m²).
- The FIB Model Code 2010 calculation method does not reflect the reality of the full-scale test results from three sets of three one-way continuous spans as tested and reported in this paper.

Therefore, an application multiplying factor (η_{det}) like that in the EN-SS 812310 design standard for steel-fiber-reinforced concrete structures, must be applied to the EN14651-derived flexion strength to obtain the design strength. Without this adjustment, the EN 14651 and the FIB Model Code act as deterrent rules for using steel fiber reinforcing as the main reinforcement in suspended slab structures.

The same TU Eindhoven slab test is further discussed in Chapter 6, which covers fire resistance.

REFERENCES

B. Espion. Test Report n° 33.396, Nov.2004, Université Libre de Bruxelles, Faculté des Sciences Appliquées, Génie civil.

B. Massicotte, A. Ing, K. Moffatt. Développements pour l'Utilisation des Fibres dans les Applications Structurales, Rapport de' Recherche, Dpt. du Génie Civil, Ecole Polytechnique de Montréal, Projet CDT-P2867/rapport ST03-15, December 2003.

X. Destree, C. Kleinman, H. Lambrechts. "Steel fibre as only reinforcing in free suspended elevated slabs: design conclusions of a tunnel formed slab and walls based upon full scale testing results" BEFIB 2012- Rilem-ACI.

3.5.10 The Klaipeda test slab in Lithuania (2011) (G-SFRC)

Primekss SIA constructed a suspended slab on piles in Klaipeda, Lithuania, in 2011. The slab is located near the shore of the River Dane in an area with weak sludge and clay-type subbase ground that has very low load-bearing capacity, resulting in plastic settlement that quickly leaves the slab in a suspended elevated condition.

The full-scale testing procedure, commissioned by the owner of the building, was performed by Pr. Dr. L. Furmanavičius of the Vilnius Technical University.

The suspended slab, reinforced with 50 kg/m^3 of Twincone fiber (straight steel wire fibers with conical anchoring ends), is constructed using C30/37 concrete with proprietary Prime DC liquid and hydraulic additives designed to generate an 800 μ-strain restrained expansion, resulting in permanent compression stress across the section. The slab features a 4 m x 4 m pile grid with 1.0 m diameter pileheads and is 210 mm thick to carry 35 kN/m^2 UDL. Each bay covers an area of 4000 m^2, with construction joints approximately 60 m apart.

The full-scale loading test consisted of imposing a total of 29 kN/m^2 UDL in steps of 5.8 kN/m^2 each. The load was imposed over 24 hours, reaching a total load of 464 kN on a pile grid section of 4 m x 4 m. The loaded area was 3 x 4 m in length and 2 x 4 m in width, totaling six fields of 16 m^2 each, or 96 m^2.

Under the 29 kN/m^2 UDL imposed for 24 hours, a deflection of 0.57 mm was recorded at mid-span between two adjacent piles. At the center of the load field (not accessible for measurement), the deflection is likely twice as much, estimated at 1.14 mm, corresponding to span/3500, which is far stiffer than the Eurocode limit of span/500 = 8 mm. Not a single crack, even a very minute one, developed.

The strain gauges embedded in the slab showed a 14 x 10^{-6} strain increase at the field center, indicating a flexion tensile stress of:

14 x 10^{-6} x 35,000 N/mm^2 = 0.49 N/mm^2, which is far below the flexion first crack limit of approximately 5 N/mm^2.

The 30 kN/m^2 UDL remained imposed on the loaded area for 3 months during which a maximum deflection of 0.9 mm was observed. This corresponds to a span-to-deflection ratio of span/0.9 = 4400, which is far stiffer than the Eurocode span/500 limit.

REFERENCE

X. Destrée, R. Cepuritis. Chemically Post-Tensioned steel Fibre Reinforced Concrete Suspended Slabs from Design to application., The Institute of concrete Technology, 2019–2020, London, UK.

3.5.11 The Friends Arena (Stockholm, 2011) proof loading (G-SFRC)

We must also mention the Swedbank/Friends Arena foundation slab in Solna-Stockholm (as shown in Figure 3.47), constructed with a mandatory thickness of 300 mm and a 4 m x 4 m span above 1 m x 1 m pile heads. The slab was designed to resist a service load of 40 kN/m² UDL using a 45 kg/m³ dosage rate of HE + 1/60 steel fiber (1 mm diameter x 60 mm length hooked-end steel wire fibers with 1500 N/mm² tensile strength).

The SFRC, constructed with 16,000 cubic meters of concrete, serves as the foundation slab of a 65,000-seat multi-purpose arena. The Friends Arena was inaugurated in 2011 and has been successful since then.

During the jobsite progress, several huge cranes were used and circulated on top of the slab to install the retractable roof structure. Each crane weighed 700 tons, as shown in Figure 3.52, imposing a 200 kN/m² pressure under the steel caterpillars—five times the maximum service loads.

There was no damage at all, nor were any claims reported.

This slab has been a real success story for the general contractor, who saved six months in the planning process by eliminating all steel cutting, bending, and placing operations.

Consider the cost of running such a large jobsite for one day—possibly 10,000 to 20,000 € when accounting for all fixed costs, such as safety, insurance, administration, office personnel, jobsite offices, electric power, cranes, and tower cranes.

The 45 kg/m³ steel fibers replaced 150 kg/m³ of traditional reinforcing steel, or at least 2,400 metric tons of rebars, which would have required a minimum of 10 workman hours per ton on-site. This resulted in a saving of at least 24,000 workman hours, especially valuable in the harsh Swedish winter conditions of freezing, thawing, icy rain, and cold winds.

It is noteworthy that the structural engineering firm initially in charge—a large and prominent firm—was unwilling to consider the fiber concrete

Figure 3.52 Overloading the Friends Arena SFRC piled raft up to 200 kN/m².

Figure 3.53 Aerial view of the Swedbank/Friends Arena.

solution, preferring instead to adhere blindly to outdated reinforced concrete standards. As a result, the firm was replaced by engineers more knowledgeable in yield line design and steel fiber-reinforced concrete slab design, who signed off on the foundation slab.

After the project was completed, it became widely acknowledged in Sweden that there was a pressing need to draft a new standard to explain and facilitate the structural design and use of steel-fiber-reinforced concrete.

A paper summarizing the challenges of this task was prepared and presented at the FIB-2012 Stockholm conference by X. Destrée, Ir., and Prof. J. Silfwerbrand of the K.T.H.

Note that the arena's initial name, "Swedbank Arena," shown in Figure 3.53 was later changed to Friends Arena.

In 2015, the SS 812310 standard, titled "Fibre Concrete - Design of Fibre concrete structures" was finalized and circulated. This standard serves as a complement to SS-EN 1992-1-1, the Swedish version of the Eurocode.

REFERENCES

SS 812310 "Fibre Concrete- Design of Fibre concrete structures"

X. Destrée, J. Silfwerbrand. "Steel fiber reinforced concrete in free suspended slabs: case study of the Swedbank arena in Stockholm." FIB 2012 – Stockholm conference.

3.5.12 The Tingstad test slab (Sweden, 2014) (G-SFRC)

The full-scale testing was conducted on a selected area of the slab by a third party (CBI Betonginstitutet—The Cement and Concrete Institute) to verify

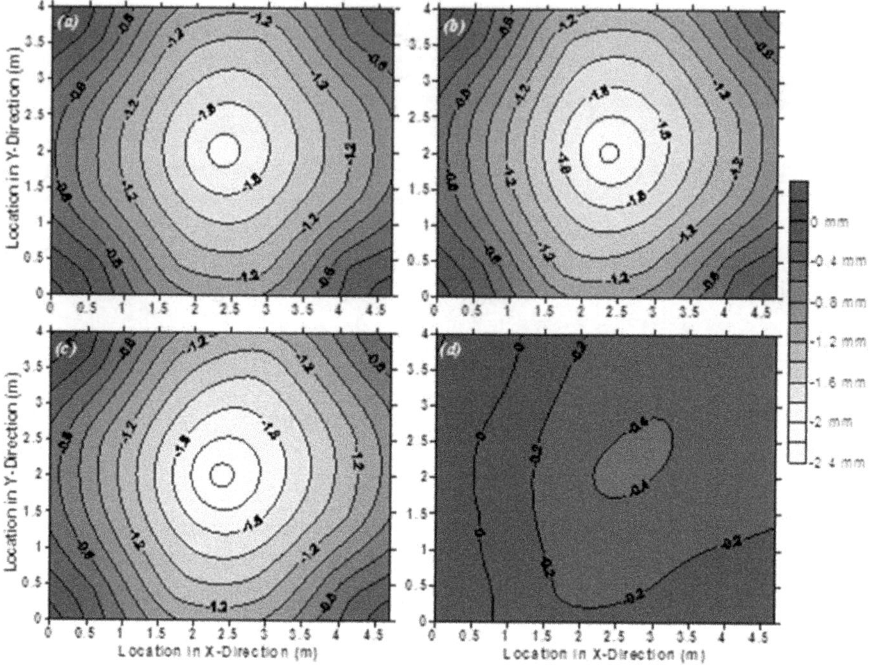

Figure 3.54 The deformed shape of the Tingstat slab.

conformity with project-specific requirements regarding allowable load-induced deflections and crack widths.

The slab is a suspended design with a 4.0 m x 4.7 m grid of piles featuring 1 m diameter pile heads. The slab thickness is 220 mm, constructed using C35-45 concrete, which includes 300 kg CEM I 42.5 N Anläggningcement, and is reinforced with 55 kg/m³ of HE+1/60 fibers from ArcelorMittal. The slab was designed to carry a service load of 40 kN/m² Uniformly Distributed Loading (UDL).

Load testing was conducted by stacking concrete blocks over a small area of the slab. The slab was loaded with 40,320 kg of concrete blocks to achieve 44.8 kN/m² over a 9 m² area and held for 8 days within a 4 m x 4.7 m (18.8 m²) field, resulting in an effective load of 21.4 kN/m².

At 45 kN/m² UDL, the deflection reached 2.1 mm at the center of the span, as shown in Figure 3.54.

This corresponds to a deflection of span/2400, which is far better than required by norms (1/500), project specifications (8 mm), or predictions by the CBI testing program.

The very high stiffness of the slab indicates that it has a load-bearing capability far exceeding the required values..

The deformed shape of the slab, with the average pile settlement removed, is shown after (a) 1 day, (b) 4 days, (c) 8 days of loading, and (d) immediately after unloading.

It demonstrates a span/deflection ratio of 4400 / 2.4 = 1833, which is significantly stiffer than any standard requirement.

At the ultimate stage, such a slab can undergo up to 25 mm deflection, indicating that the maximum slab capacity is a high multiple of the 40 kN/m^2 specified here.

The maximum SLS loading conservatively occurs at a span/deflection ratio of 1200, leading to an allowable loading capacity of: 40 kN/m^2 x 2.833 / 1500 = 75 kN/m^2.

At such high loading intensity, the pile could settle permanently, 2.4 mm x 75 / 21 = 9 mm instant settlement.

The piles are more critical than the slab. Interestingly, this opens the possibility of designing such a Primekss SFRC slab based solely on the maximum pile reaction and the SFRC post-cracking properties.

Indeed, it is a simple method, which will be reviewed further in a specific design chapter.

REFERENCES

CBI Uppdrasrapport KTH/CBI, Provbsbelastning av industrigolvet, Göteborgs hamn.07/02/2014.
P. Sia, K. Kamars. Tingstad Loading Test Report.

3.5.13 The Goteborgs test (Sweden, 2019) (G-SFRC)

This full-scale test has not yet been reported in a public paper by its author, the prominent Professor Björn Täljsten, who has unparalleled experience in bridge testing and engineering. Therefore, we will not unveil the entire test but provide only a short summary of the results, as it is highly relevant to this chapter.

Here, I will summarize the process and present the design data, calculations, test setup summary, and a brief summary of the results.

The test slab was built at approximately 1 m above ground level to ensure there was no support between the piles. The slab is 250 mm thick and rests on a 3.80 m x 3.80 m pile grid. Each pile has a 700 mm diameter and was constructed in open-sky external weather conditions.

The aim of the test was to verify the design of 40 kN/m^2 UDL in service and the factored loading at ULS.

The built test slab covers an area of approximately 320 m^2, supported by 36 piles. The first edge span lengthwise is only 1.80 m. The piles, made of traditional reinforced concrete, were built in the first step.

The next step was to install the fiber-reinforced concrete slab using only steel fiber reinforcement, without any rebars or wire mesh. The mix was C30-37 concrete (320 kg/m³ CEM I, 42.5 N), with 45 kg/m³ of HE+1/60 steel fibers (1 mm diameter, 60 mm length, with hooked ends and a 1500 N/mm² tensile strength). The test slab was installed by Primekss SIA, a Latvian concrete contractor.

The load distribution on the test slab was designed by Dr. G. Plizzari, a prominent Italian university professor in structural engineering renowned worldwide for expertise in fiber-reinforced concrete. A sophisticated elasto-plastic finite element calculation software, which included the post-cracking properties of the fiber-reinforced concrete, was used for the design.

The test UDL was applied following an L-path from the edge to the center of the slab to create the highest possible negative and positive moments.

Figure 3.55 shows an aerial view of the test slab as well as the L-shape loading pattern of SLS and ULS cases.

The test loading program included by means of big bags of heavy aggregates (ca. 50 kN/m³ density):

SLS checking in step 1 2- days loading increase to 40 kN/m² intensity followed by a 7-days loading intensity at a constant 40 kN/m² in order to check the service ability of the slab regarding deflections and crack opening. The unloading happened in1 day.

At 40 kN/ m² loading intensity, 256 tons (2560 kN) of heavy stones in big bags were placed on 64 m² of slab as shown on Figures 3.55 and 3.56.

The following steps 2,3 and 4 are for the checking of the Ultimate Limit State(ULS)

ULS-checking and final collapse:

Step 2: 52 kN/m² imposed during 7 days, then the slab is unloaded (357 tons on 67 m² area)

Step 3: 55 kN/m² imposed during 1 day, followed by the step 4

Step 4: 60 kN/m² imposed during 8 days (350 tons on 58 m² area followed by unloading

The last step, step 6, is to identify the final collapse under a higher loading intensity than ULS.

Step 5: loading increase up to the final collapse at 399 tons (3990 kN) on 51 m² area thus at 78 kN/m² as shown in Figure 3.57.

As reported, the final collapse developed gradually, so it was not a sudden brittle collapse.

In conclusion, it is reported that the test slab met all standard criteria at SLS (a serviceable slab under 7 days of loading duration with limited cracking and deflection), at ULS under 1.5 x SLS loading intensity for 21 days, and at collapse with 30% more than ULS loading.

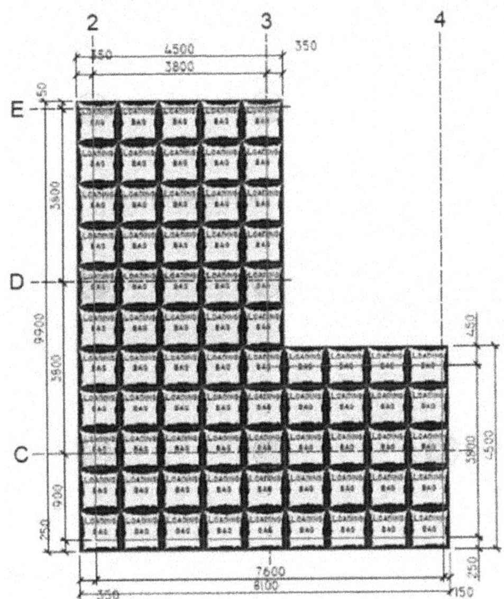

Figure 3.55 Aerial view of the tested slab and the L-shaped of the UDL pattern.

Figure 3.56 Maximum UDL service loads imposed following the most critical loading pattern, indeed in L-shape.

Figure 3.57 Test slab after the final collapse at 78 kN/m².

Summary of most critical results: *At SLS*:
The maximum long-term deflection recorded is of 6,08 mm thus the
span/625, thus better that Eurocode span/500
Crack width: Maximum top crack width recorded of 0.41 mm
Maximum bottom crack width recorded of 0,08 mm.
At ULS:

Step 2: an 8 mm deflection instantly together with 0.8 mm crack width on
top and 1.86 mm in the bottom, followed by a 18 mm deflection after
6 days. Crack depth of 97 mm at edge bottom and 32 mm in top
Step 3: 18 mm deflection observed instantly
Step 4: 22.1 mm deflection after 7 days; after unloading, a permanent
deflection of 14,9 mm
The crack width under 60 kN/m² during 7 days was limited to 2.25
mm
Step 5: Just before the collapse, the maximum deflection was of 30 mm
while the last recorded deflection was 55 mm and the last crack width
of 3 mm at collapse.

The final collapse of the slab was not brittle but resulted in a "calm" process,
as defined by the expert professor organizing the test.

The general conclusion is that the test slab met all standard requirements
for suspended slabs under a 40 kN/m² imposed SLS loading intensity.

It is interesting to observe that a 3 mm maximum crack opening, typical
of the f_{r4} (EN 14651), was observed only under the collapse loading intensity (78 kN/m²), far beyond the ULS loading, and associated with a 55 mm
deflection, resulting in a span/deflection ratio of 69. Is f_{r4} truly a relevant
design parameter?

This observation further demonstrates the limitations of TR 34 4th
Edition in its design methodology for SFRC piled slabs. The use of the f_{r4}

derived from the EN 14651 small standard beam test leads to significant overdesign, resulting in unnecessary resource use and environmental waste.

Based on the numerous successful full-scale slab tests and results exceeding expectations according to standards, it is difficult to understand why the following statement:

"Fibres may be used in place of the transverse reinforcement but may substitute the main longitudinal reinforcement only up to 30% of the total required amount," will be included in the last 2024 actual Eurocode draft version

Such a statement should be limited to specific beam types and should never be generalized to all SFRC applications, especially not to statically indeterminate slabs.

We can also see that the test slab is fully verified using the ACI 544 6R15 design method.

The calculation verification is done according ACI 544 6R15 of the American Concrete Institute.

$t := 250 \, mm$ — Slab thickness

$f_{ck} := 30$ — Cylinder strength(characteristic)

$f_{ckc} := 37$ — Cube strength

Tensile strength of plain concrete

$$f_{ctm} := 0.3 \cdot \left(f_{ck}\right)^{\frac{2}{3}} = 2.896$$

Strength ratio

$$\omega := \frac{f_{ck}}{f_{ctm}} = 10.357$$

$f_{r3} := 4.59 \; N/mm^2 \; @ \; 45 \; kg/m^3 \; HE+1/60$

ArcelorMittal

EN 14651 - average value

$\sigma_{cr} := f_{ctm}$

In ACI 5448R16, p.13

$$\mu := \frac{f_{r3}}{3.104 \cdot f_{ctm}} = 0.511$$

$$u := 1 \, \frac{N}{mm^2}$$

(Eq.6.4.1b and H.8)

$$m_{Rdaci} := \frac{3 \cdot \omega \cdot \mu}{\omega + \mu} \cdot f_{ctm} \cdot \frac{t^2 \cdot u}{6} = 44.04 \cdot kN \cdot \frac{m}{m}$$

$\gamma_c := 1.5$

$$\gamma_c := 24\,\frac{kN}{m^3}$$

Concrete volumic mass

$$q := 40\,\frac{kN}{m^2}$$

Unfactored UDL

$$W_G := \gamma_c \cdot t = 6 \cdot \frac{kN}{m^2}$$

Slab weight

$$D := 700\,mm$$

Pile diameter

$$L_x := 3.80\,m$$

spans

$$L_y := 3.80\,m$$

$$\lambda_{DL} := 1.4$$

Load factor on the own weight

$$\lambda_{LL} := 1.6$$

Variable load factor

$$\lambda_Q := \frac{(\lambda_{DL} \cdot W_G + \lambda_{LL} \cdot q)}{W_G + q} = 1.574$$

Average load factor

$$L_{rx} := L_x - D - t = 2.85\ m$$

Net spans

$$L_{ry} := L_y - D - t = 2.85\ m$$

$$\Phi_P := 0.90$$

reduction factor of resisting moment design moment Middle span (eq.7.5.3c and 6.3.1.1m)

$$M_{Px} := \left[\frac{(\lambda_{DL} \cdot W_G + \lambda_{LL} \cdot q) \cdot (L_{rx})^2}{16}\right]$$

$$= 36.754\,\frac{1}{m} \cdot kN \cdot m$$

$$\frac{M_{PX}}{\Phi_P \cdot m_{RPC}} = 0.927$$

<1, OK

---

punching shear verification:

Shear resistance at critical perimeter (eq.7.6a)

$$V_{pc} := \left[2 \cdot \pi \cdot \left(\frac{3 \cdot t + D}{2}\right) \cdot t \cdot 0.66\right] \cdot \left(\mu \cdot \sigma_{cr} + 0.175 \cdot \frac{\sigma_{cp}}{u}\right) \cdot u = 1.111 \times 10^3 \cdot kN$$

$$R := (\lambda_{DL} \cdot W_G + \lambda_{LL} \cdot q) \cdot L_x \cdot L_y = 1.045 \times 10^3 \cdot kN$$

Total factored reaction

$$R_{SLS} := \frac{R}{\lambda_Q} = 664.24 \cdot kN$$

Total unfactored reaction

$$\frac{R}{V_{pc}} = 0.941$$

<1, Ok

Deflection

$$E := 15000 \frac{N}{mm^2}$$

Long term modulus of elasticity PC

$$\delta := \frac{0.185 \cdot q \cdot L_x \cdot \left(\frac{L_{rx}}{t}\right)^3}{E} = 2.777 \cdot mm$$

Eq. J.2 (ACI 5446R15) deflection:

$$\frac{L_x}{\delta} = 1.368 \times 10^3$$

Span-to-deflection ratio > 500, OK! (Eurocode 2 limit) The slab is a lot stiffer!

$$\lambda_M := \frac{\lambda_{DL} \cdot W_G + \lambda_{LL} \cdot q}{W_G + q} = 1.574$$

Average loading factor

$$P_{ULT} := \lambda_M \cdot \frac{(W_G + q)}{M_{PX}} \cdot (\Phi_P \cdot m_{RPC}) = 78.076 \cdot \frac{kN}{m^2}$$

Proposed collapse loading intensity

$$\delta_{NCC} := \frac{0.420 \cdot q \cdot L_x \cdot \left(\frac{L_{rx}}{t}\right)^3}{E} = 6.305 \cdot mm$$

Revised expression to match the 7-day test load result

REFERENCE

Fullskaleprovning av pelarunderstödd platta – Provningsprogram 2019-11-10, Björn Täljsten

3.6 CONCLUSIONS

The test slab, 250 mm thick and constructed with C30-37 concrete reinforced with 45 kg/m³ HE+1/60 steel fibers, is designed to carry a 40 kN/m² UDL in service, following the ACI 544 6R15 design methodology.

We can also compare to moments following the EC2 2027 draft Annex L for SFRC:

$h := 250mm$ — Thickness

$L := 3.80m$ — Span

$f_{rlk} := 4.41 \dfrac{N}{mm^2}$ — EN 14651 SLS

$f_{r3k} := 4.59 \dfrac{N}{mm^2}$ — EN 14651 ULS

$f_{Fts} := 0.37 f_{rlk} = 1.632 \cdot \dfrac{N}{mm^2}$ — EN 14651 SLS tensile strength

$\gamma_f := 1.5$ — Material factor

$$M_{Ud} := \left[\left(0.57 \cdot f_{r3k} - 0.26 \cdot f_{rlk}\right) \cdot \dfrac{h^2}{2} + \left(f_{Fts} - 0.57 \cdot f_{r3k} + 0.26 \cdot f_{rlk}\right) \cdot \dfrac{h^2}{6}\right] \cdot \dfrac{1}{\gamma_f} = 31.744 \cdot kN \cdot \dfrac{m}{m}$$

$$M_{Ud} := \left(1.14 \cdot f_{r3k} - 0.15 \cdot f_{rlk}\right) \cdot \dfrac{h^2}{6 \cdot \gamma_f} = 31.744 \cdot kN \cdot \dfrac{m}{m}$$

ULS resisting moment Material only

$u := 1m$

$A_{ct} := \dfrac{h \cdot L}{u^2} = 0.95$ — A_{ct} scale factor $< 1.5 = 1 + A_{ct} \times .5 < 1.5$

$\kappa_{rd} := 1 + 0.5 \cdot A_{ct} = 1.475$ — A_{ct} in m²

$M_{Rd} := \kappa_{rd} \cdot M_{Ud} = 46.822 \cdot kN \cdot \dfrac{m}{m}$ — ULS structural resisting moment.

The EC2 2024 Annex L M_{Rd} is thus very similar to the M_{Rd} by ACI 544 6R.

The Goteborg slab is quite stiff, with K_{Rd} at almost its maximum of 1.5.

In the case of a slab 150 mm thick with a 2.5 m span, K_{Rd} = 0.5 x 0.16 x 2.50 + 1 = 1.20, which is almost 25% less than the 1.5 maximum allowed value.

Sadly, EC2 Annex L is a disadvantage for thinner slabs on piles placed in greater numbers and close to each other. This is the case when spans are between 2 m and 3 m, such as when high-speed piles are used—an economical and quick method for installing up to 60 piles by each machine in one shift.

3.6.1 A full-scale test of elevated suspended slab (SFRS type) reported in the literature

There are very few cases reported, but we can comment briefly on two important references.

The reference. [12] in Chapter 3 reports fully the repetition of the Bissen test in 2004, the number 4 test in this chapter. All sizes are the same in both (3 x 6 m - 3 x 6 m; 200 mm thickness; center point loading in center and edge span.

There were however differences as shown in Table 3.11:

Bissen:

Unlike the Bissen test, the X1 slab did not include the APC 3DIA 16 mm bottom rebars spanning from column to column.

We firmly disagreed with such a lack and decided to resign, thus not participating in the test process. Our experience, dating back to 2002 with Prof. B. Massicotte at Polytechnique de Montréal (Canada), as detailed in Reference (9), has shown that all SFRC suspended elevated slabs must include APC rebars following the Canadian Reinforced Concrete Code of 1992.

These rebars are designed to keep the slab suspended, even with significant deflections, in case a column accidentally disappears. This event was also tested in Bissen by Prof. U. Gossla when the slab was subjected to the maximum service loading intensity.

The X1 slab used a mix design including 400 kg of CEM I 52.5 and CEM V, which proved to be very prone to early cracking.

The mix was quite 'light,' with only 2290 kg/m³ density, much lower than our permanent recommendation of over 2380 kg/m³.

It is well known that water and insufficient compaction can reduce the density. Low-density concrete is weak and prone to excessive shrinkage.

During installation in a cold and windy November, the slab cracked rapidly before hardening. Consequently, the first flexion crack during the test appeared at 100 kN center-point loading, half of the 200 kN at Bissen, and the slab collapsed at 325 kN, almost 30% less than the 450 kN at Bissen.

Our design principle for suspended slabs, based on R&D findings, is that such slabs should not crack in flexion or shearing under the most onerous service loading, ensuring full serviceability throughout their lifespan.

While the report in reference (12) confirms that the slab met all FIB MC 2010 provisions, despite the poor quality of the concrete and its installation, it fails to mention the sudden collapse of the test lab when the edge was loaded last. The slab was severely cracked by shrinkage and the center-point loading of the central span.

The collapse was so sudden that a technician sustained broken bones and was out of work for weeks. It was fortunate that nobody was beneath the slab at the time, avoiding potential fatalities.

Table 3.11

	Fibers and dosage	Concrete	Traditional reinforcing	Lab test	First crack	Collapse kN -deflection
Bissen (L)	100 kg/m³ TABIX 1.3/50 f y = 900 MPa	C30 37 350 CEM I 42.5 120 PFA W/C = 0.50	APC rebars 3 dia 16 mm	fe3=4.8N/mm²	200 kN	450 kN 60 mm
X1 Limelette (B)	70 kg/m³ HE+1/60 fy = 1500 MPa	C30-37 **200 CEMI 52.5** **200 CEM V** **2290 kg/m³**	**NO APC rebars**	fib MC 2010 class 5c and 3e	100 kN	325 kN 60 mm

The lesson is very clear and of the utmost importance for SFRC elevated suspended slabs: it is not a process where the question is simply about replacing rebars with steel fibers based on questionable standard testing to define performance.

Designing and building an SFRC elevated suspended slab is a precisely defined process where the use of APC rebars is completely mandatory, along with high-quality, high-density concrete and installation by a skilled team capable of properly curing the slab.

The authors write that first cracking occurred at 8.7 kN/m² UDL (i.e., 261 kN!), which is 13% higher than the 7.7 kN/m² quasi-permanent combination with 7 mm deflection. At a UDL of 9.8 kN/m², the cracking slightly exceeded the 0.3 mm limit.

Only at ULS did cracks appear on the surface of the slab. The slab, "loaded up to 14 kN/m², didn't show signs of pre-failure damage, so the loading process continued until reaching 16.0 kN/m². During 25 hours, the maximum deflection increment was 10 mm."

Further testing was stopped due to the propagation of considerable shear cracks around the corner column, although the prototype maintained its structural integrity.

Authors conclude:

> The SFRC of the same f_{r1} f_{r3} proved to be sufficient to guarantee a suitable structural response for load representative of residential buildings in a column-supported slab with a span to depth ratio of 30.

As earlier in this chapter, we deplore that the Eurocode draft of 2024 will include that

> Fibres may be used in place of the transverse reinforcement but may substitute the main longitudinal reinforcement only up to 30% of the total required amount.

This kind of sentence is highly destructive to the SFRC structural design of slabs and is extremely regrettable, given the vast and satisfactory experience accumulated on over 30 million m² over the past 30 years.

It is very regrettable, especially when considering the advantages of SFRC over traditional reinforcement in terms of:

- Reducing direct production costs,
- Shortening the critical path in planning by weeks,
- Improving safety, as rebars and meshes are frequent causes of accidents and injuries,
- Enhancing durability, and
- Significantly reducing the carbon footprint compared to rebars and meshes.

Why such a stark difference between the FIB/EC2 Annex L approach to SFRC and other methods?

There is a notable contrast between the Eurocode/FIB committee's method of working and that of the ACI committees.

Indeed, ACI committees comprise dozens of multidisciplinary members (engineers, architects, material suppliers, contractors, code-writing bodies, and government agents) who collaborate effectively to prepare practical documents addressing well-defined questions, methods, or processes.

A structured balloting process ensures that each member can cast justified vetoes, which are then addressed through contradictory discussions in writing. The veto author alone has the authority to lift their veto or maintain it, depending on the arguments presented by other committee members.

If the veto is not resolved, the entire process halts until it is addressed.

At any time, any of the 25,000 ACI members are allowed to ask questions, each of which requires a written, justified reply.

The whole process is conducted in writing, ensuring that there are no obscure deals or decisions that go against common sense or the interests of the industry.

Finally, the ACI TAC Committee must approve or reject the first draft before it is sent back to the committee for amendments and another balloting process. The draft then returns to the TAC Committee once more for final approval.

The ACI 544 6R15 report on the design and construction of SFRC suspended slabs faced 256 vetoes across dozens of ballots!

No wonder ACI Committee documents are clear, practical, straightforward, and to the point.

Regarding SFRC, Eurocode 2 committees are mainly composed of academics, typically one or two per country, many of whom lack practical job site experience or real experience with SFRC.

Eurocode coefficients often result from negotiations among a small group of actively working members.

I have often said that with such an organization and working method, reinforced concrete would never have been developed in the late 19th century. Instead, construction would have remained reliant on bricks and stone blocks.

Many positive building solution opportunities will be lost in this rapidly changing, demanding, and challenging world.

Indeed, rigid design norms are a significant hindrance to the future of structural building.

These norms fail to contribute to the optimum future, as they discourage innovation and prevent practical discoveries from being developed.

Their primary function appears to be indemnifying norms-abiding designers, regardless of the real and practical needs of the project and how it will be built.

In Europe, there are two government bodies negatively impacting the future of construction. One of them is the German DIBT Institute, a large organization that must approve anything deviating from German standards. This process can take 5, 10 years, or more, and is often very costly before anything drastically novel becomes possible only through leapfrogging.

I have often used the example of the Holocaust Memorial in Berlin to illustrate this point.

This open-air park, located behind the Brandenburg Gate, consists of 2,400 hollow dark concrete blocks, each measuring 1 m wide, 4 m long, and 2 to 4 m high, installed 1 m apart in a rectangular pattern. The monument was designed with a 300-year life expectancy, which posed a significant challenge.

A committee of the twelve most prominent concrete academics in Germany was appointed to develop and decide how to construct the monument.

It was soon admitted that no standard reinforcing or material types could meet the requirements.

It was therefore decided to construct the monument using plain standard concrete with specific mechanical properties: very high strength and low modulus of elasticity. The mix design was developed and installed.

However, two to three years after completion, around 2007, many blocks began to crack. I visited the monument in 2018, where I observed large cracks visible from 30 m away, affecting so many blocks that it posed a hazard to pedestrians, who could be injured or crushed by falling blocks.

To mitigate this risk, many blocks were fitted with unsightly stainless steel ring belts.

One of the 12 design experts explained to me that the issue was not the mix design, which would have met the requirements, but rather that the concrete supplied on-site did not conform to their specifications.

The nonconforming concrete supply continued for two years.

My understanding is that the specification aimed for a standard solution but ultimately failed to provide a real solution capable of meeting the requirements and needs.

In France, the construction industry must overcome the CSTB, a ministerial organization touted as "le futur de la construction" (the future of construction). Like the German DIBT, it is a highly bureaucratic body employing thousands of public agents. It delivers certification stamps for all processes outside the DTU (Document Technique Unifié).

The CSTB process is very expensive, as it does not accept third-party laboratory testing results, requiring all available tests to be redone. A large upfront payment is required without any guarantee of a positive outcome.

Such a heavy bureaucratic system did not prevent the collapse on 23[d] of May 2004 of a showcase of the French engineering, the recently built terminal building at CDG airport (Paris), which had an innovative concrete structure and shape and was provided with all the necessary stamps. The collapse occurred suddenly, fortunately in the middle of the night, but tragically resulted in four casualties.

If such a bureaucratic system had existed at the end of the 19th century, the famous French entrepreneurs and engineers who made France renowned for brilliant engineering inventions in concrete structures would not have been allowed to achieve their successes.

It is sad that the two most important countries in Europe have their construction industries constrained by such restrictive systems, effectively stifling innovations in structural concrete reinforcing.

REFERENCES

1. J. Silfwerbrand. "Industrial fiber concrete floors." *Concrete International*, p. 33, May 2021.
2. X. Destree, B. Mobasher. "Industrial floors with fiber-reinforced concrete: A review of knowledge and experience." *Concrete International*, p. 28, July 2022.
3. A. Orbe, J. Cuadrado, R. Losada, E. Rojf. Framewor of the design and analysis of steel fiber reinforced self-compacting concrete structures, Construction Engineering Area, Dpt. F Mechanical Engineering Faculty of Bilbao, University of the Basque Country (UPV/EHU).
4. Technical Report N° 63: Guidance for the Design of Steel-Fiber-Reinforced Concrete. Report of a Concrete Society Working Group (March 2007).
5. ASTM C 1550: Standard Test Method for Flexural Toughness of Fiber Reinforced Concrete (Using Centrally Loaded Round Panel).
6. Design of floors on Ground, Technical Report 550, J.W.E. Chandler Cement and Concrete Association, 1988.
6. Concrete Industrial Ground floors, Technical Report 34, 4th Edition, The Concrete Society, 2013.
7. X. Destrée, R.C. Modern Developments, The University of British Columbia, 1995; pp. 76–86, "Twincone structural Steel Fiber Reinforced Concrete".
8. B. Mobasher, Mechanics of Fiber and Textile Reinforced Cement Composites, Chapter 16, pp. 295–314, CRC, Taylor and Francis Group, 2012.
9. B. Massicotte, K. Moffatt. Développements pour l'Utilisation des Fibres dans des Applications Structurales, Polytechnique de Montréal, Projet CDT-P2867/rapport ST03-15, 2003.
10. CSA Standard A23.3-94, 13.11.5.1 p.98.
11. I. U. Gossla. "Development of SFRC free suspended elevated flat slabs." Full scale Test, Research Report prepared for Trefil ARBED Bissen SA, Aachen University of Applied Sciences, Dpt. of. of civil engineering, May 2005.
12. On the reliability of the design approach for FRC structures according to fib Model Code 2010: the case of elevated slabs. M. di Prisco, P. Martinelli, B. Parmentier, Structural Concrete(2016), N°4.p.588–601.
13. I. U. Gossla. "Development of SFRC Free Suspended Elevated Flat Slabs", Full scale Test, Research Report prepared for Trefil ARBED Bissen SA, Aachen University of Applied Sciences, Dpt. of.
14. X. Destrée, B. Pease. "Reducing CO2 emissions of concrete slab constructions with the PrimeComposite slab system." ICCS 2013, Tokyo.

OTHER RELEVANT REFERENCES

ACI 544 6R-15, Report on Design and Construction of Steel fiber Reinforced Concrete Elevated Slabs; on p.11, 5.5; on p.34–35, Appendix H.

A. Maturana, J. Canales, University of the Basque Country (Spain): Experimental-theoretical study of SFRC plates supported on columns Construction and Building Materials, Conbuildmat-D-13-00550, 2013.

C. Soronakom, B. Mobasher, X. Destrée. *Modeling of center loaded round panel test using finite element simulations of rigid crack model*, Arizona State University, Dpt of Civil and Environmental Engineering, 2009.

X. Destrée. SFRC in Free Suspended Elevated Slabs: Full Scale Testing Results, Design Assisted by Testing Route, Comparison to the Latest SFRC Standard Documents. FIB 2008, Chennai, India.

C. Soronakom, B. Mobasher, X. Destrée. *Elevated Slabs with Steel Fibre Reinforced Concrete: Full scale testing and inverse Analysis of FRC Round Panel Tests*, Arizona State University, Dpt. of Civil and Environmental Engineering, 2007.

X. Destrée. Structural Applications of Steel fibers as Principal Reinforcing: Conditions, Design, Examples, Lyon, BEFIB, 2000.

C. Paciorek. *PV n°UA-05-08-94, Annexe 2*, Université d'Artois, France, 1994.

ACI 544 6R-15, Report on Design and Construction of Steel fiber Reinforced Concrete Elevated Slabs; on p.11, 5.5; on p.34–35, Appendix H.

Fullskaleprovning av pelarunderstödd platta – Provningsprogram 2019-11-10, Björn Täljsten.

X. Destrée. Structural Applications of Steel fibers as Principal Reinforcing: Conditions, Design, Examples, Lyon, BEFIB 2000.

C. Paciorek. *PV n°UA-05-08-94, Annexe 2*, Université d'Artois, France, 1994.

B. Espion. Test Report n° 33.396, Nov.2004, Université Libre de Bruxelles, Faculté des Sc Projet CDT-P2867/rapport ST03-15, 2003 Sciences Appliquées, Génie civil.

B. Espion. Test Report n° 33.396, Nov. 2004, *Université Libre de Bruxelles*, Faculté des Sciences Appliquées, Génie civil.

B. Massicotte, A. Ing, K. Moffatt. Développements pour l'Utilisation des Fibres dans les Applications Structurales, Rapport de' Recherche, Dpt. du Génie Civil, Ecole Polytechnique de Montréal, Projet CDT-P2867/rapport ST03-15, December 2003.

X. Destree, C. Kleinman, H. Lambrechts. "Steel fibre as only reinforcing in free suspended elevated slabs: design conclusions of a tunnel formed slab and walls based upon full scale testing results" BEFIB 2012- Rilem-ACI.

X. Destrée, R. Cepuritis, Chemically Post-Tensioned steel Fibre Reinforced Concrete Suspended Slabs from Design to application., The Institute of concrete Technology, 2019–2020, London, UK.

CBI Uppdrasrapport KTH/CBI, Provbsbelastning av industrigolvet, Göteborgs hamn.07/02/2014.

Chapter 4

SFRC pile supported slabs (G-SFRS)

During the last 30 years, the structural use of steel fibers as the only principal reinforcing material has been under constant technical development in order to replace (partially or totally) the traditional reinforcement of concrete.

The SFRC pile-supported slab application was certainly, from 1993, the first structural solution to be offered. The piles are used as a support when the ground in place is not stable, too much compressible and prone to plastic settlements.

Such a step further was achieved by experience we attained in steel fiber reinforcing and its shrinkage cracking control over long distances between construction joints in SFRC jointfree slabs on grade from the early 1980s.

Indeed the piled slabs are subjected to a significant restraint by friction over each pile. Thus, shrinkage and cracking are important issues.

As for joint-free floors on grade, a low shrinkage concrete has to be used. Effective curing is also essential to avoid early cracking. Early age cracks often develop further over time, exceeding the serviceability limit.

The technical conditions regarding steel fibers and concrete specifications are reviewed hereafter as well as design methods, shrinkage, and testing.

Two practical design examples are outlined, among the ca. 5,000,000 m^3 piled SFRC slabs completed so far.

4.1 INTRODUCTION

Full-scale testing up to the ultimate loading, of SFRC suspended slabs, concludes and confirms the following points:

1. The observed ultimate loading intensity results from a completely ductile rupture process where the suspended slab deforms along yield lines where all rotations are concentrated.
2. The observed ultimate loading intensity depending on the case investigated, ranges from 3 to 4 times the first flexion crack loading.
3. Regardless of the point loading intensity, the slab doesn't punch out around the piles.

DOI: 10.1201/9781003188315-4

4. Yield lines moment, as described by Johansen in Reference (1), are easily back-calculated. From the shortest pattern of yield lines, as observed in full-scale testing, the yield moment of the slab is consistently defined and repeated across tests. Quasi-full-scale testing in a laboratory when statically indeterminate circular slabs (supported along the whole perimeter of the slab) are subjected to center point loading up to rupture supplies the same slab yield moment.

4.2 TYPICAL STRUCTURAL STEEL FIBER REINFORCED MIX SPECIFICATION

It is expected that the slab section is able to develop yield lines according to the theory of Johansen. Therefore:

Saturate the matrix with closely spaced steel fibers, taking into account the concrete aggregate grading as well as the maximum size of aggregate. Typical dosage rates range from 35 kg/m³ up to 50 kg/m³ of steel fibers of 0.8 mm to 1 mm wire diameter (ASTM type I) by 50 to 60 mm length and tensile strengths of more than 1100 MPa The fibers feature efficient anchoring (a deformed-shape fiber indeed) to the matrix, as appraised by the EN 14651 post-cracking stresses in flexion f_{r1} and f_{r3}. Moreover, the mix needs to be and remain well-functioning, often pumpable and doesn't need any mechanical vibrating.

Various types of steel fibers are possible and have been used so far. They are still recommended and preferred for the effective reinforcing of concrete:

- Undulating steel fibers: 1 mm x 60 mm, 1450 MPa strength, undulation depth of 0.8 mm at the axle with a 7–8 mm pitch of undulation (brand names are TABIX, AFT, etc.)
- Hooked end steel fibers: 1 mm x 60 mm or 0.9 mm x 60 mm, 1450 MPa tensile strength brand names like HE+, D4, etc.)
- Twincone fibers: 1 mm x 54 mm length of 1100 MPa tensile strength, with a 60° angle of the conical heads at each end were used up to the year 2012.

The workability of concrete (Slump test) is only moderately affected (typically 60 mm slump loss after steel fibers are integrated into the concrete matrix) at such dosage rates of the suitable steel fibers.

HE 90/60 steel fibers, a hooked-end steel fiber with 0.9 mm diameter, 60 mm length, and 1200 N/mm² steel wire tensile strength, are also used. The type name such as HE 65/60" reflects the length over diameter ratio (which is 65) and length (60 mm), which is identical to the HE 90/60 or HE 0.9/60.

The dosage rate, by design and long experience starting in 1993, is 40 kg/m³, 45 kg/m³, or 50 kg/m³. Smaller dosage rates are possible but additional reinforcing in rebars and wiremesh is needed, depriving the system from one of its main advantages, the omission of all traditional reinforcement.

Following Romualdi's theory, the steel fibers in the concrete mix are separated by 16 to 19 mm average distance together with a 20 mm (3/4 in) maximum aggregate size.

Larger aggregates cause smaller post-cracking strengths, while smaller aggregate sizes such as pea gravel (up to 12 mm) increase the shrinkage as they require more mortar in the mix design. Larger aggregate sizes can also increase the fiber showing on the concrete surface.

To summarize, the mix design should be as follows at minimum:

-Continuous aggregate grading from 0 to 20 mm. A small fraction of a larger aggregate is sometimes acceptable. The aggregate grading curve should fall within the shaded envelope, as shown in Figure 4.1:

Cement type CEM I, II, or III at 330–350 kg/m³ content: CEM II shall be examined carefully to ensure no inclusion of limestone fillers exceeding 12% of the cementitious content unless it is proven that such a cement is not prone to cracking under site conditions. Inclusion of 35% ground granulated blast furnace slags (GGBS) or mixtures with fly ashes (Class E, ASTM) are acceptable.

In case of a very thick slab design, it is advised to use low-heat cement.

Follow the standard guidelines for the concrete temperature to be at most 25°C and at least 10°C at arrival on the job site.

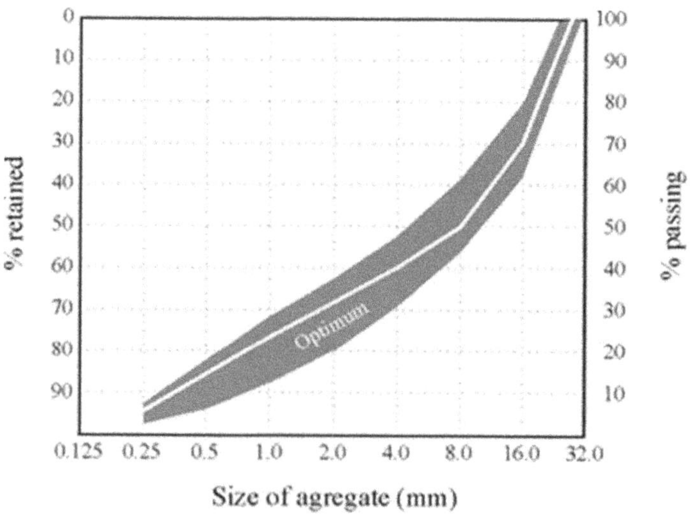

Figure 4.1 Aggregate grading envelope.

- W/C ratio should be less than 0.50 (W: free water), with a 50–75 mm slump of plain concrete before any additions of fibers and workability admixtures.
- The required workability is obtained using an HRWRA (superplasticizer), taking into account a slump loss of 40–80 mm after the introduction of steel fibers. A final slump of 180 mm is advisable so that no poker vibrating is needed. When poker vibrating is used for flowing SFRC around starter rebars and details, extensive attention is needed to avoid segregation. The clear distance between these local rebars should be no less than 120 mm. Otherwise, the compacting of the SFRC becomes difficult, and moderate poker vibrating is needed.
- The experience on-site indicates that the structural fiber reinforced concrete (SFRC) of this kind is pumpable over 100 m distance provided that the pump and the pipes are of 125 mm (5 in.) minimum diameter. Careful attention is needed to verify that the fine particles (smaller than 200 microns, cement included) content of concrete is at least of 450 kg/m^3. Pipes of 4 in. diameter requires more mortar content per cubic meter, which will increase shrinkage and make the mix more sticky.
- The plain concrete supplied should be C30/37 as a minimum. Most often the final hardened concrete will prove to be closer to C 35–45, if not even a C40–50 at 90 days of maturity. Its fresh unit weight shall be close to 24 kN/m^3, as a smaller value results from additional free water in the mix.
- The installation of the fiber concrete on site requires only light surface vibration.
- Floating the concrete surface is recommended to improve flatness and help prevent fiber from showing on the surface.

A laser screed machine helps to minimize fiber visibility, while other finishing tools, such as bull float as shown in Figure 4.2, are also helpful to push the fibers below the very surface of concrete, to achieve an almost fiber-free slab appearance. Mechanical power troweling should start at the right time—not earlier than when the heel of footprint is at a maximum of 3 mm; otherwise, the power trowels will pull back fibers to the surface. Similarly, it should not start too late, as the power trowels will no longer be able to push the surface fibers back down.

Slender steel fibers, with aspect ratios (length-to-diameter ratio) ranging from 65 to 100, are more likely to show on the surface, together with more slump loss so that the water content together with the superplasticizer content will have to be increased and become the cause of more shrinkage, more construction joint opening, more cracking, and more fiber visibility on the surface. The pumping can become a real burden on site when too slender steel fibers are used.

Figure 4.2 A typical bull float used for better flatness and less fiber showing on the surface.

Note that some concrete pumps with too narrow pipe diameter or with an elbow at the bottom of the hopper could prove the pumping of SFRC impossible.

The possible segregation of the SFRC in the pump hopper makes it impossible to pump. Vibrating hoppers should be avoided as they can cause SFRC to segregate.

4.3 STANDARD FLEXION STRENGTHS

The flexion strength is defined by the EN 14651 standard where a 600 mm x 150 mm x 150 mm beam size is used, for a span of 500 mm subjected to a mid-span concentrated loading above a 25 mm notch in the bottom. The notch acts as a crack localizer so that the samples can crack only once to form a plastic hinge. The loading intensity is recorded in function of the net deflection or the crack mouth opening displacement (CMOD). An "elastic" flexion stress is calculated at CMOD 1 = 0.5 mm to determine f_{r1} (N/mm^2) and at CMOD 3 = 2.5 mm to determine f_{r3}.

The piled slabs SFRC show the average values of f_{rim} in the range of f_{r1m} = 4,5 to 5.5 N/mm^2 and $f_{r3\,m}$ = 4 to 6,5 N/mm^2. The characteristic values of the post-cracking strength are unreliable since the small prismatic specimens exhibit very high variations attributable to their small size, their manufacturing, and the testing method itself. Such high variations are typical to the test method only unlike the real SFRC on site. The standard itself should be blamed, as it lacks the definition of a precise methodology to ensure a consistent low coefficient of variation, up to 10%, as obtained in quasi-full-scale and full-scale tests.

Indeed the standard doesn't provide information such as the stiffness of the form (e.g., what constitutive material it is made of?), how it is clamped to the vibrating table, the vibration amplitude and frequency, the duration, or the total energy spent.

Before conducting valid EN 14651 tests, any laboratory should develop its own procedure with a specific SFRC and the standard test hardware to achieve the smallest possible and the most consistent coefficient of variations.

Otherwise, the results can vary by 50% or more, meaning that sometimes a mix of 30 kg/m³ steel fibers shows better results than one with 50 kg/m³ of the same steel fiber.

4.4 DESIGN

The design method presented here is rather simple, making it a quick and easy method that is also quite understandable even to nonspecialized engineers in the matter.

Finite elements methods (FEM) are time-consuming and cannot be used for everyday applications involving several cases. FEM is also restricted to highly specialized engineers who can use the right parameters and calculation grids. Otherwise, there is a risk of ending up with grossly overdesigned solutions. If elastoplastic FEM is involved, only highly specialized engineers can perform it reasonably.

Based on the analysis of the results from full-scale tests and quasi-full-scale tests (specifically round indeterminate panel tests), a plastic design is adopted according to the yield lines theory of Johansen. The theory is outlined by Johansen in reference (1) including other authors Hognestad in reference (2) or Craemer in reference (3) or the ACI 544 6R15 in reference (4)

The Johansen yield-line theory for plates assumes the plate in flexion behaves as a deformed plane mechanism. Indeed, the slab is seen as a mechanism consisting of plane plate elements, separated by yield lines, such that the initial undeformed slab becomes a mechanism of folded plates.

All deformations are concentrated in rotations along the yield lines only, meaning there is no bending of the plate between the yield lines or any bending moment.

The elastic deformation is neglected.

The moment of flexion M_L under the ultimate loading is by definition of the yield line theory, constant along the whole yield line. The maximum and constant moments are located in the yield lines also, so that at no section of the slab does the moment exceed M_L.

Under these conditions, it can be concluded that the "reinforcement percentage" must be sufficient to ensure the yielding of the section without allowing a brittle failure of the concrete at ULS.

A plain concrete slab or an under-reinforced concrete slab doesn't show yield lines.

The yield lines are straight lines and must end at the slab boundaries. A yield line cannot start and stop midway a few meters further.

Sometimes, inexperienced engineers believe that shrinkage cracks are yield lines!

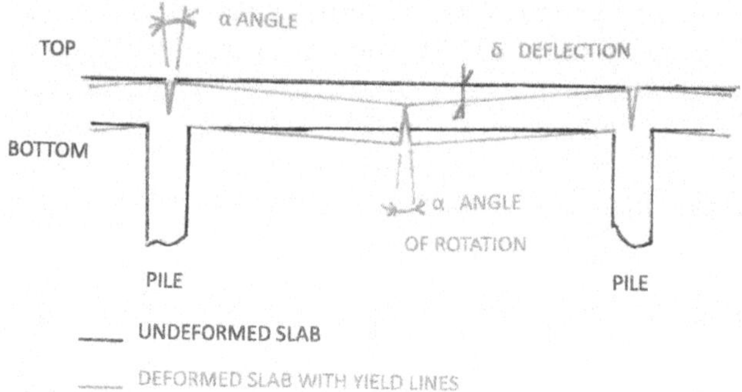

Figure 4.3 Top yield line from pile to pile as well as bottom mid-span yield line.

The "yield lines" must end at the edge of the slab, extending from edge to edge. The yield lines are subjected to a significant angle of rotation of one plate against its adjacent plate. There are always yield lines passing over the supports such as piles, columns, or walls.

The yield lines passing over the piles are visible on the surface of the slab, where the slab descends from the pile to the mid-span without bending the plate itself between two adjacent yield lines.

All deformations are concentrated in rotations along the yield lines. A yield line must also exist at half the distance between the piles; otherwise, the adjacent plates cannot rotate on top of the piles. As shown in Figure 4.3, the yield lines are represented by the angles opening successively on top of the piles and at the bottom at mid-span.

Indeed, without a hinge angle in the bottom at mid-span, no hinge angle can open on top of the piles.

The external moments are equilibrated by the rotation moments along the yield lines.

For one set of loadings the yield mechanism to use the one with the shortest total length of yield lines. Each meter run of the yield line supplies a fraction of the total resisting moment of which the minimum is taken to be safe.

For uniformly distributed loadings (UDL), one span is loaded and the only possible yield lines are parallel, passing over the piles at the top (negative moment) and at mid-span at the bottom (positive moment).

4.4.1 Moment equilibrium expression

Span L, with a total UDL = Q (kN/m²), between one top and one bottom yield line, the external moment for a 1 m width strip of slab, is:

$$M_{ext} = Q \times L / 2 \times L / 4$$

The resisting moment M_R is the sum of the rotation moment along each yield line; thus, $m_L \times 1\ m + m_L \times 1\ m = 2\ M_L$ (kN m/m)

so that the equilibrium equation is : $Q \times L^2/8 = 2\ M_L$ or $Q \times L^2 = 16\ M_L$

where M_L is the ultimate moment that must be divided by 1.5, the concrete material factor, to obtain the design resisting moment.

In the case of point loadings, the sum of all point loading in the same span $\Sigma\ P_i = P$ are located at mid-span, so that:

$$P / 2 \times L / 2 = 2\,M_L\ or\ P = 8\,M_L$$

where M_L is the resisting moment of the yield line per unit of length of the yield line.

This also shows that, under the same total loading intensity in one span, the point loads cause twice (or 16/8) as many moments as the UDL does.

To account for the effect of the slab thickness and pile diameter, a net span is defined. From the lateral face of the pile at the bottom of the slab, a 45° angle line meets the mid-depth of slab. The net span is taken between these two mid-depth points.

For a 4 m center-to-center span between the 400 mm diameter piles underneath a 230 mm thick slab, the net span is: $L_N = L - 400\ mm - 230\ mm = 3370\ mm$.

4.4.2 Rupture mechanism

The rupture mechanism to be considered is the least favorable for the proposed load and support, that is, the one giving minimum ultimate loading intensity Q_{ULT}. Equation and figures are taken from Destrée in reference (5), presented at the BEFIB 2000 Conference.

4.4.2.1 Central panel

Uniform thickness of slab (flat bottom slab) plastic rotation is shown in Figure 4.4: $Q_{ULT} \times L_N^2 = 16\,M_L$

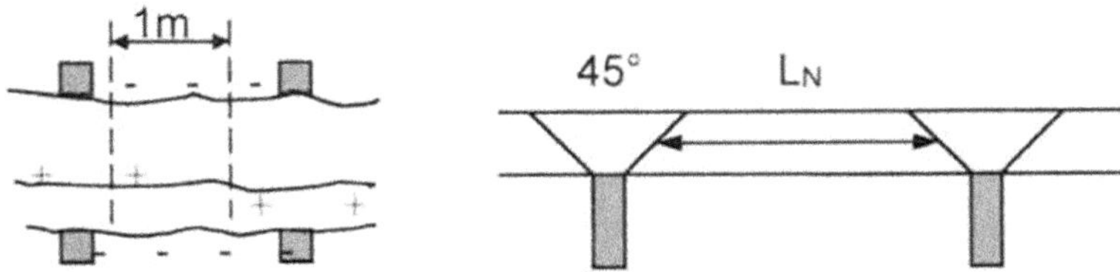

Figure 4.4 Typical yield line pattern and location.

4.4.2.2 What about pile heads?

There is a slab thickening above each pile, ensuring that the slab and the pile head are monolithic. In that case the pile heads are installed together with the slab itself, constructed in SFRC.

When the pile heads are installed together with the piles to be monolithic, the yielding mechanism remains as described above, since the pile head supports the slab, as shown in Figure 4.5.

If L_{N^*} is the net span, h is the slab thickness, H is the total thickness of the slab and the thickening, then:

$$L_{N^*} = L - p - 2H + h$$
$$Q_{ULT} \times L_N^2 = 16\,M_L$$

4.4.2.3 Rupture of head according to a fan pattern

The case of punching-out is shown in Figure 4.6:

This mechanism is, however, less critical than those described in the above-mentioned case of piled slabs.

No punching-out case of a pile has been observed in the 30-year experience of SFRC piled slabs as defined in this chapter.

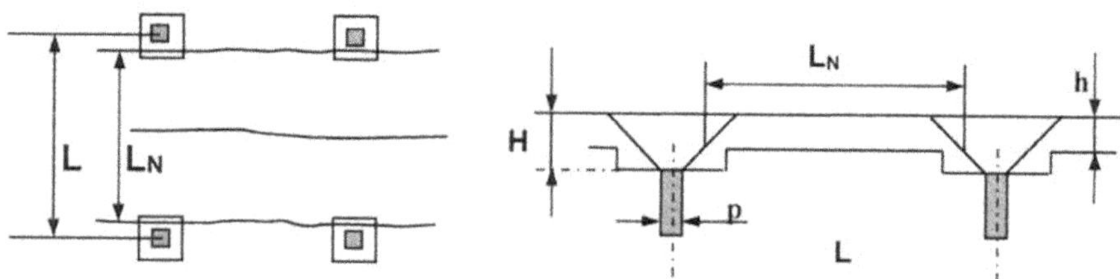

Figure 4.5 Yield line location of slab with pile heads.

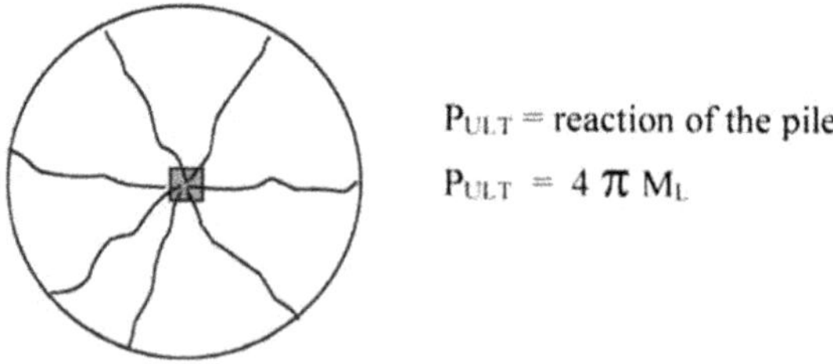

Figure 4.6 Punching out pattern of yield lines.

where P_{ULT} denotes the reaction of the pile

4.4.2.4 Simply supported edge panel case

$$L_{N^*} = L - p/2 - H + h/2$$

with an edge beam restraint as shown in figure 4.7: $Q_{ULT}\, L_N^2 = 16 * M_L$, thus same as for the continuous slab in Figure 4.4 with single support: $Q_{ULT}\, L_N^{*2} = 12 * M_L$ Generally, for a constant slab thickness, the edge span is reduced.

4.4.2.5 Rupture pattern under leg loading

Central panel: racking loads in a back-to-back case as shown in Figure 4.8.

$P_{ULT} * L_N = 8\, M_L * b$ where $P_{ULT} = \Sigma\, P_i = 4 \times P/4 = P$ and b is the aisle center to center distance.

Typical rackings are 2.30–2.60 m wide and 3–3.6 m long. Only four leg loads are summed and located at mid-span when the pile grid size exceeds the racking width.

In the direction of the aisle, the next four legs are situated in the adjacent span, so they don't influence the yield line designs.

4.4.3 Point loadings and uniformly distributed loadings are combined

With: G, own weight of the slab; Q: variable **UDL**; P(Σ 4 legs) = P_{rack} ; M_R = Ultimate moment

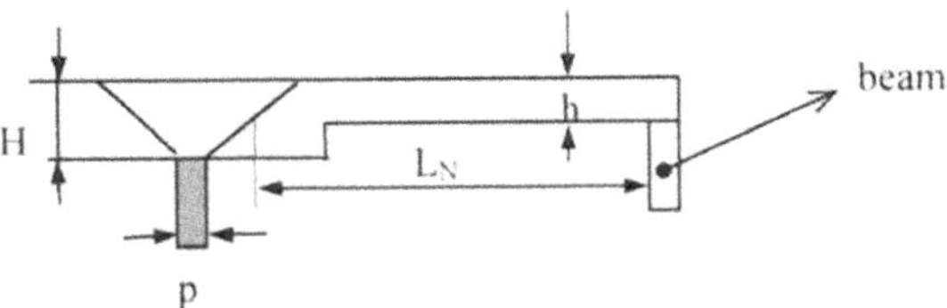

Figure 4.7 The edge span panel case.

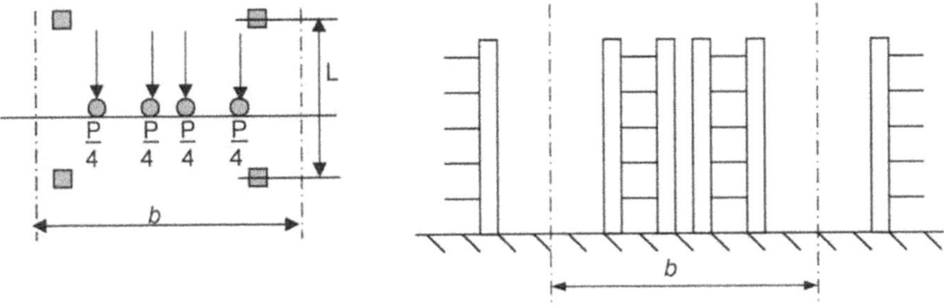

Figure 4.8 Yield lines in case racking legs loading.

$$\Upsilon_G = 1.35 \text{ (Europe) for the slab own weight and permanent loads}$$
$$= 1.2 \text{ (USA)}$$
$$\Upsilon_Q = 1.5 \text{ (Europe)}$$
$$= 1.6 \text{ (USA)}$$
$$\Upsilon_f = 1.5 \text{ is the material factor for steel fibre reinforced concrete}$$

In case of UDL only, the moment equilibrium equation is:

$$\Upsilon_f^* \left(\Upsilon_G^* G + \Upsilon_Q^* Q_L \right)^* L_N^2 < 16\, M_L \text{ (central panel or restrained edge span)}$$

$$\Upsilon_f^* \left(\Upsilon_G^* G + \Upsilon_Q^* Q_L \right)^* L_N^2 < 12\, M_L \text{ (single support under edge panel)}$$

In case of UDL and Point loadings combined in a middle panel:

$$\gamma_p \cdot P \cdot \frac{L_n}{8 \cdot b} + \left(\gamma_g \cdot G + \gamma_p \cdot Q \right) \cdot \frac{L_n^2}{16} \leq \frac{M_R}{\gamma_M}$$

Where b is the distance between two center lines of aisles between rackings.

In case of UDL and Point loadings combined in an edge panel:

$$\gamma_p \cdot P \cdot \frac{L_n}{6 \cdot b} + \left(\gamma_g \cdot G + \gamma_p \cdot Q \right) \cdot \frac{L_n^2}{12} \leq \frac{M_R}{\gamma_M}$$

At this point, we still have to determine M_R or M_L, what we'll do later in this chapter.

4.4.4 Shear and punching-out

In 30 years of experience and over 30 million square meters of SFRC piled slabs completed, we have not encountered a single case of punching-out. This result speaks for itself.

The highest shear develops around the piles or the pile head. The maximum pile reaction causes the highest shear stress, and each pile type, in a given ground, has a maximum unfactored bearing capacity

Different design standards stipulate different critical shear perimeters at a distance d from the side face of the pile or pile head, where:

$$d = h/2 \text{ or } 1.5\,d \text{ or even } 2\,d.$$

To be practical, during the last 30 years, we adopted an h/2 distance (a 45° spread from the face of the pile).

In the case of a minimum C30–37 concrete, the shear strength τ_{rs} under unfactored loading is 1.1 N/mm², 1.3 N/mm², and 1.45 N/mm² for 40 kg/m³, 45 kg/m³, and 50 kg/m³ of HE+1/60 or HE 90/60 steel fibers, respectively.

Unfactored loadings are used here, as the bearing capacity of the piles is often unfactored.

For a D = 200 mm pile diameter and h = 200 mm slab thickness, the lateral surface of shear is:

$$\pi^* \left(D + h \right) * h = 6.28 \times 400 \times 200 = 251.200\,\text{mm}^2.$$

The minimum pile capacity is calculated as:

R max = 251.200 mm² x 1.1 N/mm² = 276 kN at 40 kg/m³ dosage rate; 326 kN at 45 kg/m³ and 364 kN at 50 kg/m³. Under 364 kN, most 200 mm diameter pile are already exceed their bearing capacity.

Under 364 kN unfactored reaction, the 200 mm pile undergoes a compression stress of 11,6 N/mm², so that the crushing strength of concrete should be at least: 11,6 N/mm² x 1.5 x 1.5 /0.85 = 31 N/mm². Auger cast-in-place piles generally do not attain such a level of strength.

As far as codes for shear/punching-out of slabs are concerned, most codes today remain silent or extremely limited, including the Swedish Standard SS 812310 (2014—Design of Fiber Concrete Structures), which does not accept shear strength for fiber concrete without traditional reinforcement, such as rebars or wire meshes. However, the ACI544-6R15 (2015) is an exception, as it confirms the evidence that SFRC slabs possess considerable shear strength and resistance. All other documents still treat fiber concrete in shear/punching-out as plain concrete.

In our opinion and based on extensive practical experience, this is a completely negative and incorrect approach. Most FIB and Eurocodes-related publications fail to report full-scale slab tests or quasi-full-scale SFRC slab tests because these tests are structural tests rather than material tests. SFRC as a material is represented by standard small laboratory beam tests with spans of 450 mm to 600 mm, with or without a notch at mid-span..

Due to their small size, the manufacturing processes for standard samples—from mixing to pouring and compacting the "material" in the form—produces results biased by an experimental unpredictable scatter, often exceeding 50%. Only a single crack develops above the notch in the mid-span region, regardless of the steel fiber type and dosage rate, neglecting the possibility of multiple cracks.

Moreover, flexion stresses in these tests are highly concentrated, unlike in any structure, and cannot redistribute further, as would occur in a statically indeterminate structure.

Steel fibers are rarely used to reinforce beams for this reason and due to the limitations of the small beams test method. However, in statically determinate lintels design, similar to the small prismatic beams of the standard test method, it makes sense to use a design derived from these standard prismatic specimen tests in bending.

4.4.5 Determination of M_L, the yield moment

The yield moment of slabs cannot be derived from small beam tests but only from slab tests. In an earlier chapter, both quasi-full-scale (round indeterminate slabs with 1.50 m diameter) and full-scale tests (multiple spans in both OX–OY directions) led to the same post-cracking strength values in flexion, with scatter reduced to a 5–10% range. This reduction in scatter is due to the multiple cracking, which increases in number as the dosage rate of steel fibers increases—unlike in small prismatic specimens.

The indeterminate round panel test is shown in Figure 4.9.

Deriving from the round slab test as follows, where:

P_{ULT} is the maximum load recorded
c is the radius of the load application cylinder (0.15 m)
D is the diametral span of the slab (1.50 m), R = D/2
M_L is the yield moment also noted M_R
h is slab thickness
f_{tu} is the section plastic tensile strength, constant over 90% of the total
   thickness. This behavior is observed at the maximum point loading intensity. When the deflection increases further beyond the peak,
   cracking progresses with additional opening of the yield lines, reaching 15–20 mm deflection, while the residual load remains 2–2.5 times
   the first crack loading intensity. Multiple cracking develops, resulting

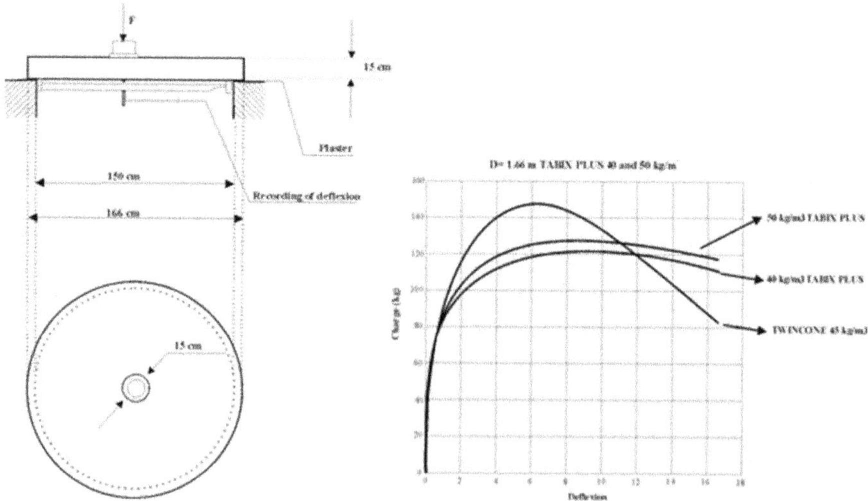

Figure 4.9 Typical diagrams with TABIX and Twincone fibers (45 and 50 kg/m³).

in 6–8 radial cracks and 11–20 circumference cracks as described by Espion reference (6) and shown in Figure 4.10.

The SLS loading intensity is attained at a deflection of less than 1500 mm/500 = 3 mm and a crack opening of less than 0.5 mm (70 kN). The ULS loading intensity is of 1.5 x SLS loading (105 kN), with collapse occurring at 160 kN, which is significantly higher than the ULS loading intensity, as shown in reference (10).

At SLS and beyond, a high number of minute cracks develop, some of which enlarge to form yield line, as shown in Figure 4.11, taken from Reference (7).

Example for 45 kg/m³ of HE+1/60 steel fibers in a C30 37 mix:

$h := 150mm$

$c := 150mm$

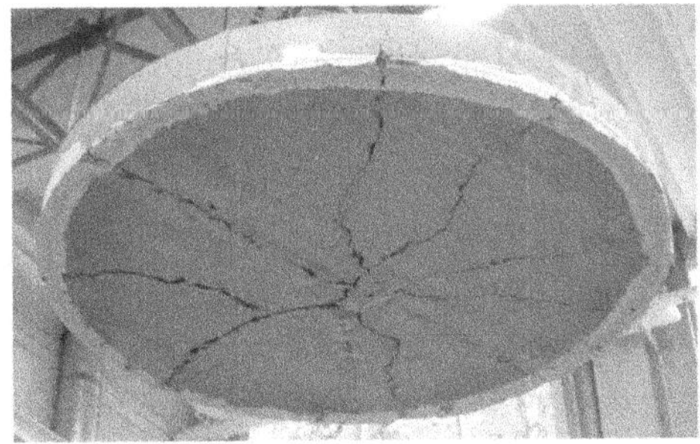

Figure 4.10 Statically indeterminate round panel test slab after testing.

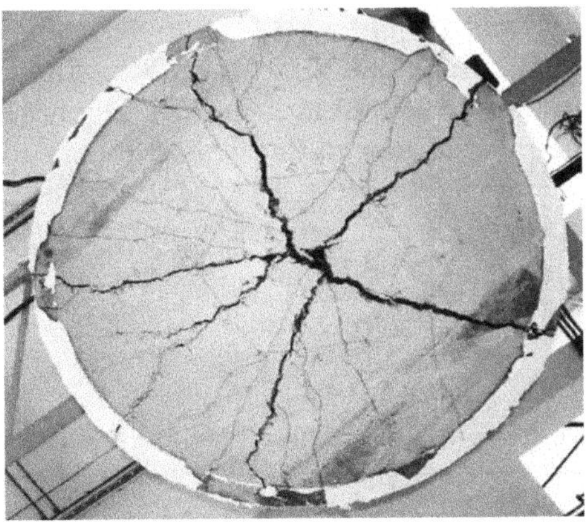

Figure 4.11 Yield lines and multiple cracking of a round indeterminate panel test at the University of Sheffield at 45 kg/m³ dosage rate of HE+1/60 steel fibers.

$$R := 1500\,\text{mm}$$

$$P_{ULT} := 155.1\,\text{kN}$$

$$M_L := \frac{P_{ULT}}{2 \cdot \pi} \cdot \left(1 - \frac{2 \cdot c}{3 \cdot R}\right) = 23.039\ \text{kN} \cdot \frac{\text{m}}{\text{m}}$$

$$f_{tu} := \frac{M_L}{0.45 \cdot h^2} = 2.275\ \frac{\text{N}}{\text{mm}^2}$$

Table 4.1 shows two different fiber types and has been used for over 25 years with complete satisfaction.

A 200 mm slab with a dosage rate of 45 kg/m³ in a C30-37 mix provides a resisting moment of $M_L/1.5$ = 0.45 x 2.35 x 200²/1.5 = 28,20 kNm/m, equivalent to an elastic equivalent of f_{ult} = 6 x 27240/200² = 6.075 N/mm².

The maximum service moment is: M_S = 28.20/(Load factor = 1.5) = 28.20/1.5 = 18.80 kNm/m

We'll show in another chapter that the resisting moment in flexion is also derived from f_{r3} standard residual flexion strength (EN14651):

$$M_{Rd} = f_{R3} \times h^2 / 6 \times 1.5.\ \text{The C30} - 37\ \text{with}\ 45\,\text{kg}/\text{m}^3\ \text{HE} + 1/60\ \text{is typical of}$$
$$f_{R3}\ \text{average} = 4.5\,\text{N}/\text{mm}^2$$

$$M_{Rd} = 4.5 \times 200^2 / 9 = 20\,\text{kNm}/\text{m}\ \text{that shows a}\ 28.20/20 = 1.41\ \text{factor of}$$
$$\text{overdesigning thus a} 19\%\ \text{thicker slab becomes needed!}$$

4.4.6 Calculation example

Carlsberg Tetley Depot of 10,500 m² (UK, completion in year 1998) and job site images are shown in Figure 4.13, extracted from reference 5

It is shown that pile grid = 3.60 m x 3.60 m. Pile head diameter = 0,600 m being part of the piles so that both pile and pile head are monolithic in traditional reinforced concrete. Here, UDL = 50 kN/m²

It is also shown that slab thickness: 250 mm with 45 kg/m³ Twincone steel fiber reinforcing of a C30-37 mix.

Table 4.1 Plastic tensile strengths value derived from round indeterminate panel tests

Fiber / dosage rate	40 kg/m³	45 kg/m³	50 kg/m³
HE+1/60.	2.2 N/mm²	2.35 N/mm²	2.50 N/mm²
HE 90/60	2.15	2.30	2.45

The Twincone steel fibers of 1 mm diameter x 54 mm length are shown in Figure 4.12, which is provided with two conical ends of 60° angle in an 1100 N/mm² steel wire. The Twincone fibers have been produced between (by Arcelor Mittal mostly and also later by Severstal—CEI) 1992 and 2012.

The production stopped as the f_{r3} values were not as high as with the HE +1/60 and TABIX +1/60 steel fibers, although the cost of production was higher.

The Twincone showed quite high f_{r1} values, quite beneficial to the cracking control which surpassed all other fiber types.

4.4.6.1 Flexion design

The design equation in flexion is:

$$\gamma_p \cdot P \cdot \frac{L_n}{8 \cdot b} + \left(\gamma_g \cdot G + \gamma_p \cdot Q\right) \cdot \frac{L_n^{\,2}}{16} \leq \frac{M_R}{\gamma_M}$$

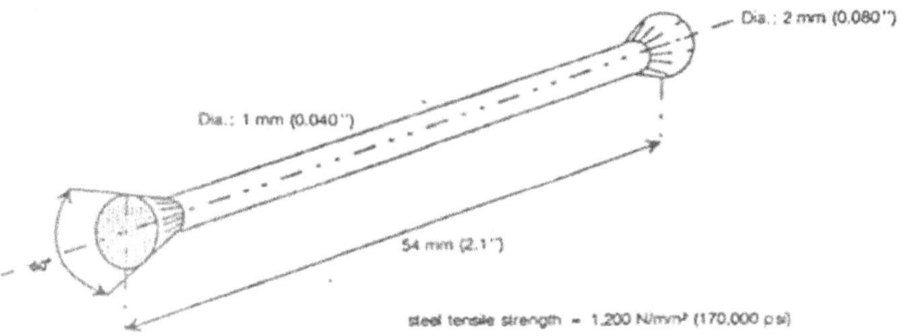

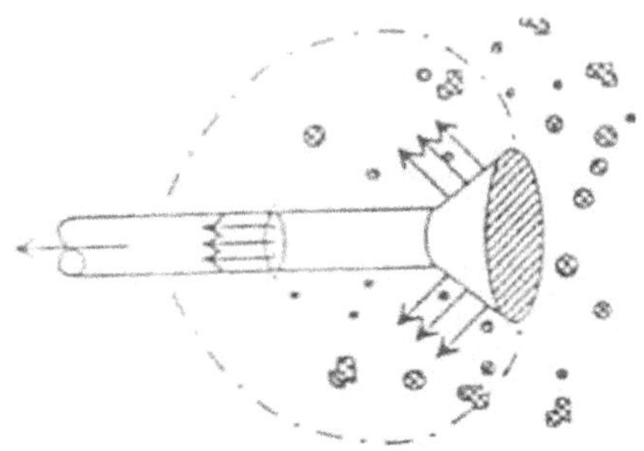

Figure 4.12 Twincone fiber geometry.

Figure 4.13 Carlsberg–Tetley jobsite with enlarged flared pile heads.

where $P = 0$; $\Upsilon_g = 1.35$; $\Upsilon_P = 1.5$; $G = 24$ kN/m³ x 0.25 m = 6 kN/m² ; $Q = 50$ kN/m²; $\Upsilon_M = 1.50$

$M_L = 0.45$ x 2.27 x 250² = 63.843 kNm/m
$M_{Rf} = 63.843$ kNm/m /1.5 = 42.56 kNm/m, the design resisting moment.
$L_N = 3.60$ m - 0.60 m - 0.25 m = 2.75 m, the net span
(1.35 x 6 kN/m²+ 1.5 x 50 kN/m²) x 2.75² m²/16 = 39.28 kNm/m <
   42.56 kNm/m

Let's calculate what type of racking such a slab could carry: the racking is of 2.30 m width long in a 1 m – 0.3 m – 1m, back-to-back case. The aisle width is 3 m, so that b = 3 m + 2,30 m = 5.30 m.

The first term of the equation (X) here above is: 1.5 x P x 2.75 m/8 x 5.30m = 0.0973 P, thus:

$\Sigma Pi = P < 42.56 / 0.0973 = 437$ kN

In case of a forklift truck of 70 kN and a dynamic factor of $\Upsilon_{dyn} = 1.6$, thus a dynamic load is 112 kN.

The racking loads are limited to 437 kN -112 kN = 325 kN or 81 kN per racking leg (4 in total).

4.4.6.2 Shear/punching-out

The shear/punching-out around the pile: the pile reaction is 437 kN (unfactored)
The shear perimeter is of (0.60 m + 0.25 m) x π x 0.25 m = 0.66759 m² = 667,590 mm²
The shear stress $\tau_{Rs} = 437,000$ N/ 667,590 mm² = 0.654 N/mm² < 1.3 N/mm².

4.4.6.3 Flexion stress control

Let's say h = slab thickness; D = pile diameter; L = span center to center; Q = U.D.L.; G = slab own weight L_N = net span; σ_s = flexion strength under unfactored loading. It is an "envelope" type of stress so that it is a maximum possible intensity under the UDL and own weight.

The piled slab is not supposed to "moment" crack beyond the service limit under the unfactored loading.

$$h := 250\,\text{mm} \qquad D := 0.6\,\text{m} \qquad L := 3.60\,\text{m}$$

$$Q := 50\,\frac{\text{kN}}{\text{m}^2}$$

$$\gamma_c := 24\,\frac{\text{kN}}{\text{m}^3} \qquad G_c := \gamma_c \cdot h = 6\,\frac{\text{kN}}{\text{m}^2}$$

$$L_N := L - D - h = 2.75 \ \text{m}$$

$$u := 1 \ \text{m}$$

$$q := sG + Q = 56\,\frac{\text{kN}}{\text{m}^2}$$

$$n := \max\left[8, \ \left(1 + 0.85 \cdot \log\left(\frac{h}{0.18 \cdot u}\right)\right) + \left(11.57 - 1.21\frac{L_N}{u}\right)\right] = 9.364$$

$$\sigma_s := \frac{6 \cdot q \cdot L_N^{\,2}}{n \cdot h^2} = 4.342\,\frac{\text{N}}{\text{mm}^2}$$

The first flexion crack could develop around approx. 5N/mm². This slab will however crack by drying shrinkage restraint. At 4.5 N/mm² flexion stress, the slab remains elastic and very stiff with little deflection of as small as span/1000 to 3000.

4.4.6.4 Deflection

The deflection δ is estimated by an "envelope" expression that has been developed by mean of the use FEM software in a number of cases.

The deflection to span ratio χ is given by the ACI 544 6R document in p.37 (J.2) and is used here on the 4.d case: the E-modulus is taken at 25.000 N/mm² as the steel fibers at 40 kg/m³ dosage rate and higher limit the cracking to shrinkage micro-cracking of 20 to 40 % of the slab section.

$$\chi := \frac{0.185 \cdot Q}{E} \cdot \left(\frac{L}{h}\right)^3 = 1.105 \times 10^{-3}$$

$$\delta := \chi \cdot L = 3.977 \, \text{mm}$$

which is better than $\chi = 1/500$, the general standard limit.

In a full-scale test during the year 2019 in Sweden where a 325 m² slab of 250 mm thickness, with 45 kg/m³ of HE+1/60 steel fibers, supported by a 3.80 m x 3.80 m grid of piles provided with an 800 mm x 800 mm square pile head, a 7 day loading at 40 kN/m² gave a 4.5 mm maximum deflection that we can compare to the 5 mm obtained with the same ACI 544 6R expression.

4.4.6.5 Shrinkage

As shown in 4.E.c here above, flexural cracking beyond the stringent limit of f_{r1} under service loading is excluded, as the slab is somehow cracked with a closely spaced pattern of minute cracks caused by restrained shrinkage..

The contraction movement restraint is generated by the contact with pile heads, the large bay size (2000 or 3000m²) and the length of the bay, which can be up to 60 m or even more in some cases. The random pattern of closely spaced cracks is quite helpful as it creates an overall shrinkage stress range of opening, with a maximum of 0.80 mm.

We can compare this to the theory derived from laboratory testing on free and restrained shrinkage specimen of SFRC.

According to Grzybowski and Shah in Reference (8), the restrained condition of a 0.5% volume percentage or 40 kg/m³ dosage rate steel fiber reinforcing reduces the total crack opening by a factor 4 (a reduction of 75 %) compared to a plain mix.

According to Mangat and Azari in Reference (9), in free shrinkage contraction movement, 0.5% steel fiber reinforcing reduces the free contraction by 10%, compared to a plain mix.

We can now try to predict, as Destrée suggested in Reference (5), the structural fiber reinforced slab cracking by comparison to a simplistic prediction for plain concrete.

The plain concrete cracking can be estimated as follows:

$$E = 10.000 \, \text{N} \, / \, \text{mm}^2 \, (\text{at long term and in tension})$$

Thus, Tensile strength = 3 N/mm²

The total drying shrinkage contraction is assumed of 3.10^{-4} strain, quite a realistic value in case of W/C< 0.5 of a carefully cured slab subjected to usual internal conditions of temperature and humidity.

A plain concrete, as a reference, could show one single crack of 3 mm opening every 10 m distance, indeed

$$3.10^{-4} \times 10.000 \, \text{mm} = 3 \, \text{mm}.$$

The fiber-reinforced concrete should then crack as follows:

3 mm x 90 % (reduction of free contraction) x ¼ (total cracking reduction) = 0.67 mm at 3,0 m distance apart.

The experience shows that 0,67 mm opening is exceptional and that the observed cracking is of less opening at a closer spacing at the end of shrinkage.

-SFRC cracks of less than 0,5 mm width, never degrade, irrespective of traffic frequency and intensity. These cracks will not cause the slightest hindrance to the proper use of the floor.

Cracks between 0,5 mm and 0,8 mm width, could degrade under heavy traffic.

Cracks wider than 0,8 mm width often degrade under heavy traffic.

Cracks that have reached the critical width, that is, 1 mm (Garber), showing degradation, must be sealed and stabilized to prevent them from degrading further.

Piled slab's shrinkage cracks start mainly above pile locations and extend from these.

It is above the pile head that the shrinkage contraction restraint takes place. Between the piles, the ground underneath will settle down so that it doesn't stay in contact with the bottom of the slab. Later as the whole slab under shrinkage shortens, more cracks can also develop from re-entrant corners like pits, dock levelers, manholes, column footprints, and also from nowhere to nowhere just like for slabs on grade.

Shrinkage cracks always start in the top surface of the slab where the evaporation of the water from inside the slab happens. The concrete shrinkage is maximum at the top surface of the slab.

Some engineers think wrongly that these shrinkage cracks are yield lines so they send the wrong message that the slab is nearing collapse!

Indeed yield lines by definition are lines of constant maximal plastic moment development so that yield lines go from edge to edge of slabs. Plastic rotations take place along yield lines so that the user of the slab should observe ups and downs from up to the piles and down to midspan all over the length of the yield lines, which indeed never happens. It is easy to control quickly by using an optical level.

Yield lines divide the slab in a number of adjacent rigid plates so that the slab becomes an articulated mechanism in rotation similar to the folding meter of the carpenter. Repeated deflections from piles down to the midspan should be observed in case of yield lines; however, this never happens.

It is impossible to see a yield line on top of the piles and nothing at midspan. Yield lines, like f_{r3} or f_{r4} CMOD of 3 or 3.5 mm, should be quite visible as it is long and repeated over the whole area of the slab.

Yield lines are also sensitive to moving vehicles as its crack opening is significant (CMOD = 2.5 mm) and such a crack section is by far not as stiff as a shrinkage cracked section.

In over 30 years over more than 20 million m² of completed slabs, I have never seen or been aware of a G-SFRS with yield lines, but I have seen a number of them with shrinkage cracks that need to be fixed when it is beyond serviceability.

The shrinkage cracks opening is now calculated using the method given by Destrée in reference (10).

$L := 5000$	50 m	Length of slab between joints (mm)
$V_f := 0.58$	45 kg/m³	Volume percentage of fiber concentration
$l := 60$	60 mm	length of steel fiber (mm)
$d := 1.0$	1.0mm	diameter of steel fiber (mm)
$\mu := 0.12$	HE+1/60	Fiber shape factor by Azari and Shah.
	$f_y = 1500$	
$k := 0.1$	N/mm²	Westergaard coefficient (N/mm³) shrinkage factor piled slab (the piled slab undergoes a significant restraint similar to a slab on grade laid on a rigid base of $k = 0.1$ N/mm³

$\beta := 2$

$\varepsilon_R := \beta \cdot 10^{-4}$ Total hydraulic shrinkage

$\lambda := 1.05$ Superplasticizer/water factor

$$O := (L) \cdot \lambda \cdot \varepsilon_R \cdot \frac{\left(4.60 + \ln(k)\right)}{V_f \cdot \dfrac{l}{d} \cdot (1+\mu)^{2.5}} = 0.522 \qquad \text{Crack opening (mm)}$$

The validity of the calculation and applicable conditions are as follows:

> $k > 0.02$
> $\mu = 0.08$ with single hooked end at 1200 MPa steel wire strength;
> $\mu = 0.12$ undulated and total anchorage steel fiber made of high strength steel wire

The same calculation is now done starting from a given water-to-cement ratio (WCR) to obtain a same crack opening:

$\beta = 1$ in case of W/C< 0.45 $\lambda = 1$

$\beta = 1,5$ W/C< 0.48 $\lambda = 1,02$

$\beta = 2$ in case of W/C<0.50 $\lambda = 1.05$

| = 3 | W/C<0.55 with Superplasticizer λ = 1.15 |
| = 5 | W/C<0.65 with superplasticizer λ = 1.35 |

β could still be increased to more than 5 in the case of concrete malpractice in mixing, transporting, installing, or lack of suitable curing.

μ = 0.08 in case of hooked end fiber

μ = 0.012 in case of undulated and total anchorage fiber

Fiber volume concentration and dosage rate

Vf = 0.38	30 kg/m^3
Vf = 0.51	40 kg/m^3
Vf = 0 64	50 kg/m^3
Vf = 0.70	55 kg/m^3
Vf = 0.77	60 kg/m^3

In case of a piled slab, use k = 0.06 to 0.1, depending on the restraint.

All slabs are supposed to be installed according to the state of the art and are provided with free contraction movement at the edges, around columns, re-entrant comers, etc. The calculation here is not applicable to thermal cracks and crack localization due to poor detailing. However, plastic settlement of piles is excluded here.

The crack opening can be written as a function of the WCR as shown below, where G is a scalar mixture function:

$$\text{WCR} := 0.505 \qquad\qquad 0\ 45 < \text{WCR} < 0.70$$

$$u := 100\,\text{mm} \qquad\qquad \text{unit of length}$$

$$G : \left[35\,(\text{WCR})^2 - 9.75\,(\text{WCR}) - 1.7 \right] \cdot 10^{-16} = 2.302 \times 10^{-6}$$

$$Ol := G \cdot L \cdot u \cdot \frac{\left(4.60 + \ln(k)\right)}{V_f \cdot \dfrac{l}{d} \cdot (1+\mu)^{2.5}} = 0.572 \cdot \text{mm}$$

crack opening as a function of W/C, λ and β are not used here.

Indeed such a crack opening is well according to the practical observed values.

4.5 AREA OF APPLICATION

The process is used for internal structural ground floors laid on piles, sheltered and protected from external conditions (wind, air currents, sun, and temperature variations) during installation. External applications are also possible with additional considerations regarding the mix design, slab installation, and distance between construction joints.

The process is particularly recommended when high standards are imposed, especially in relation to:
-Control of shrinkage cracks.

- Large area between construction joints (3.000 m² bay size together with Length/Width < 1,5).
- Total elimination of traditional reinforcement.
- Planning/speed of performance.

4.6 EXECUTION

The construction joints are located at ¼ of the span between piles and are reinforced by a 5 mm steel form to protect the arris under traffic.

As being zero-moment joints, these joints must provide full shear transfer together for a minimum of curling, with free horizontal movement of shrinkage in order to reduce the shrinkage restraint.

The polythene sheets under the slab don't play any positive role since reentering folds of the sheet are bottom cracks initiators. This fact was clearly observed during the Bissen full-scale test of a suspended elevated SFRC slab, where the forming woodchip board was covered by a polythene sheet carefully nailed to ensure that this issue would not appear. The opposite happened, and the slab's first positive moment crack was exactly in that area. Figure 4.14, showing a core with a crack, illustrates the polythene sheet inside the crack!

Figure 4.15 shows the negative effect of a plastic sheet underneath the slab: a high probability of being the cause of initiation of rupture.

Traditional designers recommend the plastic sheet as it is supposed to reduce the friction against the grade, even in the case of piled slabs, as the grade is the form of the bottom of the slab. After some weeks, the grade can settle plastically of several centimeters or more, but initially, it is restrained by the early age concrete shrinkage contraction movement and hence causes slight cracking.

In my opinion, this is preferable to the initiation of a rupture line in the slab.

I have observed, however, that the plastic sheet could be the cause of more construction joint openings and slightly fewer shrinkage cracks.

In my opinion, slight cracking doesn't affect the service condition of the slab, unlike excessive joint opening, which significantly affects it and is difficult to repair.

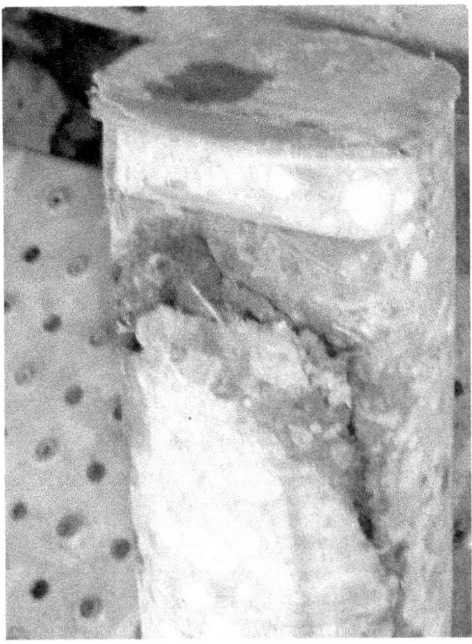

Figure 4.14 The polythene sheet with inward fold, is included in a crack that has been cored across the full depth of the slab.

Figure 4.15 The slab thickness is shown with an inward fold of the plastic sheet, indeed a severe section weakening.

The piled slab are not restrained along the free edges that can be a ground perimeter beam as support.

The slab thickness is never reduced above the piles: No re-intrant piles into the slab thickness. It is important to check the case of prefabricated piles to be cut off on site.

4.7 ADVANTAGES OF THE SFRC PILED SLABS OVER THE TRADITIONAL REINFORCED CONCRETE

The benefits from using steel fiber piled slab solutions are proven by the experience and the repeated usage by owners. The advantages are shown below.

4.7.1 Quality improvements

- Better Floor Flatness and Levelness control: Achieved by the total elimination of reinforcement bars, allowing the use of a full-sized laser screed for the floor construction. The laser screed machines produce floors with an improved surface regularity, which in turn increases the efficiency and speed of operations in logistics facilities.
- Better Crack Control: The total saturation of the concrete matrix with high tensile strength steel fibers, specially designed with hooked ends or an undulating shape for better anchorage within the concrete, provides three-dimensional reinforcement for a better distribution of shrinkage stresses and improved crack control.

  In addition, the slab is designed to be separated from the piles and pile heads, beams, and columns. This allows the concrete to shrink with less restraint to minimize cracking. This leads to significant reductions in long-term maintenance costs for facility floors and improvements in operating efficiencies.
- Reduced Maintenance at Joints: Fewer joints are needed, with greater spacing between them.
- Improved Durability: The use of SFRC also provides considerable benefits thanks to a significant amount of ductility, fatigue and impact resistance, and resistance to Spalling.

4.7.2 Easier and less cumbersome slab installation

It is done by the following methods :

- Fast-track construction method by the total elimination of reinforcement bars and the use of a Laser Screed machines for floor construction to provide significant savings in time, often many weeks are saved on the planning thanks to the shortening of the critical path.
- The elimination of reinforcement bars also reduces the risk of defects resulting from incorrect detailing, steel fixing or lack of supervision.
- Plant and labor reductions as the elimination of reinforcement bars also reduce the reliance on labor and supervision. Also time savings for detailing, drafting, and easier procurement. No risk of errors at the installation of rebars, and no need to check the reinforcing in diameter, spacing and coverage. Less inspection is needed. No risk of personnel

injury by cutting, bending, and placing the rebars, unloading them out of the truck, and handling them on site.
- No need for concrete pumps as the slab can be laid directly being dumped from the truck mixer, or by using specific dumpers on low-pressure tires

4.7.3 Sustainability

- Reduction in material usage: Both concrete and steel experience a significant reduction in tons and cubic meters, thanks to the SFRC. The potential reduction in slab thickness is often between 10% and 15 %. The blinding concrete, which is needed to properly install the rebars layers, is no longer required with SFRC. The 40 to 50 kg/m^3 dosage rate of steel fibers compares to 80–150 kg/m^3 of traditional steel.

  It results that the "Carbon Footprint" is improved quite, as addressed by Destree in Reference (11) and by de la Fuente in Reference (12)
- Better site safety is already outlined above.

4.8 COMPARISON TO A TRADITIONAL REINFORCED CONCRETE PILED SLAB

The comparison shown above is prepared by the Cogri company (Singapore) for a Rolls Royce job.

It is in very good accordance with the wide experience gained over the last 25 years with the application.

Item	Traditional R.C. Piled Slab	Steel Fiber only Reinforced slab
Design Load:	40 KN/m2	40 KN/m²
Pile Spacing:	4.33 m x 4.33 m	4.33 x 4.33 m
Reinforcement:	100 kg/m^3	45 kg/m^3
Slab thickness:	350 mm	270 mm
Construction Method:	Long wide strip method	Laser screed, wide bays
Surface regularity Specification achievable:	TR-34 Table 4.2 FM3	TR-34 Table 4.2 FM2 Special
Concreting Days:	15 days	10 days
Steel placing in man-days (on site in 2000 m² bay size)	105	18
Concrete Volume	5250 m³	4050 m³
Steel weight	525 Tons	182 Tons

The savings in cost, time, and CO_2 are very significant, making the SFRC application very attractive.

Figure 4.16 The fibers integration machines on site at Carlsberg–Tetley site.

The fiber integration is carried out on site by steel fibers blast machines, which blow the fibers at a high speed of 30 m/s into the truck mixer set at high revolution speed, covering the entire surface of fresh concrete. Figure 4.16 shows the Carlsberg Tetley slab fiber integration team with two blowers.

4.9 VERY RAPID SIMPLIFIED DESIGN METHOD

The method here is only based upon the pile reaction intensity and f_{R3} the EN14651 flexion strength:

$$f_{r3k} := 4.5 \frac{N}{mm^2}$$

EN 14651 flexion strength of the SFRC, characteristic and average value

$$f_{r3m} := 5.62 \frac{N}{mm^2}$$

$$UDL := 50 \frac{kN}{m^2}$$

The UDL intensity

$$w := 6 \frac{kN}{m^2}$$

The own weight of slab

$$L_x := 3.60 m$$

The span center to center

$$L_y := 3.60 m$$

$$R_P := \left[(UDL + w) \cdot L_x \cdot L_y \right] = 725.76 kN$$

The pile reaction (unfactored)

$$H := 0.62 \cdot \left(\frac{R_P}{f_{r3k}} \right)^{\frac{1}{2}} = 0.249 m$$

The needed slab thickness based on f_{r3k}

$$H := 0.69 \cdot \left(\frac{R_P}{f_{r3m}} \right)^{\frac{1}{2}} = 0.248 m$$

The needed thickness based upon f_{r3m}

The formula applies when there are pile heads above each pile.

Without pile heads, the slab should be 20 mm thicker up to 200 mm and 30 mm thicker up to 300 mm.

Without pile heads, the Carlsberg–Tetley slab in the UK should have been 240 mm + 30 mm = 270 mm thickness.

4.10 MORE DESIGN EXAMPLES

Design of a piled slab with a 4 m x 4 m (L = 4 m) grid of piles, each provided with 900 mm diameter pile heads (P=0,9m). Both piles and pile heads are made of traditional reinforced concrete. The edge span is reduced to 3.00 m, and the loading intensity is 40 kN/m².

We'll first determine the design moment based on the yield line pattern shown in Figures 4.2 and 4.3, according to reference (4.7).

The net span length L_N is used to account for the difference between a slender slab with a span-to-depth ratio of 20, and a thick one with a span-to-depth ratio of 12.

Real-scale piled slabs and round indeterminate laboratory slabs benefit from a considerable arching effect. In the case of small span-to-depth ratios below 24, using the reduced span length instead of the center-to-center span, is a simple way to take the arching effect into account.

The arching effect is especially visible when real piled slabs in suspended conditions are loaded beyond their service limit, showing very small deflections, such as span/2000 to 4000, as outlined here further in this paper under "Real piled slabs loading tests".

The design moment is calculated when the slab thickness is h = 210 mm; the yield lines pattern is shown in Figure 4.17:

$$\lambda_w := 1.35$$

$$\lambda_q := 1.5$$

$$q_d := \lambda_w \cdot w + \lambda_q \cdot q = 66.804 \, \frac{kN}{m^2}$$

Internal span
L= 4 m center to center span
P = .9 m square pile head side
Net internal span,

$$L_{NI} := L - P - h = 2.87 m$$

As shown in the Figures 4.17a and 4.17b

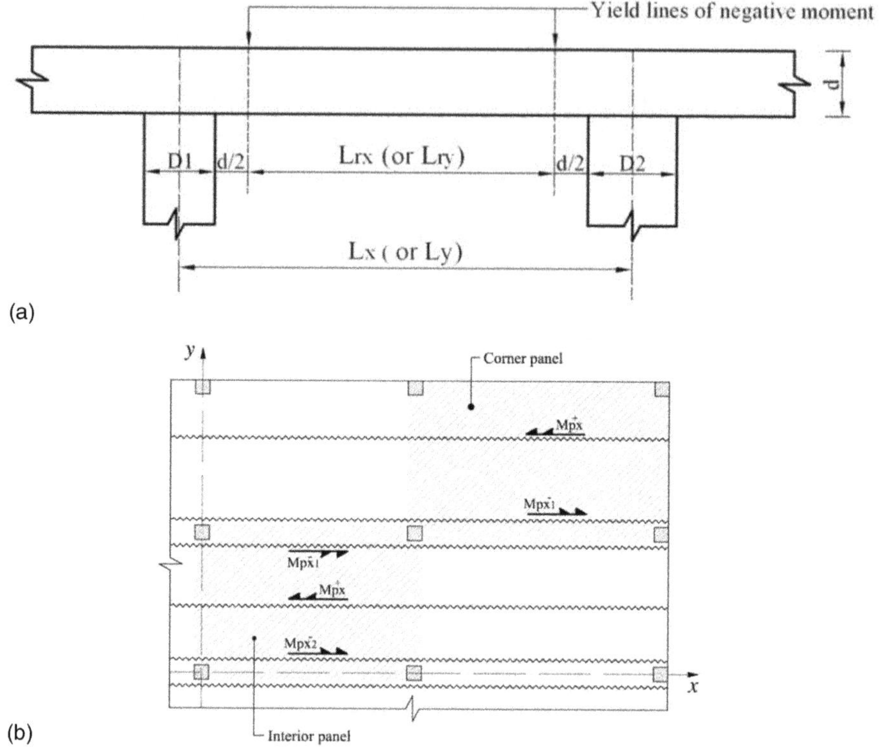

Figure 4.17 (a): definition of the net span. (b): typical yield lines pattern.

4.10.1 Design moment of an internal span

$$M_{dI} := \frac{q_d \cdot L_{NI}^2}{16} = 34.872 \text{kN} \cdot \frac{m}{m}$$

Edge span

$$L_{NE} := L_E - P - h = 1.87 \text{m}$$

4.10.2 Design moment of an edge span

$$M_{dE} := \frac{q_d \cdot L_{NE}^2}{12} = 19.886 \text{kN} \cdot \frac{m}{m}$$

4.10.3 Resisting moment

A typical C30–37 SFRC section with 45 kg/m³ HE+1/60 or TABIX+1/60 steel fibers, indeed of 1mm diameter, 60 mm long, 1500 MPa steel wire strength, respectively with hooked end or undulated shape shows resisting moment according to EN14651 and the FIB-Modelcode 2010:

$$f_{r1k} := 3.70 \frac{N}{mm^2}$$

$$f_{r3k} := 3.50 \frac{N}{mm^2}$$

$$M_u := \left(0.5 \cdot f_{r3k} - 0.2 \cdot f_{r1k}\right) \cdot \frac{h^2}{2} + \left(f_{FTs} - 0.5 \cdot f_{r3k} + 0.2 \cdot f_{r1k}\right) \cdot$$
$$\frac{h^2}{6} = 27.085 \cdot kN \cdot \frac{m}{m}$$

Simplified, it becomes,

$$M_u := \left(f_{r3k} + 0.05 \cdot f_{r1k}\right) \cdot \frac{h^2}{6} = 27.085 \cdot kN \cdot \frac{m}{m}$$

Furthermore, we can approximate that f_{r3} and f_{r1} are almost equal in suspended slab application:

$$M_u := 1.05 \cdot f_{r3k} \cdot \frac{h^2}{6} = 27.011 kN \cdot \frac{m}{m}$$

$$m_r := \frac{\eta_{det} \cdot M_u}{\gamma_f} = 36.015 kN \cdot \frac{m}{m}$$

where η_{det} = 2 in case of two-way slabs, and is the statical indeterminacy factor as defined in the SS 812310 design standard

$$m_r > M_d$$

hence, the design is okay!

4.10.4 Service condition

We check that the slab is not cracked under the most onerous flexion condition, ensuring that the flexion stress of the slab does not exceed σ = 4.5N/mm².

A formula has been developed to account for the net span, slab thickness, and loading intensity such that the flexion stress is: $\sigma = f(Q\ L^2/n)$ where $n > 8$, Q is the total unfactored loading intensity, and n is shown as follows:

$$n := \left(1 + 0.85 \cdot \log\left(\frac{h}{180\text{mm}}\right)\right) \cdot \left(11.57 - 1.21 \cdot \frac{L_{NI}}{1\text{m}}\right) = 8.558$$

$$\sigma := \frac{6 \cdot (q + w) \cdot L_{.NI}^2}{n \cdot h^2} = 5.898\,\frac{N}{\text{mm}^2}$$

This level of stress is above the cracking limit so that the loading intensity of 40kN/m² is too high. The maximum permissible service loading is given as follows:

$$q_{max} := \frac{q \cdot \sigma_{lim}}{\sigma} = 30.519\,\frac{kN}{m^2}$$

The plastic design was not the critical one.

4.10.5 Deflection

The calculated deflection is derived from an envelope diagram of different deflection case.

The parameters involved are the net spans,

loading intensity, slab thickness, and the modulus of elasticity E of concrete.

$$L_{NX} := 2.89\text{m}$$

$$L_{NY} := 2.89\text{m}$$

$$h := 210\text{mm}$$

$$E := 30000\,\frac{N}{\text{mm}^2}$$

$$w := \gamma_C \cdot h = 5.04\,\frac{kN}{m^2}$$

$$q := 30\,\frac{kN}{m^2}$$

$$\delta := 0.185 \cdot (w + q) \cdot \frac{L_{NX}{}^2 \cdot L_{NY}{}^2}{E \cdot h^3} = 1.628 \text{mm}$$

4.11 MORE LOADING TESTS

An identical slab of the same thickness, the same pile heads has been recently test loaded in Sweden. The ground underneath had settled down by 70 mm so that the slab was under suspended elevated condition (no contact with the ground underneath).

It has been loaded up to 40 kN/m² for 7 days and did not show more than 1.0 mm deflection, a figure that includes some pile settlement and no visible cracking.

The same observation was made at the full-scale test of Nieuw-Vennep, Holland, where the 180 mm thick slab was also under suspended elevated condition, with a 3.0 m x 3.60 m pile grid and 0.8 m x 0.8 m pile head support. It was subjected to 29 kN/m² UDL and 1.5 mm deflection during 150 hours, as shown in Figure 4.18, instead of the 15 kN/m² most onerous service loading intensity.

The slab was still un-cracked after the full load test, as confirmed by the rapid calculation below, which shows that the flexion stress is still smaller than 4.5N/mm², the cracking limit:

$$n := \left(1 + 0.85 \cdot \log\left(\frac{h}{180\text{mm}}\right)\right) \cdot \left(11.57 - 1.21 \cdot \frac{L_{NI}}{1\text{m}}\right) = 9.126$$

$$\sigma := \frac{6 \cdot (q + w) \cdot L_{NI} \cdot L_{NIy}}{n \cdot h^2} = 3.578 \frac{N}{\text{mm}^2}$$

$$L_{NX} := 2.02\text{m}$$

$$L_{NY} := 2.62\text{m}$$

$$h := 180\text{mm}$$

$$E := 20000 \frac{N}{\text{mm}^2}$$

$$w := \gamma_c \cdot h = 4.32 \frac{kN}{m^2}$$

$$q := 29 \frac{kN}{m^2}$$

$$\delta := 0.185 \cdot (w + q) \cdot \frac{L_{NX}^2 \cdot L_{NY}^2}{E \cdot h^3} = 1.48\text{mm}$$

The rapid design method shows:

H = 0.62 x ((29000 + 0.18 x 24000) x 3.0 x 3.6)/4.5) $^{0.5}$= **175 (mm)** thus close to 180 mm as tested.

The parametric equation to estimate the deflection of the slab in elevated condition shows quite good accuracy.

The same observation of 0.9 mm deflection was noted for a 210 mm thick Klaipeda (Lithuania) piled slab spanning 4m x 4m with 1m x 1m pile heads, loaded for 3 months at 29.7kN/m² UDL over a 200 m² area, as shown in Figure 4.19.

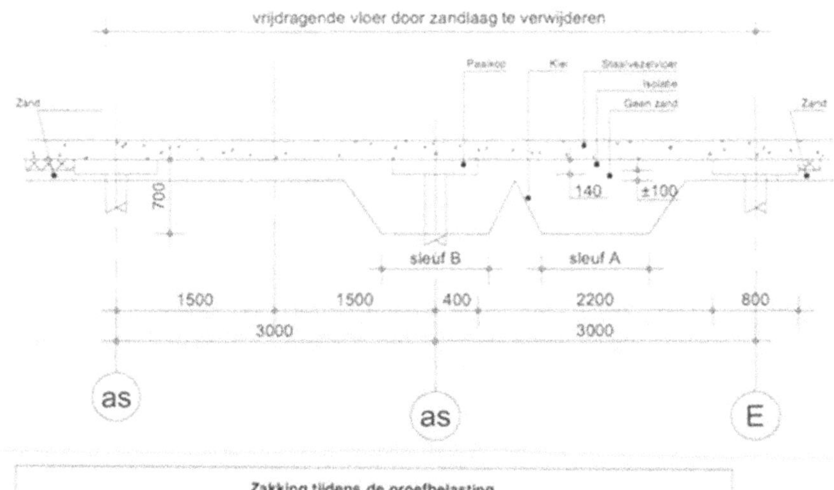

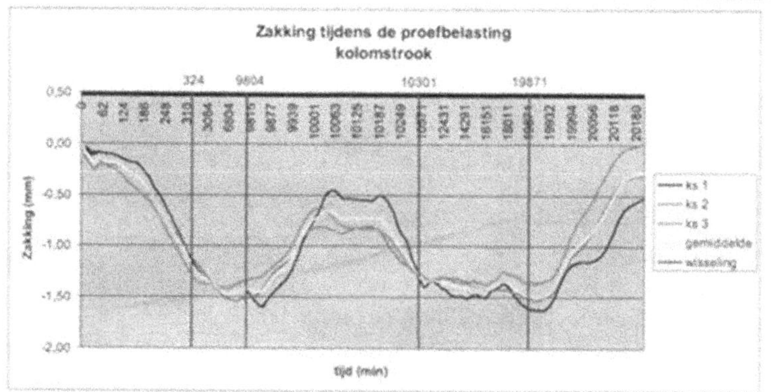

Figuur 5.3: Verloop van de zakking van de kolomstrook tussen de belaste randvelden tijdens de gehele duur van de proefbelasting

Figure 4.18 Nieuw Vennep slab section and deflections.

4.12 DESIGN TABLE

Assuming a 40 kg/m³ dosage rate of HE+1/60 steel fibers (Hooked ends, 1 mm diameter, 60 mm length, 1500 N/m² tensile strength) in a C30 -37 mix design (typical of a f_{r3k} = 4 N/mm²), Table 4.2 gives the permissible UDL as a function of the pile spacing for a flat bottom slab, that is, without pile heads on top of the piles.

The edge span is limited to 67% of the internal span, otherwise, extra traditional reinforcing is required in the bottom at midspan and in the top over the first row of piles.

When pile heads of 500 mm diameter (2.50 m span) to 1000 mm (4 m span), the thickness can be reduced by 20–30 mm (as already explained here above in Section 4.9).

Figure 4.19 Long-duration loading on the Klaipeda (Lithuania) test slab.

Table 4.2 Design table of ground piled slabs, thickness vs. pile grids and UDL. in case of 40 kg/m³ dosage rate of HE+1/60 steel fibers:

Thickness/ Span	2.50 m	3.0 m	3.50 m	4.00 m
180 mm	40 kN/m²	23	15	10
200 mm	42	27	19	12
220 mm	50	37	26	16
240 mm	63	42	29	20

In case of using 45 kg/m³ or 50 kg/m³ dosage rate of the same steel fiber, the permissible UDL of the Table 4.2 increases respectively by 6 % and 13 %.

4.13 DESIGN BASED UPON EN 14651 FLEXION TEST RESULT

4.13.1 Background of the design strengths

The small notched prismatic specimen is subjected to a mid-span (EN 14651) line loading while the loading intensity is recorded in the function of the mid-span deflection or the crack mouth opening displacement, noted $CMOD_i$, the subscript I to indicate the displacement at f_{ri} flexion strength.

It was observed early that these designs are adversely affected by an important scatter of the post-cracking strengths, resulting in up to 25 to 40% so that in many occasions, a smaller fiber volume fraction of reinforcing can show higher strengths than a higher volume fraction.

This is attributed to an erratic single crack formation that is not located at mid-span.

The single crack becomes a plastic hinge since it remains impossible to develop a multiple cracking.

The filling and compacting of the concrete has also quite a negative and unpredictable influence on the fiber distribution in the section. A consistent method that eliminates the scatter was never defined.

The notched beam specimen test as in EN 14651, was invented (from the rock fracture theory) in order to impose the crack location in the notch at the mid-span of the specimen.

The testing experience and the literature later showed that the scatter of results is even larger when a notched beam is used. Indeed the deviation observed is not typical of the SFRC material but typical of the testing method. When wide beam specimens or slab specimens are used, it has been shown that the deviation observed can be up to halved. When the SFRC becomes tested as a structure like in the case of a large square or round slab, the scatter is reduced even further. Indeed in these slab tests, multiple cracking can take place systematically so that a more consistent and reproductible post-cracking behavior is observed. The scatter observed of round indeterminate steel fiber reinforced slabs in flexion is reduced to 10%. When real scale tests are organized on multiple-span slabs, the scatter disappears completely. The increase in ductility is function of the number of cracks developed and further also a function of the number of active yield lines. Unlike with small prismatic specimens, an increasing fiber volume fraction shall always produce an increasing ductility.

The full-scale slab and quasi-full-scale testing methods, as described in the literature, exhibit structure testing instead of material testing, and are meaningful to distinguish between the material and the structural ductility. Despite the evidence and the literature, no standard method has been agreed upon for these round indeterminate slab structural testing methods although they extensively helped to understand and assist the design.

The round indeterminate slab test in flexion test method has only been admitted as reliable and structural in several documents such as the CUR 111 (Holland, 2007), and the T.R.63 (UK, 2007) but also recommended in the ACI 544 6R-15.

In the EN-SS 812310 Swedish standard for designing fiber-reinforced structures, the strengths derived from the notched (EN14651) prismatic specimen in flexion could be increased by a statical indeterminacy factor η_{det} in order to take into account the structural ductility, that is, the possible moment redistribution, and the multiple cracking of the fiber reinforced concrete that the EN 14651 never reveals.

In the case of a slab, the statical indeterminacy factor value depends on its type:

η_{det} = 1,4 when only parallel yield lines are observed in the real structure, which happens in continuous 1-way slabs,

η_{det} = 2,0 when x and y directions of yield lines are observed, and that is applicable in the case of two-way slabs such as E- and G-SFRS (ACI 544-6R) η det = 1 in case of statically determinate structures.

The formation of a multiple cracking pattern in these slabs together with the development of yield lines at ULS need a minimum volume concentration of steel fibers of 0.5% of volume concentration (40 kg/m³ of steel fiber dosage or more) typical of β > 2400 (see in chapter 1, SFRC Basics).

In the case of SFRC elevated suspended slabs, E-SFRS -ACI 544 6R15, the minimum volume concentration is 0.65% (50 kg/m³ dosage rate of steel fibers) typical of β > 3000 (see in Chapter 1) together with minimum structural integrity reinforcing, as described in CSA23-13-1 , ACI 352.1R-2011 and ACI 544-6R 15; indeed, a set of rebars is needed in the bottom of the slab from column to column. These specific rebars are also called APC rebars for Anti-Progressive Collapse.

4.13.2 Design and design strengths

4.13.2.1 Flexion

f_{r1} and f_{r3} (average values) flexion strength are derived from EN14651 standard SFRC flexion strengths.

Always use average values only instead as the characteristic values with the experimental scatter of the small notched beam specimen, indeed completely irrelevant in a real slab structure.

The maximum flexion moment at SLS is : $M_{SLS} < f_{R1} \times h^2/6$

The maximum flexion moment at ULS is : $M_d < \gamma_{LT} \times \eta_{det} \times f_{R3} \times h^2 / \left(6 \times \gamma_{Cf}\right)$

where η_{det}, a statical in determinacy factor, is defined in the EN-SS 812310,(for the design of fiber concrete structures) h is the slab thickness, $\gamma_{LT} = 0,85$ indeed eventually used as done a long-term factor, and $\gamma_{Cf} = 1,5$ is the material factor of concrete.

4.13.2.2 Shear/punching-out

$$\tau_{rf} < k / 2 \times C \times f_{R3} / \gamma_{Cf} \, (SS812310 - 6.4)$$

where C = 0.45
  and k = 1 +√(200/h) < 2 as a maximum value.
  (h =200mm, k = 2; h = 250 mm, k = 1,89; h = 300 mm, k = 1.82; h = 400 mm, k = 1.71)
The critical shear perimeter is at a distance of 2 x h from the face of the column.

Note that the SS 812310 allows a shear strength of SFRC only when the element includes also main reinforcing rebars in flexion.

We repeat here that such a provision is a non-sense as never ever a single punching-out has happened around the pile top in 30 years of experience.

The academic collection of papers about SFRC shear strength based upon the testing of statically determinate beams, shows always, at least, a set of bottom rebars to ensure that the beams will not crack and break under flexure instead of shear. Hence the wrong understanding so far, is that SFRC only doesn't show any shear resistance. Further, slabs are assimilated as being like beams!

This is how to summarize the standard non-sense for SFRC shear strength.

4.14 DESIGN EXAMPLE OF A G-SFRS

The C30-37 with 40 kg/m³ of HE+1/60 steel fibers shows f_{r1} = 4.0 N/mm² and f_{r3} = 3.7 N/mm² (EN 14651). The HE+1/60 are 1 mm diameter steel fibers of 60 mm length provided with hooked ends and a steel wire strength of 1450 N/mm².

Note that each mix design with the same fiber type and dosage rate can show a different set of f **ri** average values. The characteristic values are almost unpredictable due to extreme experimental variations attributable to the small notched beams fabrication in the lab and even more on-site, the test set-up, portal frame stiffness, deflection measurement (from mixing to filling the form, to the form material type, to the mechanical vibrations in intensity-frequencies -duration, to the way the form is placed on the vibrating table. All of these factors have a very significant effect on the fiber distribution in the specimen, which is far from a random distribution.

The resisting moment are calculated as follows:

$$\text{at ULS}: M_d < \left((0.85 \times 2 \times 3.7)/1{,}5\right)h^2/6 = 4{,}19 \times h^2/6 = 27.93\,\text{kNm/m}$$
$$\text{at SLS}: M_{sls} < 4.0 \times 0.85 \times h^2/6 = 3{,}40 \times h^2/6 = 22.67\,\text{kNm/m}$$

the shear strength is calculated without bottom rebars:

$\tau_r < 0{,}45 \times 2/2 \times 3{,}7/1.5 = 1{,}11$ N/mm² in case of a 200 mm slab where $\eta_{det} = 2$ (from SS 812310)

$V_{rd} = (200 + 4 \times 250) \times 3.14 \times 200 \times 1.11 = 836$ kN in case of a 200 mm slab onto a 250 mm diameter pile.

4.14.1 Slab data and calculations

The SFRC slab is supported by 4 m x 4 m pile grid, so the s span between centers L = 4 m

L_n is the reduced span = L - pile diameter - thickness of the slab

UDL = 10 kN/m² (Uniformly Distributed Loading intensity)

Slab own weight: 0.2 x 24 = 4.8 kN/m²

M_{max} (SLS): (10 + 4.8) x (4 - 0.25)²/10 = 20.81 kNm/m

Max f_{sls} = 6 x 20810 /200² = 3.13 N/mm² thus < f r1 = 4.0 N/mm² the maximum service flexion stress.

Pile reaction intensity in service: 4 m x 4 m x 14.8 kN/m² = 237 kN

The ULS pile reaction :Max V_d = 237 x 1.5 = 356 kN < 836 kN, the punching/shear is not critical in G-SFRS

From this we can also find the maximum possible service UDL that could be imposed:

It is: (10 + 4,8) kN/m² x 4.0/ 3.7 = 16 kN/m² total intensity minus 4.8 kN/m² own weight, thus 11.2 kN/m².

In case of the calculation is done in the linear elasticity theory, for piles provided with a pile head, the elastic moment are as follows, when the slab is loaded by a total unfactored uniformly distributed loading intensity q:

$M_i = qL_n^2/\alpha$, moments location as shown in Figures 4.19 and 4.20.

Table 4.3 Moment intensity at i location

I =	0	1	2	3	4
α =	-10	22	-50	40	-15

Note that q L_n^2/40 - (- q L_n^2/10) = q L_n^2/8, is also the maximum total moment envelope intensity.

Without pile heads, the M4 at 0.2 L_n distance to the center of adjacent pile, changes into q L_n^2/30.

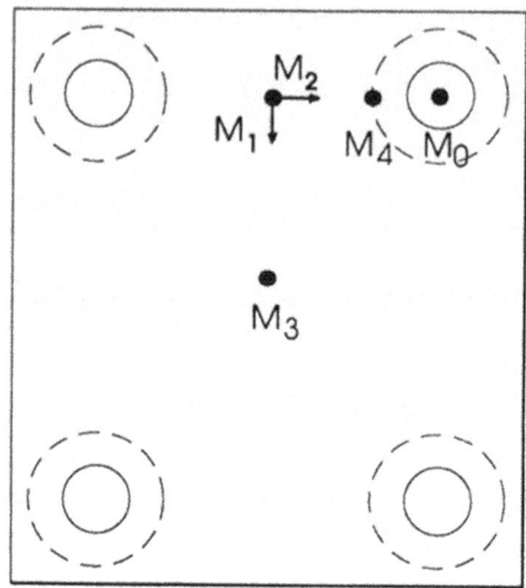

Figure 4.20 Localization of Mi.

If a construction joint is needed somewhere, it should be located at ¼ distance between two adjacent piles as it is a rather zero moment /High shear location.

In that case, the construction joint is a zero-moment joint and has to provide full shear transfer with a free shrinkage contraction movement.

The first span along the free edge is generally 25 to 33 % shorter to keep the section design identical, otherwise, this first edge span has to be thicker to resist 50% more moment.

The ULS verification with 10 kN/m² UDL:

$$10^3 \text{x} \, (1.35 \text{ x } 4.8 + 1.5 \text{ x } 10) \text{ x } (4 - 0.25 - 0.2)^2/16 < 3.7 \text{ x } 200^2/6 \text{ x } 1.5$$

16,918 Nm < 16 440 Nm thus missing shortly the target.

The solution is to increase the dosage rate of steel fibers by 5 kg/m³ up to 45 kg/m³.

In that case, f_{r1} and f_{r3} are likely to be 4.4 and 4.0 N/mm².

Only the experience with the local concrete type available and the expertise in SFRC of the testing laboratory, f_{ri} experimental values will be reliable.

A good piece of advice is to never accept the contractual responsibility on f_{ri} values if it is required to show it from jobsite manufactured standard samples and tested by an unqualified laboratory that has not developed its own expertise in the EN 14651 application, so that two different batches of the same mix design won't supply the same f_{ri} values. Up to 50 % scatter of the results is not reassuring!

To not follow such advice is like in a lottery or at the casino: completely unpredictable numbers!

4.14.2 Span-to-depth ratio

As long as the flexion stress under the most onerous service remains smaller than f_{r1}, the deflection is very limited to between span/1000 and span/3000.

G-SFRS slabs are generally subjected to much higher loading intensities than E-SFRS.

Hence, the span-to-depth ratio of G-SFRC is generally smaller than 22, as compared to E-SFRS ratio to be smaller than 30.

REFERENCES

1. K. Johansen. "Pladeformler" 2. Udgave, Polyteknisk Forening, Copenhagen, 1949, 172 pp.
2. E. Hognestad. "Yield line theory for the ultimate flexural strength of reinforced concrete slabs." *Journal of the American Concrete institute*, 24, 637–656.
3. H. Craemer. "Slabs spanning in two directions analyzed by consideration of pattern of fractures." *Concrete and construction Engineering (London)*, 45(8), 279–282, 1950.
4. ACI 544-6R 15: Report on design and construction of SFRC suspended elevated slabs.
5. X. Destrée. "Structural application of steel fibre as principal reinforcing: conditions - design – examples", BEFIB 2000, Lyon, France, RILEM PRO 15, 2000.
6. B. Espion. Test report N° 33410, Aug. 09-2006, Faculty of Applied Science, ULB.
7. KTP report N°001454, University of Sheffield, Aug. 31-2009.
8. S. Grzibowski. "Shrinkage of Fiber Reinforced Concrete." *ACI Materials Journal*, 101, 138, March-April 1990.
9. P. S. Mangat and M. Motamedi Azari. "Shrinkage of steel fiber cement composites." *Materials and Structures*, 21, 163–171, 1998.
10. X. Destrée. "A crack opening model of steel fibre only reinforced concrete slab, both ground bearing and pile supported slabs are considered." *BEFIB 2016*, Vancouver, 9 th Rilem Symposium on Fiber Reinforced Concrete.
11. X. Destrée, B. Pease. "Reducing CO2 emissions of concrete slab constructions with the Primecomposite slab system." *International Conference on Concrete Sustainability*, ICCS13, Tokyo.
12. A. Blanco, A. de la Fuente, A. Aguado. "Sustainability of steel fiber reinforced concrete slabs." *International Conference on Concrete Sustainability*, ICCS16, Madrid, Spain.

OTHER REFERENCES TO RECOMMEND FOR READING

X. Destrée, "Free Suspended Elevated Slabs of Steel Fiber-reinforced Concrete: Full-Scale Test Results and Design," *7th International Symposium of Fiber-reinforced 28 Concrete: Design and Applications BEFIB*, R. Gettu, ed., Sept., pp. 941–950, 2008.

X. Destrée, "Structural Application of Steel Fibers as Only Reinforcing in Free Suspended Elevated Slabs: Conditions—Design Examples," *Sixth RILEM PRO 39 Symposium on Fiber-reinforced Concrete*, Varenna, Italy, V. 2, Sept., pp. 1073–1082,2004.

X. Destrée, J. Mandl. "Steel fiber only reinforced concrete in free suspended elevated slabs: Case studies, design assisted by testing route, comparison to the latest SFRC standard documents.", Tailor Made Structure, J. Walraven and D. Stoelhorst, eds, 2008.

D. Mitchell, W. D. Cook "Preventing progressive collapse of slab structures." *Journal of Structural Engineering*, V. 110, No. 7 July. International FIB 2008 Symposium, Amsterdam, CRC Press, 1984.

C. Soranakom, B. Mobasher, X. Destrée "Numerical Simulation of FRC Round Panel Tests and Full-Scale Elevated Slabs." In *Deflection and Stiffness Issues In FRC And Thin Structural Elements, ACI SP-248*, American Concrete Institute, Farmington 48 Hills, MI, 2008.

Chapter 5

SFRC suspended elevated slabs (E-SFRS)

5.1 INTRODUCTION AND BACKGROUND

This led to a number of SFRC suspended slabs being full-scale tested in elevated conditions—10 in total, as outlined in Chapter 3 of this book, from 1994 to 2024: Ternat (Belgium, 1994), Townsville (Australia, 2000), Bissen (Luxembourg, 2004), Tallinn (Estonia, 2007), Eindhoven (Holland, 2010), Limelette (Belgium, 2011), Göteborg (Sweden, 2019), Poprad under Euro and ISO standard fire test (Slovakia, 2019), and four more slabs in Jelgava (Latvia, 2023 and 2024) organized by the Primekss concrete contractor.

All these tests have concluded the robustness of these E-SFRS, capable of carrying SLS loadings with almost no cracking or with cracks of less than 0.5 mm opening, remaining stable but deformed beyond SLS at ULS loadings, and collapsing under loading intensities well above the ULS.

Dozens of buildings, ranging from 2 to 17 floors, of residential, commercial, and industrial use, have been built using Elevated Suspended Steel Fiber Reinforced Concrete (E-SFRS) in eight different countries. This is detailed in the ACI 544-6R15 report on "Design and Construction of Steel Fiber-Reinforced Concrete Elevated Slabs."

A number of peer-reviewed papers have been produced to detail the results, show design methods, or provide back-calculations. In the literature, you'll find X. Destrée in references (1–15), and Mobasher and Destrée in reference (16).

The back-calculation theory of the section is explained in depth in reference (23) by B. Mobasher in Chapters 16 and 17, pp.295 -338. The book is a bible on the mechanics of fiber-cement composites.

Other researchers and authors have conducted full-scale tests of SFRC-free suspended elevated slabs, where steel fibers are the main flexural and shear reinforcement, concluding the reliability of the concept.

Aidarov, Mena and de la Fuente in reference (17) (pp. 12), conclude in the paper: "*The results obtained in this experimental program represent a new evidence of the structural suitability of the SFRC in elevated flat slabs, even in total substitution of the longitudinal steel bars.*" Or with more details, it is concluded that" *a SFRC with fr1m =7.2 N/mm² (CoV.= 35.3%)*

DOI: 10.1201/9781003188315-5

and fr3m = 7.7 N/mm² (CoV. = 24.7%) proved to be sufficient to guarantee a suitable structural response for loads representative of residential buildings in a column-supported slab with a span to depth ratio of 30."

Di Prisco, Martinelli, and Parmentier in reference (18) about the full-scale test organized in Belgium concluded that *"the limit analysis gives a reliable prediction even if the failure mechanism does not coincide with the experimental one and the role of arch and membrane effects is not negligible in this redundant structure, despite the lack of conventional reinforcement"*

Parmentier, Van Itterbeek, Skowron in Reference (19) compared the experimental results to the conclusions for design.

We remind the reader that the lack of conventional reinforcement—specifically, the (anti-progressive collapse (APC) rebars—was only the choice of Prof. M. di Prisco for this full-scale test, and that we disagreed and stepped out of the testing process. His work showed that the APC rebars could hide or dilute the contribution of the steel fibers.

In my definition and experience, the purpose of such an E-SFRS full-scale test is to demonstrate the safety, robustness, and viability of an elevated slab system, rather than the performance of steel fiber reinforcement in a structure.

The test organized in Limelette (Belgium) was not of such a system, as its mix design was incorrect, and the absence of APC rebars led to the complete collapse of the edge span, unlike any other similar full-scale test performed before.

In Reference (17), the authors report an SFRC full-scale elevated slab test with a span-to-depth ratio of 30 and a mix design similar to the authors of reference (18), with a 70 kg/m³ dosage rate of 1 mm x 60 mm 1500 MPa hooked end steel fibers. It was also decided to omit all APC bottom rebars going from column to column in X and Y directions, unlike what is specified as mandatory in the ACI 544-6R15 Report on Design and Construction of SFRC elevated slabs.

The real concrete as supplied and installed for references (18) and (19) was found to be rather poor; the characteristic residual tensile strengths reported were as low as 1.25, 1.44, and 0.77 N/mm² (with an average of 1.15N/mm²). In the reference (17), the authors report f_{r3m} = 7.7 N/mm², which we estimate to be about 5.6 N/mm² "characteristic" and thus, for a characteristic residual tensile strength of 0.37 x 5.8 N/mm² = 2.15 N/mm², which is 87% higher than in the reference (18)! As per reference (18), the behavior of slab provided a safety factor, as shown in the authors' conclusion, γR =1.28.

The slab shown in Reference (18) could not carry more than 320 kN point loading at mid-span, unlike over 500 kN reported in References (4–6, 8–11).

The reader should now understand why it is so important to design, prepare, and build the E-SFRS, or SFRC elevated suspended slabs, in full compliance with the ACI 544-6R15 provisions, including the inclusion of APC rebars, which should never be omitted, and the $f_{r\,m}$ (EN 14651) values of at least 5 to 6 N/mm², with a low coefficient of variation of less than 15%.

E-SFRS slabs do not result from the simple addition of steel fibers at a nominal dosage rate, but rather from a complete process that includes design, preparation of the mix design, detailing the drawings, following the method statement during installation, and real-time control to ensure the optimum quality is achieved.

APC rebars are also referred to as continuity rebars because they "tie the whole structure together," ensuring that a local collapse does not lead to a global collapse. Today, according to most design standards, including CSA (Canada Standard Association, which was the first standard in 1992 to make their use mandatory), ACI, Eurocodes, and others, all slabs in reinforced or pre/post-stressed cast-in-place construction must include them.

When I started the R&D activities for ArcelorMittal on E-SFRC in 2001 at the Polytechnic of Montréal with Professor B. Massicotte, we decided to include the APC rebars and make them mandatory.

The complete omission of traditional standard flexion and shear rebars or reinforcement mesh in E-SFRS has now become a reality.

The design method takes advantage of both the material and structural ductilities. It is applicable only in the case of two-way statically indeterminate and one-way indeterminate slabs with a minimum of two adjacent continuous spans, where SFRC benefits from full moment redistribution, and with a span-to-depth ratio limited to 30.

E-SFRS is not applicable for statically determinate slabs, where full-bottom traditional reinforcement remains necessary.

However, a minimum continuity reinforcing is mandatory, consisting of at least a set of bottom rebars from column to column in both the X and Y directions, to prevent a local column collapse from becoming a global collapse of the building.

Such continuity reinforcement is also defined as APC reinforcing because it prevents the slab from detaching from a column when a fatal action strikes the building structure.

The APC rebars are calculated to act like a cable, carrying the dead weight and the service loading of the slab. In the Canadian Standards Association (CSA), this provision was introduced in the early 1990s to apply to all types of suspended elevated slabs.

These APC rebars are also useful if, unfortunately, delays in concrete supply, breakdowns of concrete pumps, or any other accidental issues cause a cold joint to form in the slab. In this case, the APC rebars do not only eliminate the need for introducing wire mesh at the top and bottom across the cold joint before the concrete hardens but also provide a slab continuity.

The installation of E-SFRS is made easier because this structural E-SFRS concrete is a flowing mix that does not require mechanical vibration, allowing for more uniform fiber distribution throughout the depth of the element.

However, it is not classified as self-compacting concrete, which typically has a high mortar content that leads to excessive shrinkage.

It is counterproductive to believe that higher steel fiber concentrations in concrete will offset overall shrinkage. Early-age and thermal cracking can result from too high a cement and mortar content in the mix.

As with any SFRC slabs, whether thinner or thicker, it is highly recommended to start with a plain mix that has the lowest possible shrinkage content, and a slump test of 50 mm to 75 mm natural workability at jobsite arrival, without superplasticizer admixtures and without steel fibers.

To identify a good mix design, the flow table can help: check that, after the test, no free water seeps at the perimeter of the cake and that the fiber concentration is evenly spread from the center of the cake to the perimeter, as shown in Figure 5.1. An SFRC of 60 kg/m³ HE+1/60 steel fibers is suitable for the pumping. If the fibers are more concentrated at the center of the cake, the mix is unstable and needs to be corrected. In the case of pumping, it could create clumps and cause blockages.

For each job site, trial mixes have to be organized at the concrete plant lab so that the real nature of available aggregates and cements will be taken into account in the mix design.

Eliminating the vibration is a significant advantage as it eliminates the noise as a nuisance, and this is even more when the jobsite is in urban or residential areas.

Typical E-SRFS mix design included a continuous aggregate grading up to 20 mm maximum size when $\beta = V_m \times L_f/d_f$ is smaller than 3000, and a larger fraction of 16 mm if it is larger than 3300 (see Chapter 1). This mix also includes a 350 kg/m³ cement content, W/C < 0.50, and high workability F5 resulting from the addition of superplasticizers so the mix can be pumped and placed without poker vibrating. Generally, a 475 kg/m³ fine content

Figure 5.1 Slump flow cake of 60 kg/m³ HE+1/60 steel fiber ready for the pumping.

smaller than 200 microns, including all cementitious constituent materials, is used to pump the mix even to high floors.

Tall E-SFRS were successfully in repeated commercial and residential applications for clients and end users' satisfaction in The span-to-depth ratio is limited to a maximum of 30 as higher values have not been tested at full scale.

As reported in the earlier chapter on "Full-Scale Tests of SFRC Suspended Slabs," no fewer than 9 full-scale SFRC suspended slabs have been tested following our specifications from 1994 to 2018. All these tests showed that the test slabs do not crack at the service load intensity.

In one case of a full-scale test in elevated suspended conditions, a minute crack with a 0.4 mm opening occurred during weeks of imposed UDL at maximum service conditions under the most unfavorable load distribution.

The FIB Model Code 2010 and the draft of Eurocode 2-2024, define f_{r1} post-cracking flexion strength together with a 0.5 mm crack mouth opening displacement (CMOD) as a SLS crack opening limit.

Based on all the experience with real structures, a rule of thumb for design is that the slab should not crack due to flexion or shearing under the most onerous service load. However, it can undergo some shrinkage cracks, as this is almost inevitable when concrete is used.

I have always told clients who don't accept any cracks that they should not accept concrete as a material—maybe they know of a better material?

What they certainly need and must expect is that the slab remains serviceable and functional.

The slab contractor should always offer a guarantee of serviceability.

It is possible for a slab to meet the ULS conditions but no longer be serviceable (fit for SLS), as in the case of excessive deflection, construction joints stepping, wide cracks with several millimeters of opening, chipping at edges, delamination, or vertical movements along construction joints.

These issues can occur regardless of the type of reinforcement, whether traditional reinforcement, post-tensioning, or steel fiber reinforcement.

Avoiding such cases requires proper design, the use of low-shrinkage structural concrete, and installation in accordance with the code of good practice.

A simple design method based on the yield line theory of Johansen has been used by me since 1994 for G-SFRS applications, and in 2004, it was extended to E-SFRS, later endorsed in the ACI 544 6R15 "Report on Design and Construction of SFRC Suspended Elevated Slabs."

All design equations are used as already shown in the "SFRC pile supported slabs," Chapter 4 in this book and in the references (13, 16).

The main difference in design, however, is that the SFRC elevated suspended slabs (noted E-SRFS in ACI 544 6R15) always include the "Anti-Progressive-Collapse" set of rebars running from column to column at the bottom of the slab, following the X and Y main directions.

APC rebars are designed to prevent the slab from detaching from the columns during extreme events such as gross overloading, earthquakes, poor concrete quality, fire, etc., in order to avoid the sudden fall of the slab, which could crash to a lower level and continue collapsing further down, resulting in a pancake-type collapse of the building.

The reader must understand and accept that these APC rebars in SFRC suspended elevated slabs are mandatory and should never be omitted under any circumstances.

All our full-scale SFRC and E-SFRS tests, as outlined in an earlier chapter of this book "The need of full-scale tests of real structures", have included the APC rebars.

These APC rebars act like a chain capable to carry the entire slab under its maximum service load, from column to column. Each direction (X or Y) carries 50% of the total unfactored loading intensity while the APC steel is not stressed more than $0.85\ f_y$.

The APC rebars also contribute to meeting the fire resistance requirement of 120 minutes and more, as shown and explained in the chapter 6 "Fire resistance of SFRC suspended elevated slabs (E-SFRS) and walls" later in this book, also as reported in reference (15).

The purpose of these APC rebars was first proposed and explained by Mitchell and Cook in reference 20. Later, in 1993, it has been included in the Canadian reinforced concrete code in reference (21) and more recently in the ACI 318 Code.

Indeed, we wish to clearly repeat that all types of suspended slabs, regardless of the reinforcing type, are required to include these APC rebars, which are now also referred to as "minimum continuity reinforcing" as shown by Mitchell and Cook in Reference (20).

The section of the APC rebars is calculated like a cable in tension, capable of carrying the unfactored own weight, permanent, and service loading of a single span between two adjacent columns in both directions of the two-way system.

5.2 A E-SFRS CASE STUDY

We provide a case study addressing the structural design and use of steel fiber reinforcement for E-SFRS elevated slabs building as built at LKS site, in Mondragon, Spain.

Indeed, the LKS office building located in Mondragon near to Bilbao, Spain, as shown in Figures 5.2a and b, were constructed by Galdiano contractors and completed in 2010 to serve as the headquarters of a highly visible Spanish engineering firm.

A typical floor is shown in Figure 5.3. The X direction spans consist of four bays, each 7.80 m, while the three Y direction spans measure 8.00 m, 4.90 m, and 6.40 m, respectively.

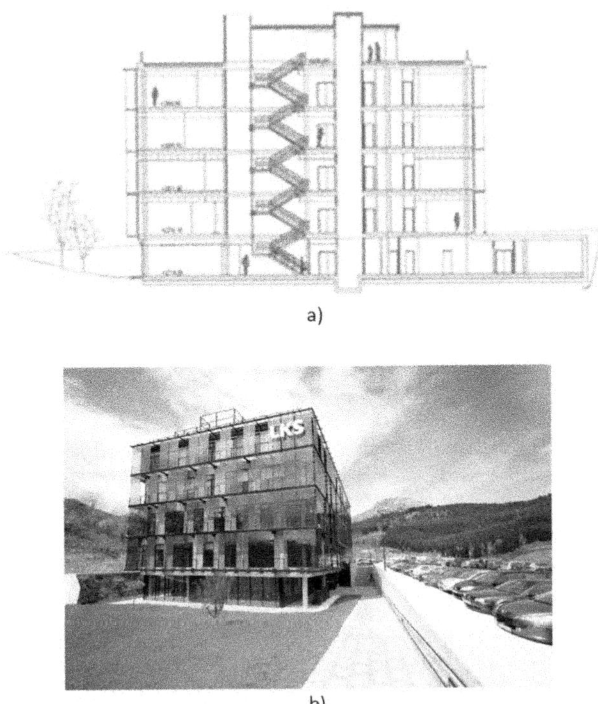

a)

b)

Figure 5.2 (a) View of the LKS building and (b) view of LKS building showing four levels plus roof.

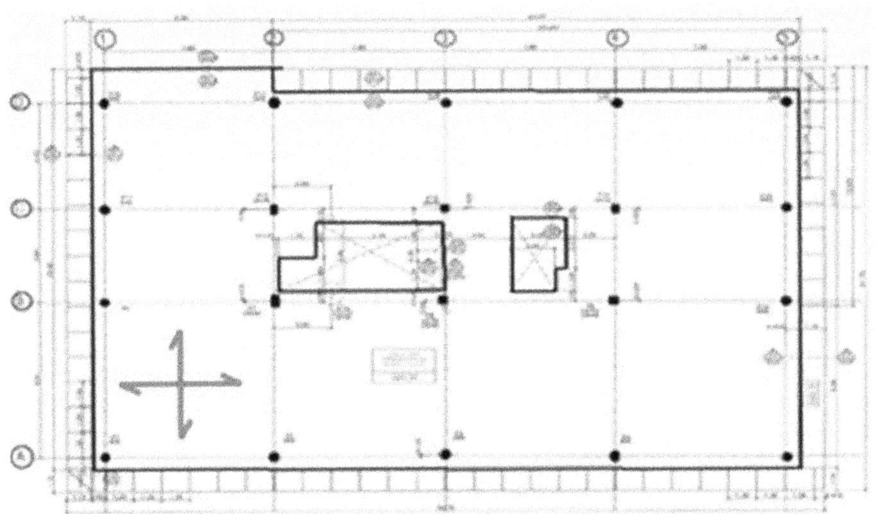

Figure 5.3 A typical floor.

The owners' requirements were based on the new building to take advantage of the most modern technologies and therefore wished to use SFRC in all cast in situ elevated slabs. This included the first floor to the fourth and the roof floor level totaling an area of about 4200 m² (45000 sq.ft.). SFRC suspended elevated slab offered significant advantages in terms of cost savings and short length, streamlining the schedule when compared to traditional rebar reinforcing.

Since the fiber reinforcing of the slab solution was not, at that time, covered by any valid structural design standard, the whole process from design to installation was evaluated and controlled by the Polytechnic School of Bilbao, Spain, under the leadership of Prof. A. Maturana, as reported in Reference (22).

The load-bearing structure consists of a set of 20 reinforced concrete columns with a section of 450 mm x 450 mm (17 in 3/4 x 17 in 3/4). The latter set supports a t = 300 mm thick flat slab without beams or drop panels, also named E-SFRC (ACI, 544-6R 2015). The elevated suspended floors for the five stories are 32.45 m (106 ft 5 in) long x 19.45 m (63 ft 9 in) wide, including 4 and 3 consecutive continuous spans in x and y directions, respectively.

The layout of the building (Figure 5.2) and (Figure 5.3) extends from a central lift shaft facing a stairwell so that the typical slab includes two large holes separated by a single one-way span slab. The column grid varies from its maximum of a corner panel of 8.00 m (26 ft 2 in 3/4) x 7.80 m (26 ft 6 in 4/5), to 7.80 m (26 ft 6 in 4/5) x 8,00 m (26 ft 2in 3/4) of an edge panel to 4.59 m (15 ft) x 7.80 m (26 ft 6 in 4/5) of an interior panel.

All these slabs are subjected to a total live loading intensity of 4 kN/m² (80 psf), which includes partition walls and the floor covering. The C 30 (30N/mm² or 4500 psi) mix design included 400 kg/m³ (675 lbs/cu.yd) CEMII 42.5 with 16 mm (0.63 in) maximum aggregate size in which 100 kg/m³ (165 lbs/cu.yd) steel fiber was added at the batching plant. The steel fibers were made out of a 850 N/mm² (120000 psi) steel wire of 1.3 mm (1/2 in) diameter and 50 mm (2 in) length, indeed TABIX 1.3/50 type of ArcelorMittal.

The TABIX 1.3/50 provided with an undulated shape and 850 MPa wire tensile strength, is no longer produced and is replaced today by a minimum 50 kg/m³ of HE+1/60 steel fibers of 1500 MPa tensile strength.

This size and geometry of steel fiber were selected for their very user-friendly nature, as they do not tend to ball at all during dry or wet stages of mixing and transporting operations. Their disadvantage is that they are not the best-performing type of fibers in hardened concrete, as can be seen below in the μ coefficient of 0.6 in references (21–23). (Mobasher, 2008, 2014).

More slender steel fibers of 0.9 mm or 1 mm diameter are also best used and more economical in fine aggregates despite the extra cost for the higher strength (1500 MPa) of steel and more specific attention to introducing and mixing in concrete.

For example, 100 kg/m³ of TABIX 1.3/50 (undulated, 1.3 mm diameter x 50 mm length—850 MPa tensile strength) steel fibers are—in suspended slabs applications—equivalent to 50 kg/m³ of HE +1/60 (1 mm diameter x 60 mm length—1450 MPa tensile strength)

Back to our design example, here is a reprint from the FRC 2014 Joint ACI-FIB International Workshop in Montréal, referenced as (14), following the ACI 544-6R15 design report method:

5.2.1 Flexion

The final slump flow as F5 for pumping purposes and a live load intensity q_L is:

$$q_L = 4kN\,/\,m^2 = 80psf$$

The design loading intensity q is calculated as follows:

$$\lambda_{DL} = 1.2, \quad \lambda_{LL} = 1.6, \quad t = 300mm, \quad w_g = 24kN\,/\,m^3$$
$$G = w_g t = 7.2kN\,/\,m^2$$
$$q = \lambda_{DL}G + \lambda_{LL}q_L = 15.04kN\,/\,m^2 \tag{5.1}$$

All design moments are obtained by the application of the yield line theory (ACI, 2013). In the case of the edge panel, the formula 6.3.1.1.o (*ACI, 2013*) gives the following when the longest span L_x is of 8.00 m while the column thickness D is of 450 mm (17.72 in) and thickness of 300 mm so that a net span length is calculated as follows:

$$L_{rx} = L_x - D - t = 7.25m\ \left(23.79ft\right) \tag{5.2}$$

And the positive design moment M_{Px} becomes regarding the edge panel:

$$M_{px} = \frac{qL_{rx}^2}{2\left(\sqrt{2}+1\right)^2} = 67.82\ kN\frac{m}{m}\left(15.25Kip\frac{ft}{ft}\right) \tag{5.3}$$

Note that the denominator is calculated as 11.657 which is almost equal to 12. The longest corner panel is of 8.00 m so that the net span is calculated:

$$L_{xe} = 8.0m \qquad L_{rxe} = L_{xe} - D - t = 7.25m\ \left(23.79ft\right)$$

$$M_{pxe} = \frac{qL_{rxe}^2}{8} = 98.82 \ kN\frac{m}{m} \left(22.21Kip\frac{ft}{ft}\right)$$

The interior panel is of 7.80 m so that the net span length is 7.05 m. The positive moment of the interior panel M_{Pxi}:

$$M_{pxi} = \frac{qL_{rxi}^2}{16} = 46.72 \ kN\frac{m}{m} \left(10.5Kip\frac{ft}{ft}\right)$$

The design resisting moment M_{Rd} of the section is calculated as per *Soronakom, Mobasher, (2008) and ACI (2013)*:

$$M_{Rd} = \frac{3\mu\omega}{\mu+\omega}M_{cr} \tag{5.4}$$

μ represents the residual tensile strength parameter as a fraction of the tensile strength, and ω represents the compressive strength to tensile strength ratio of the concrete matrix.

$$f_{ck} = 30N/m^2 \qquad \omega = 1.52\sqrt{f_{ck}} = 8.325 \qquad \sigma_{cr} = \frac{f_{ck}}{\omega} = 3.6$$

$$\sigma_{cr} = \frac{f_{ck}}{\omega} = 3.6 \qquad \mu = 0.6 \qquad \sigma_{cr} = \mu\sigma_{cr} = 2.16 \ N/mm^2 \tag{5.5}$$
$$\left(Mobasher, \ 2008, \ 2014\right)$$

Hence, we obtain:

$$M_{Rd} = \frac{3\mu\omega}{\mu+\omega}M_{cr} = 50.37kN\frac{m}{m} = 11.324kip\frac{ft}{ft} \tag{5.6}$$

The corner panel shall need extra-reinforcing steel rebars as the fiber reinforcement of concrete does not offer a sufficient resisting moment:

$$\Delta M = M_{pxe} - M_{RD} = 948.45 \ kN\frac{m}{m} \left(10.89Kip\frac{ft}{ft}\right) \tag{5.7}$$

If c, the coverage of steel bars, is of 30 mm, f_{ys} is of 435 N/mm², and b is the unit width of 1000mm, we can determine the section of additional bottom steel S needed in the corner panels:

$$S = \frac{\Delta Mb}{0.9f_{ys}(h-c)} = 458.33mm^2 = 0.216in^2/ft \tag{5.8}$$

A bottom wiremesh of 10 mm (0.40 in) diameter and 150 mm (6 in) spacing in both directions offers 523 mm²/m (0.247 in²/ft) reinforcing section per meter run thus more than 458 mm²/m (0.216 in²/ft) is needed. This wiremesh shall be placed in all four corner panels. One can also simplify the calculations of the design moments of all types of panels in observing that $M_{Rd} = q \times L_{rx}^2 / (16, 12, \text{ or } 8)$ respectively in the case of an interior panel, an edge panel and a corner panel reference (2).

Additional steel reinforcement (i.e., other than steel fibers) is needed in various places where a moment redistribution is not available: between the stairwell and the lift shaft, where we can identify an almost statically determinate one-way slab with an $L_s = 4.80$ m (16ft) span. Therefore, the installation of a bottom mesh of S section per meter is required:

$$S = \frac{M_{sd}b}{0.9f_{ys}(h-c)} = 409.77mm^2 = 0.194in^2 \, / \, ft$$

Therefore, a bottom mesh of 10 mm diameter wires and of 150 mm (6 in) spacing shall be placed. Along the perimeter of the building, the slab is cantilever over 500 mm distance where, although the moment is very insignificant and limited to $15 \times 0.50 \times 0.25 = 1.88$ kNm/m. It is recommended to specify the installation of a 8 mm (0.315 in) diameter wiremesh with 150 mm (6 in) wire spacing over 1m width at the top, as safety reinforcement in case a cold joint forms on site for any reason.

At last, as shown in Figures 5.3 and 5.4, from column to column in the bottom in both x and y directions, minimum continuity reinforcing bars (which are also called anti-progressive collapse rebars) of A_{sb} are needed. The A_{sb} section of steel is calculated as follows:

$$w_s = \max(G + q_L, 2G) = 14.4kN \, / \, m^2 \qquad (300.75psf) \qquad (5.9)$$

$$A_{sb} = \frac{0.5w_sL_xL_{xi}}{0.85f_{ys}} = 1.215x10^3mm^2$$

Three rebars of 25 mm (1 in) diameter, 1472 mm² (0.695 in²/ft) of section and larger than 1215 mm² (0.574 in²) shall be installed continuously in the bottom from column to column in both x and y directions. These rebars need to cover the footprint of each column, as shown in Figure 5.4.

Note that the Figures 5.4, 5.5 and 5.6 are extracted from the reference (22)

Figure 5.5 shows the complementary bottom of 10 mm x 150 mm wire mesh in two larger corner spans.

In Figure 5.6, shows complimentary 16 mm diameter bottom rebars next to the stairwell and lift shafts where the slab is, indeed, a one-way slab.

Note that the corner spans are typical of higher moments, so a 10 mm × 150 mm bottom wire mesh is necessary, along with the fiber reinforcing.

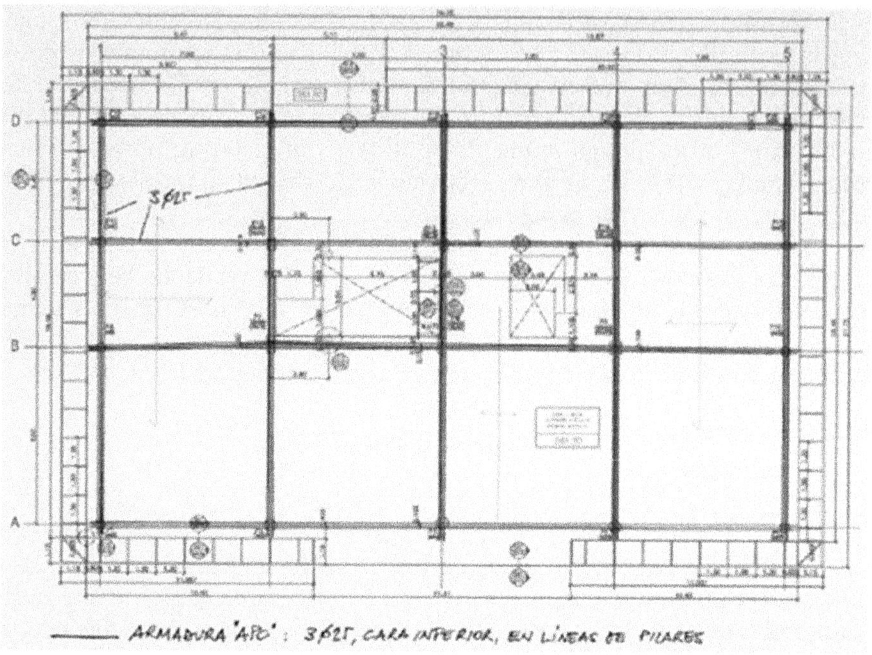

Figure 5.4 A set of three APC rebars of 25 mm diameter, in the bottom from column to column.

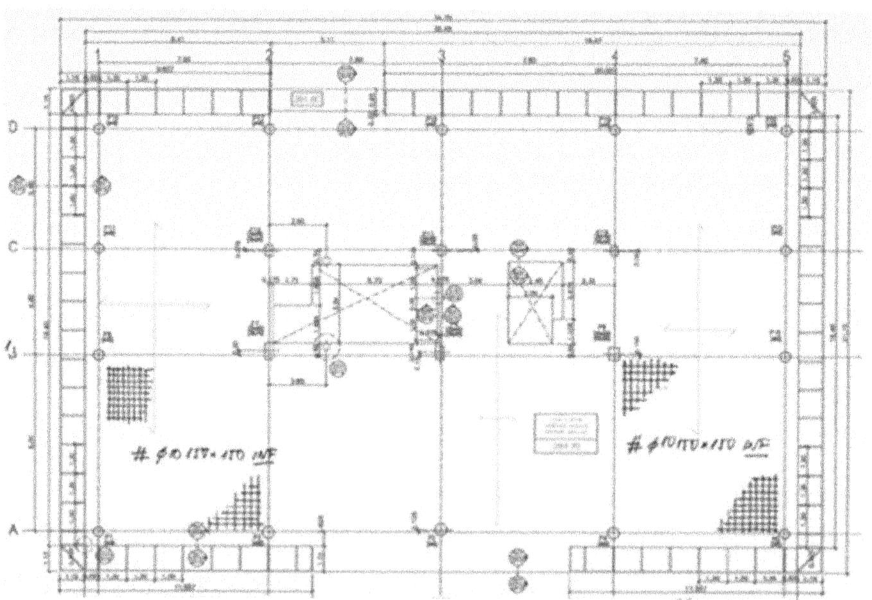

Figure 5.5 Complementary wire mesh in two larger size corner spans.

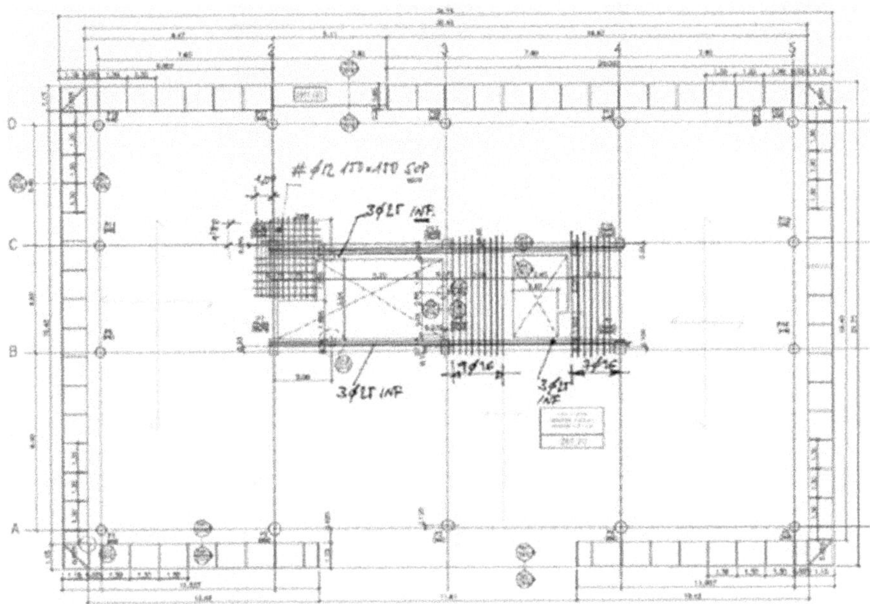

Figure 5.6 Complementary rebars of 16 mm diameter in one-way slab section between stairwell and lift shaft.

In cases where the suspended SFRC is supported by walls instead of columns, the APC bottom rebars are spaced every 1.5 m or up to 2 m apart, as shown in Figure 5.7 from a jobsite in 2005 in the UK, where the 100 kg/m³ TABIX 1.3/50 steel fiber concrete is being pumped for the first floor of row houses.

In this case, each rebar has to carry, like a cable, its own weight and the service loads of a strip of 1.5 m or 2 m width if the slab is a one-way slab.

In the case of two-way slabs, with both sides supported by walls, each APC rebar in each direction carries only the half of the total load of the strip.

5.2.2 Shearing and punching

A verification of V_{Rd} punching shear resistance of columns has also to be done.

The design shear strength is *(ACI, 2013)*:

$$\tau_{Rd} = 0.66\,\mu\sigma_{cr}u = 1.427 \ N\,/\,mm^2 = 207\,psi$$

$$t_c = \sqrt{\frac{a^2}{\pi}} = 0.254m \ (10 \ in) \tag{5.10}$$

where t_c is the equivalent radius of the column

Figure 5.7 Typical continuity bottom and pumping of the SFRC.

$t = 300mm(1 \text{ ft})$ where t is the slab thickness

$$V_{Rd} = 2\pi\left(\frac{t_c + 3t}{2}\right)t\,0.66\,\mu\sigma_{cr}u = 1.552 x 10^3 kN = 348.9 kip$$
$$(DIN\ 1045-1,\ ACI,\ 2013)$$

(5.11)

The factored column reaction is calculated:

$$R_c = qL_{rx}L_{xy} = 0.760 x 10^3 kN = 172.88\ \ kip$$

where L_{rx} and L_{ry} are the net spans in x and y directions, so that it is verified:

$$\frac{R_c}{V_{Rd}} = 0.495$$

A rapid deflection control is done as follows as the span to depth ratio needs to be smaller than 30, $\dfrac{L_x}{t} = 26.667$:

$$\delta = \frac{0.185(G+q_L)L_x}{E}\left(\frac{L_x}{t}\right)^3 = 10.48mm,(0.413in)$$
$$(Destr\ddot{e}e,\ 2006)\ \text{and}\ (ACI,\ 2013)$$

(5.12)

Figure 5.8 A 2000 mm diameter with 200 mm thickness round panel test in flexion.

And the span-to-deflection ratio needs to be larger than 500, $\dfrac{L_x}{\delta} = 763.53$

At the LKS job site, a tight quality control procedure was implemented and among other test specimens, six round panels of 2000 mm diameter x 200 mm thickness, as shown in Figure 5.8, were cast in order to record a 200 mm diameter center point loading diagram while the round panels were simply supported along their perimeter.

The average ultimate loading intensity was 260.25 kN, with a variation of plus or minus a maximum of 4%, confirming once again that the considerable variation of up to 30–50% recorded on all prismatic type of specimens tested (EN 14651) is typical of these small specimens and not characteristics of the steel fiber reinforced concrete (SFRC).

From the round panel test, the post-cracking strength $\sigma_p = \mu\,\sigma_{cr}$ can be verified by back-calculation as follows, where R is the radius of the panel, a is the radius of loading plate, and g is the panel unit area weight:

$$M_u = \frac{gR^2}{6} + \frac{P_{max}}{2\pi}\left(1 - \frac{2a}{3R}\right) = 39.46 kN\frac{m}{m}, (8.87\ kip) \qquad (5.13)$$

Hence, *(Destrée,1995, 2000, 2004 and 2008)*

$$\sigma_P = \frac{M_u}{0.45 b^2} = 2.19 MPa, (317.92\ psi) \qquad (5.14)$$

Instead of $\mu\,\sigma_{cr} = 2.162$ N/mm² (317 psi) as in the design calculation here above, the back-calculation shows a very similar but somewhat larger value.

The 0.45 coefficient is based on the observation that the round indeterminate panel test, with a 1.50 m diameter and 150 mm slab thickness, subjected to a center-point loading, shows 90% of depth crack penetration at maximal load intensity while the crack opening remains limited to 0.5 mm.

5.2.3 Advantages of E-SFRS

The elimination of the traditional double layer of reinforcing steel is such that it reduces in construction time and shortens the job site planning critical path.

Because the SFRC is pumped to each level, crane time for lifting the rebars to each level is eliminated. This adds significant freedom for using the cranes for other applications, which further speeds up the whole jobsite. Everybody is aware of that the everyday cost of using tower cranes or mobile cranes is quite high.

The reduction in the number of workers on-site improves safety.

There is also the elimination of all risks of injuries and fatalities associated with rebar and mesh handling, cutting, and placement.

In addition, there is no need to store tons of mesh and rebars on-site, which becomes increasingly important as more and more sites are tight and congested.

A reduction in the volume of concrete is noticed, as no concrete coverage is needed with steel fibers.

Increased architectural freedom, as "strange" floor shapes become economical since they no longer require cutting rebars to different lengths.

Other examples of E-SFRS are shown in Figures 5.9 to 5.11.

As shown in Figure 5.9, a balcony preparation before the E-SFRS concrete installation. We observe the APC rebars from column to column in the bottom as well as top rebars along cantilever edges.

As shown in Figure 5.10 the pumping and placing of the E-SFRS concrete in Estonia in 2008 are shown without any mechanical vibrating like pokers or screeds.

In Figure 5.11, the multistory triangle office building with E-SFRC slabs, we can notice that the cantilever edge is provided with a top mesh of 8 mm x 150 mm size as there is no structural ductility in such an area of the slab.

In Figure 5.12, the preparation of an E-SFRS slab at the 17 floors high residence at the Rocca Tower (2008) in Tallinn can be seen.

Figure 5.13 shows a typical renovation in Belgium, of an old four-floor high building in downtown where all 19th century timber slabs were replaced by E-SFRS concrete slabs. It was indeed overly complicated and time-consuming to handle the rebars across doors and windows then install the traditional reinforcing. As slabs are one-way, the APC rebars are parallel at 3 m distance apart, with 7 m spans and 220 mm slab thickness.

Figure 5.9 APC bottom rebars.

Figure 5.10 Installation of the SFRC suspended slab.

Figure 5.14 shows the E-SFRS installation of the first floor of a row of houses in the Birmingham region at Single Hayes Rd,. The structure is wall-supported so that the APC rebars are distributed evenly in both directions.

Figure 5.15 shows the preparation of an E-SFRS in Hannut, Belgium, for a light commercial building.

Note the APC rebars and in the bottom with the electricity wiring included in the concrete slab.

Figure 5.11 The Triangle office multi-storey building with SFRC slabs in Tallinn, Estonia.

Figure 5.12 Suspended and elevated E-SFRC slab preparation.

Figure 5.16 shows the installation of concrete by pumping and without mechanical vibrating.

Figures 5.17–5.19 show an old multistory building in Riga, fully emptied of all old slabs and some walls.. The finished E-SFRS includes steel beams instead of old walls.

Holes are also possible in E-SFRS such that the resisting moment of the missing part of the slab is replaced by bottom rebars at the perimeter of the hole.

Figure 5.13 Old 19th-century mansion speedy rehabilitation in Belgium by E-SFRS installations.

Figure 5.14 Single Hayes Rd E-SFRS on-way slab installation.

Finally, in Figures 5.17, 5.18, and 5.19, a 19th-century old building in Riga is being renovated, where only the façades are kept pristine. A series of cores drilled in the hardened E-SFRS to allow the placement of PVC pipes, and a large rectangular opening was created where the slab has been reinforced with additional perimeter rebars in the bottom. The cored holes are indeed close to a long wall support. Coring next and close to a column will require extra reinforcing for both flexion and shear.

Figure 5.15 APC rebars in preparation of a light commercial building.

Figure 5.16 Pumping an E-SFRC at 100 kg/m³ dosage rate of TABIX 1.3/50 steel fibers.

Figure 5.17 The old building façade remains unlike the replacement of the original structure inside.

Figure 5.18 Inside view of the completed E-SFRS slab structure with a large rectangle opening.

Figure 5.20 shows an office building in Tallinn where all slabs are E-SFRS of a design identical to the 2004 full-scale test slab of Bissen (ArcelorMittal, Luxembourg): 200 mm thickness, 6 m x 6 m span with the same set of APC rebars of 5 diameters 16 mm in the bottom from one column to another.

Both Figures 5.20 and 5.21 are provided by the ArcelorMittal steel fiber team.

For those who wish to study in depth the progressive collapse of reinforced concrete buildings, I recommend to read in detail (the Ph.D. thesis of Engg. A.R.I. Ali El-Gamal).

Figure 5.19, shows the pumping and the installation of a one-way statically determinate 180 mm thickness SFRC slab of 5.5 m span, in C30-37 with 100 kg/m³ dosage rate of the TABIX1.3/50 of 1. 3 mm diameter and

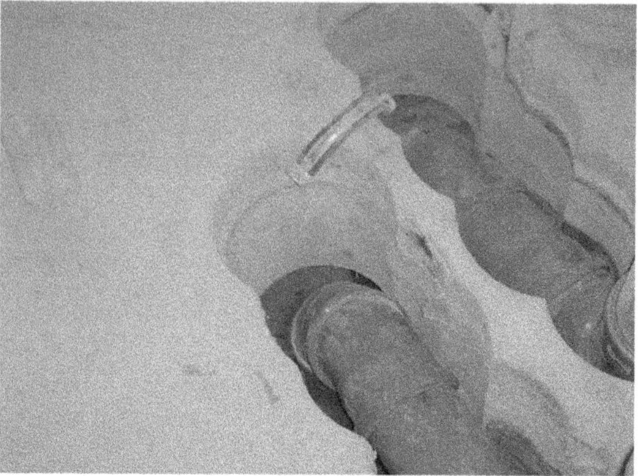

Figure 5.19 Holes drilled in the E-SFRS to insert PVC pipes next to a wall support of the slab.

Figure 5.20 E-SFR Slabs in this this Ericsson office building in Tallinn (Estonia).

50 mmp length- 850 MPa wire tensile strength. The mix free falling out of the pump hose is self-compacted thus installed without poker vibrating. The slab includes the APC rebars in the bottom at 1 m distance apart to absorb akas well the moment peak at mid-span.

Note the perfection of the mix design to ensure homogeneity and high workability without segregation or bleeding. It is a stable mix in which the aggregates are still visible, and there is a total absence of any water leaking

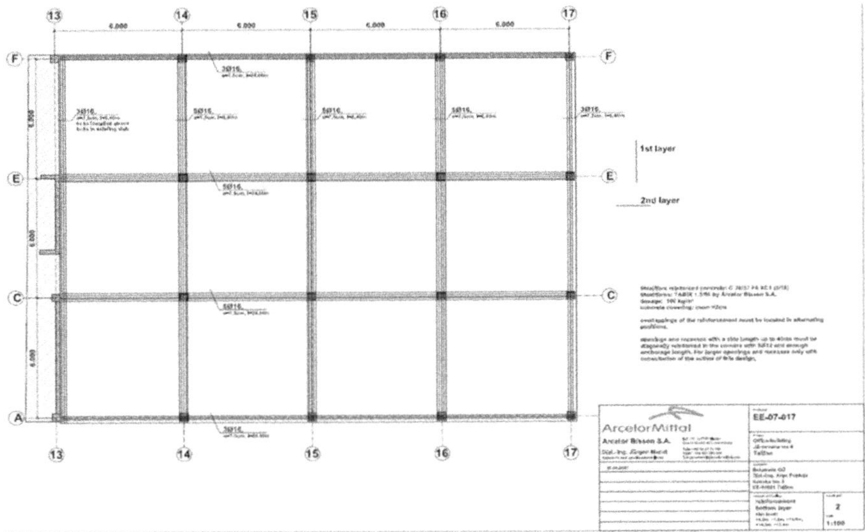

Figure 5.21 Typical E-SFRS slab of 6 m × 6 m span and APC rebars, in an Ericsson building.

Figure 5.22 100 kg/m³ TABIX 1.3/50 mix pumped and installed without poker vibrating on a smooth plywood form for a statically determinate suspended elevated slab.

out onto the plywood form. The undulating TABIX 1.3mm diameter × 50 mm length and 850 MPa wire tensile strength has been used as a low aspect ratio fiber of 38, which is very user-friendly even at a 100 kg/m³ dosage rate, without no tendency to ball together whatever the mixing method used.

An equivalent standard flexion performance is achieved by 50–60 kg/m³ of HE+1/60 1500@MPa wire tensile strength (Figure 5.22).

I hope that the reader understands how the E-SFRC slabs process is conservative, as it offers a quite high global safety coefficient (the collapse loading to service loading ratio) no less than 4 and often 7 or higher. It is also much more economical compared to traditional methods, easier to design and place, and more sustainable than traditional reinforced slabs.

E-SFRC slab is an innovation that perfectly fits the needs of the 21st century, being lean, effective, economical, and durable.

The only barriers it encounters are psychological and traditional construction practices.

In the next chapter, I'll show the readers that E-SFRC slabs also offer high fire and temperature resistance, up to 3 and 4 hours, much more than traditional slabs.

REFERENCES

1. X. Destrée. "Planchers structurels en béton de fibres métalliques; justification du modèle de calcul » Troisième colloque Internationnal Francophone sur les Bétons Renforcés de Fibres Métalliques. Université Laval, CRIB, 1998.
2. X. Destrée. "Structural application of steel fiber as principal reinforcing: Conditions, design examples." RILEM PRO 15, Fiber Reinforced concrete(FRC) BEFIB, 2000.
3. X. Destrée. "Steel fibre reinforcement for suspended slabs." UK, Concrete, September 2001, pp. 58–59.
4. X. Destrée. "Structural Application of steel Fibre as only reinforcing in free suspended elevated slabs: Conditions-Design-Examples", *Rilem Pro39, 6th International RILEM Symposium on Fibres-Reinforced concretes*, Vol. 39, 2004, pp. 1073–1082.
5. X. Destrée. "Concrete Free suspended elevated slabs reinforced with only steel fibres: full -scale testing results and conclusions - Design examples" *RILEM PRO 49, Workshop on High Performance Fiber Reinforced Cementitious composites in structural Applications*, 2005.
6. X. Destrée. "Steel fibre reinforced self-compacting concrete in free suspended elevated slabs; Full-scale testing conclusions, design and examples". *SCC 2005, second North American conference in the Design and Use of Self-Consolidating Concrete and Fourth International RILEM symposium on Self-Compacting Concrete*, Chicago, 2005, pp. 417.
7. X. Destrée, "Structural steel-fibre-reinforced concrete construction." *Concrete*, 41 (8): 23–24, September 2007.
8. X. Destrée, J. Mandl. "Steel fibre only reinforced concrete in free suspended elevated slabs: case studies, design assited by testing route, comparison to the latest SFRC standard documents" FIB 2008, Amsterdam, *Tailor Made Concrete Structures: New Solutions For Our Society*, CRC Press, p.111.
9. X. Destrée. "Free suspended elevated slabs of steel fibre reinforced concrete: full scale test results and design." pp. 941–950, RILEM Pro 60, 7th Rilem International Symposium-BEFIB 2008, Chennai, India.
10. X. Destrée:"Steel- fibre-only-reinforced concrete in free suspended elevated slabs", *Concrete Engineering International*, 30, 47–49, Spring 2009.
11. X. Destrée, "Steel Fiber Reinforced concrete in Free Suspended Elevated Slabs." *ACI SP 30*, 268–300, October 2009.

12. C. Kleinman, X. Destrée, A. Lambrechts, A. Hoekstra. "Steel fibre as only reinforcing in free suspended one way elevated slabs: design conclusions of a tunnel formed slab and walls based upon full scale testing results." *8th Rilem International Symposium-BEFIB 2012*, 2012, Guimaraes, Portugal.
13. X. Destrée, J. Silwerbrand. "Steel Fibre Reinforced Concrete in Free-Suspended slabs: a case study of the Swedbank arena in Stockholm." *FIB Symposium 2012, concrete Structures for Sustainable Community*, 2012, pp. 97–101.
14. X. Destrée. "Steel fibre reinforced concrete elevated suspended slabs: design cases in Europe and in the USA." *FRC 2014, Joint ACI-FIB International Workshop.*
15. X. Destrée, A. Krasnikovs. "Fire resistance of steel fibre reinforced concrete elevated suspended slabs: ISO fire Tests and conclusions for design." FIB, 2020, *Shanghai, 17th FIB Symposium, Concrete Structures for Resilient Society.*
16. B. Mobasher, X. Destrée. "Design and construction aspects of steel fiber-reinforced concrete elevated slabs." American Concrete Institute, ACI special Publications SP274, pp. 95–107, 2010.
17. S. Aidarov, F. Mena, A. de la Fuente. "Structural response of a fibre reinforced concrete pile-supported slab: full-scale test." Engineering Structures-229 (2021) 112292, Elsevier.
18. M. di Prisco, P. Martinelli, B. Parmentier. "On the reliability of the design approach for FRC structures according to FIB Model Code 2010: the case of elevated slabs." *Structural Concrete*, 4, 588–602, 2016.
19. B. Parmentier, P. Van Itterbeek, A. Skowron: The behaviour of SFRC slabs: The Limelette full-scale experiments to support design model codes. FRC 2014 *Joint ACI-FIB International Workshop. Fibre-reinforced concrete: From design to structural applications.*
20. W. D. Cook Mitchell. "Preventing progressive collapse of slab structures." *Journal of Structural Engineering*, 1100(7), 1513–1532, July 1984.
21. CSA Standard (Canadian) A23.3-94, 13.11.5.1, p. 98, 1994.
22. A. Maturana, R. Canales, E. Ansola. Vegueria: Experimental study of SFRC Plates Supported on Columns. Dpt. of Mechanical Engineering, Escuela Superior de Engineering, University of the Basque country, 2013.
23. B. Mobasher. *Mechanics of Fiber and Textile Reinforced Cement Composites.* Chapters 16 and 17, pp. 295–338. CRC Press, Taylor and Francis Group, 2012, ISBN978-1-4398-0660-9.

Chapter 6

Fire resistance of elevated suspended SFRC slabs (E-SFRS) and walls

6.1 INTRODUCTION

Steel fiber-reinforced concrete (SFRC) as a material has been investigated by many authors over the last 35 years when subjected to high temperatures ranging from 200°C to 1100°C.

The ACI. report 544.5R-13 on "Physical Properties and Durability of fiber-reinforced concrete" concludes that "Reinforcement in the form of steel fibers only... generally improves the performance of structural concrete members under extreme temperature and fire" and further states that "... fibers have been successful in extending the safe time of fire exposure for many practical and proven applications. Extending the safe time of fire exposure allows fire-fighters more time to evacuate structures and extinguish the fire safely." The door is indeed open for many different applications of SFRC under fire.

Among these applications, elevated slabs are an important and successful one, as we will show in detail in this chapter.

Kodur (1998) in reference (1) concluded that "The mechanical properties of fiber-reinforced concrete are more beneficial to fire resistance than those of plain concrete," and that steel fibers prevented early cracking and contributed to the compressive strength of concrete at elevated temperatures.

To summarize the benefits of SFRC in the fire resistance of concrete, it is preferable to cite V. Kodur, a prominent figure in fire resistance of structures at the National Research Council of Canada, as well as a Professor at the University of Toronto and Fellow of the A.C.I., who wrote the following on the thermal properties of SFRC:

> "The effect of aggregate type on the thermal conductivity of steel-fiber reinforced concrete is similar *to approximately 400°C. This is in contrast of the plain concrete which decreases slightly with increasing temperature up to 400*C."*
>
> *"The tensile strength of both plain (made of calcareous and silicious aggregates) decreases with temperature. However, the strength of fiber concrete decreases at a lower rate than that of plain concrete throughout the temperature range."*

DOI: 10.1201/9781003188315-6

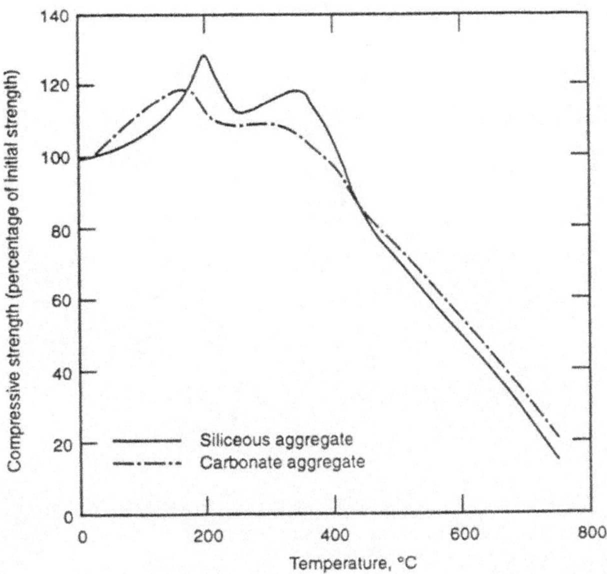

Figure 6.1 Compressive strength of steel fiber-reinforced concrete as a function of temperature.

> *"For temperatures up to about 360°C, the strength of fiber-reinforced concrete is significantly higher than that of plain concrete" as shown in Figure 6.1 in Reference (1) further below.*
>
> *"The increased tensile strength delays the propagation of cracks in fiber reinforced concrete structural members and, is highly beneficial when the member is subjected to bending stresses."*
>
> *"Steel fiber reinforced concrete attains higher ultimate strains than plain concrete at elevated temperatures. Thus, the presence of steel fibres increases the ultimate strain and improve the ductility of the concrete".* Indeed, this is quite visible in Figure 6.2, extracted from Reference (1), where we see that at 600 °C, the SFRC strain is six times higher than the plain concrete.

It is also concluded by Nurchasanan, Massoud and Solikin in Reference (2) that *"future concrete having steel fiber will act considerably as fire protective"*.

6.2 SFRC ELEVATED SUSPENDED SLAB APPLICATION

In addition to the investigation of SFRC as a material, real-scale loading tests on a real structure were conducted at ambient temperature (20°C) and compared to the same tests under fire at high temperatures for several hours.

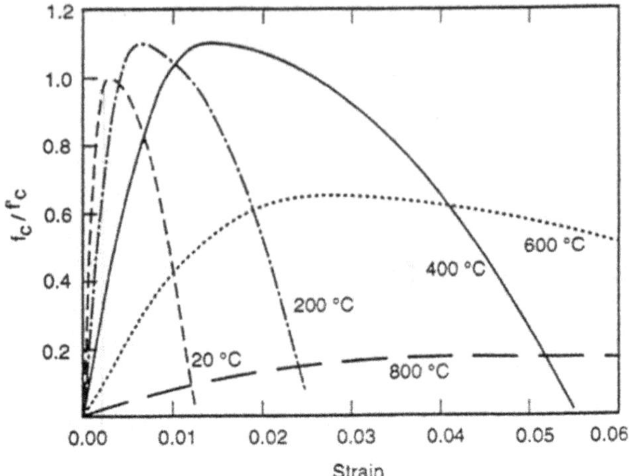

Figure 6.2 Stress–strain relationships of steel-fibre reinforced concrete as a function of temperature. Indeed the ratio of stress at a given temperature on the stress at cold (20°C) is used in the diagram.

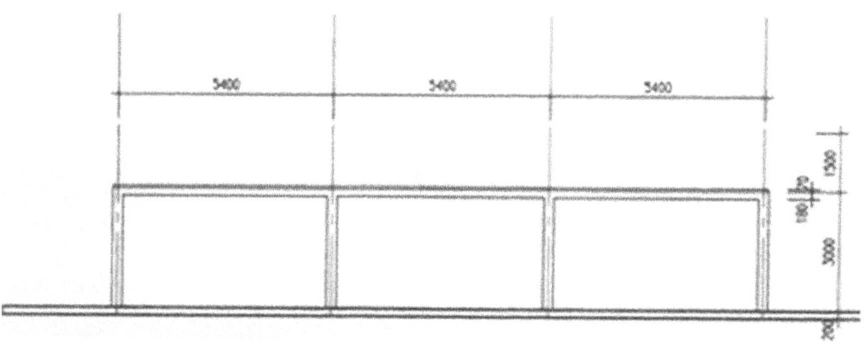

Figure 6.3 Tunnel formed slab and walls full-scale test.

6.2.1 Full-scale test at cold temperature (20°C) of a free suspended elevated one-way slab

The slab was cast over a tunnel-form installation, where the slab and the supporting walls are cast together. The tunnel forms are released the next day for reuse at the next location, as shown in Figure 6.3.

The technique of tunnel forming is frequently used in many countries for condos and hotel buildings.

Several consecutive inverted U steel forms, firmly attached and braced together, are installed with a camber of up to Span/300 on a bottom slab. Three or four spans are assembled together, allowing an area of 20 m long by up to 10 m wide to be installed, including walls and slabs.

Since the form is quite expensive, it must be reused multiple times. Therefore, once the entire typical area of 20 m x 10 m has been filled with steel fiber-reinforced concrete and allowed to harden to a compressive strength of 14 N/mm² after 16 hours, the U-forms are released and taken by a large crane to be installed one level higher, as seen in, as shown in Figure 6.4.

In Figures 6.3 and 6.4, the test here at cold temperature has been back-calculated according to the FIB Model Code provisions.

The test slab, organized at the Technical University of Eindhoven (Holland), was 180 mm thick to bridge three spans continuously, each 5.40 m, divided into three 2 m wide sections to repeat the loading tests. At one end, there was an attached balcony, 250 mm thick with a 2 m overhang. The walls, also 250 mm thick, were installed together with the slab. The phases of installation are shown in Figure 6.5.

Placing the inverted U-forms, including holes for vertical shafts (see at 7 in Figure 6.5), fixing, and shoring to maintain an 18 mm camber, bracing (see 10, 11, and 12), installation of all electrical piping, including heating pipes (9), pouring the fiber concrete from the bottom of the wall to the top of the slab (5), with concrete falling from a hopper (6), or alternatively, it can be pumped. The concrete is compacted naturally or with mechanical vibration, finished to the required level, and carefully cured (14). A guard rail for safety is also essential (13).

The testing committee sadly decided not to include the set of APC rebars as defined in the chapter 5.

The mix design consisted of 350 kg of CEM I (EN) and CEM III (EN) with a W/C ratio < 0.50, 0–16 mm to 32 mm aggregate grading, and included 50 kg/m³ of type I hooked-end steel fibers with a 0.9 mm diameter, 60 mm

Figure 6.4 A tunnel form is being installed.

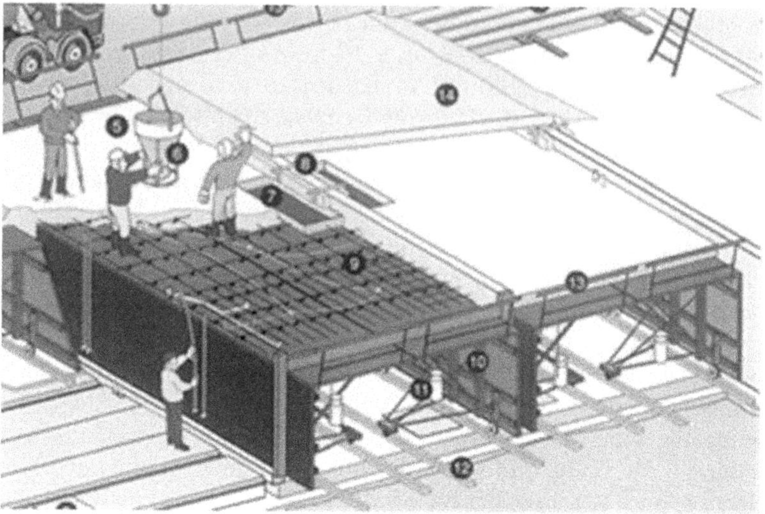

Figure 6.5 Installation of multiple tunnel form elements.

Figure 6.6 Pouring the SFRC with a flying hopper.

length, and 1200 N/mm² tensile strength. The fibers were introduced via a conveyor belt, which discharged them into the truck mixer at the batching plant. Both loose and collated fibers were used and mixed together, with no fiber balling observed. The mix had an F4 fluidity, ensuring that no mechanical vibration was needed during installation. Figure 6.6 shows the installation of the test slabs.

The completed structure is shown in Figure 6.7.

Figure 6.7 The hardened test slab and walls after the form release.

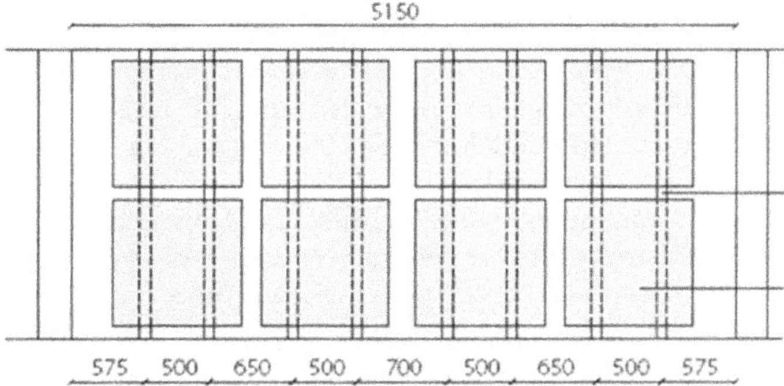

Figure 6.8 Loading schema of each six spans with the concrete blocks loading.

The compressive strength was f_c (14 hours) > 14 N/mm², enabling the early-age release of the inverted U-forms at 14 hours for reuse at the next level.

The hardened concrete at 28 days, showed an average compressive strength of 66.6 N/mm² (150 mm cube strength), a maximum average flexion strength (EN 14651) of f_{max} = 6.5 N/mm²; the average limit of proportionality LOP: 6.0 N/mm²; average f_{r1} = 5.5 N/mm², average f_{r3} = 5.0 N/mm² and average f_{r4} = 4.5 N/mm² (all following EN 14651 flexion testing procedure) with an average volume weight of 2384 kg/m³. The average modulus of elasticity was 27.184 N/mm².

The loading consisted of a distributed intensity, applied in incremental steps with up to 16 concrete blocks of different sizes as shown in Figure 6.8 extracted from ref.6.4 in Befib 2012), to achieve a final surface-imposed load of up to 12.18 kN/m², as shown in Figure 6.9. The blocks were placed on the slab by crane, on lumber, with 500 mm spacing between them.

During the test, each loading intensity step increase was applied after stabilization of the former.

Figure 6.9 Loading test of an internal span (same as Figure 3.50).

The average deflection due to the applied load, just before collapse at the last stable observed step of loading, was 20.8 mm at a load of 12.18 kN/m², excluding the slab's own weight.

The 20.8 mm was the total average deflection from demolding until just before collapse, representing the deflection caused only by the superimposed loading intensity. This agrees well with the predictions.

Under 6.5 kN/m² uniformly distributed loading intensity, the first negative moment minute crack was observed.

Under an 8 kN/m² uniformly distributed loading intensity applied by means of the heavy concrete blocks stacked as described above, the deflection of the central span was 8 mm, which increased to 10 mm after 24 hours. At this point, five parallel minute bottom cracks, with openings of up to 0.4 mm, were visible, as shown in Figure 6.10.

At 8 kN/m² loading intensity, the slabs remained stable with no further increase of deflection with time.

A long-term loading during 6 months at 7 kN/m² intensity of an edge span confirmed no significant visible creep.

By comparison, the housing specification of variable loading intensity requires a long-term loading intensity of 2.4 kN/m² in Holland, excluding the slab's own weight.

Practically, we can design the slab so that it doesn't show flexion cracking under both negative and positive moment under the most onerous service (Service Limit State) load.

This sets the SLS imposed load limit to 6.0 kN/m² for our TU Eindhoven full-scale slabs, with a deflection of approx. 6 mm, resulting in a span-to-deflection ratio of 900 well above the SLS limit of 500.

Including the partition walls, floor coverings, and ceiling, a total imposed load never exceeds 6.5 kN/m² but typically ranges from 4.5 to 5 kN/m² in residential and light commercial buildings.

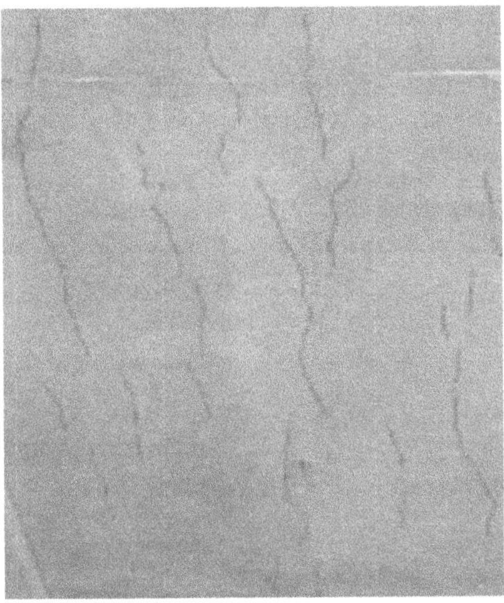

Figure 6.10 Minute bottom cracking at mid-span under 8 kN/m² uniformly distributed loading.

6.2.2 Conclusions of the full-scale test carried out at cold temperature

- All nine spans tested showed a quite ductile behavior resulting from a significant moment redistribution together with a multiple cracking pattern under sagging moments.
- The deviations between the nine spans tested are significantly smaller than those recorded during EN14651 beam tests.
- EN14651 test method shows a deviation that is typical to the test method only and not to the real behavior of the structure as tested.
- The first crack experimental loading intensity in all nine spans occurred at more than 300% of the standard housing service loading intensity in Holland (2.5kN/m²)
- The ultimate loading experimental intensity of all nine spans occurred at more than 500% of the standard housing service loading intensity in Holland (2.5kN/m²).
- The FIB Model Code 2010 calculation method does not reflect the reality of the full-scale test results of three sets of three one-way continuous spans as tested and reported in this chapter.

Therefore, an application multiplying factor, such as η_{det}, the structural indeterminacy factor defined in the EN-SS 812310, "Design Standard for steel fiber reinforced concrete structures", must be applied to the EN 14651-derived flexion strength. Otherwise, the EN 14651 flexion strengths, as used as in the FIB Model Code, can lead to uneconomical SFRC solutions due to increased costs and a lack of competitiveness, making both documents a deterrent to the use of SFRC structures.

All experimental values and test constatations here are from reference (3) by Herremans and reference (4) by Kleinman, Destrée, Lambrechts.

6.2.2.1 Back calculations

$h := 0.180 \text{m}$ Slab thickness

$w := 24 \cdot \dfrac{\text{kN}}{\text{m}^3}$ Volume weight of slab

$p := h \cdot w = 4.32 \cdot \dfrac{\text{kN}}{\text{m}^2}$ Slab's own weight

$l := 5.20 \cdot \text{m}$ Net span taking into account the wall thicknesses

$s := 6.000 \cdot \dfrac{\text{kN}}{\text{m}^2}$ Imposed loading intensity

A three consecutive spans l beam of supports 0, 1, 2, and 3
M01 is the max 0–1 positive span moment
M1 is the minimum negative moment above 1
M12 is the minimum positive moment in the 1–2 span
M01, M1, and M12 are the maximum envelope moment with the most onerous combination of s and p

$$M01 := p \cdot \frac{l^2}{12.5} + s \cdot \frac{l^2}{10.} = 25.569 \cdot \frac{\text{kN} \cdot \text{m}}{\text{m}}$$

$$M1 := -p \cdot \frac{l^2}{10.0} - s \cdot \frac{l^2}{8.6} = -30.546 \cdot \text{kN} \cdot \frac{\text{m}}{\text{m}}$$

$$M12 := \frac{p \cdot l^2}{40} + \frac{s \cdot l^2}{13.3} = 15.119 \cdot \text{kN} \cdot \frac{\text{m}}{\text{m}}$$

$f01 := 6 \cdot \dfrac{M01}{h^2} = 4.735 \cdot \dfrac{\text{N}}{\text{mm}^2}$ Maximum SLS flexion stress under positive moment in the edge span

$f1 := 6 \cdot \dfrac{M1}{h^2} = -5.657 \cdot \dfrac{\text{N}}{\text{mm}^2}$ Maximum SLS flexion stress under negative moment over the first support

$f12 := 6 \cdot \dfrac{M12}{h^2} = 2.8 \cdot \dfrac{\text{N}}{\text{mm}^2}$ Idem over the second support

Conclusion 1

At 6 kN/m² imposed variable loading, the first crack negative moment strength f1 = 5, 6 N/mm² is attained. The ULS (Ultimate Limit Stage) loading intensity is of 6 kN/m² x 1.5 = 9 kN/m²

This is as observed during the full-scale test

$$f_{R3} := 5.0 \, \frac{N}{mm^2}$$

-EN 14651 flexion test at 50 kg/m³
HE+1/60 C30-37

$$\eta_{det} := 1.4$$

-Statical indeterminacy factor SS
812310 Swedish standard 1-way
continuous slabs

$$\gamma_m := 1.5$$

-material factor

$$M_{Rd} := \frac{\eta_{det} \cdot f_{R3} \cdot h^2}{6 \cdot \gamma_m} = 25.2 \cdot kN \cdot \frac{m}{m}$$

Resisting moment SFRC

$$\gamma_{Q1} := 1.35$$

Permanent Load factor

$$\gamma_{Q2} := 1.5$$

Variable Load factor

$$s_{max} := \frac{\left(16 \cdot \eta_{det} \cdot f_{R3} \cdot h^2\right)}{6 \cdot l^2 \cdot \gamma_{Q2}} - \frac{\gamma_{Q1} \cdot P}{\gamma_{Q2}} = 11.023 \cdot \frac{kN}{m^2}$$

Maximum Uniformly Distributed
Loading intensity at collapse

Conclusion 2
The maximum loading intensity at collapse is **11 kN/m², excluding the
slab's own weight, to be compared with the 12.8 kN/m² average experimental imposed loading intensity just before collapse at full-scale tests**

$$E := 28000 \, \frac{N}{mm^2}$$

Elasticity modulus

$$f_{max} := \frac{5 \left(p + s_{max}\right) \cdot l^4}{384 \cdot E \cdot \left(\dfrac{h^3}{12}\right)} = 10.734 \cdot mm$$

Deflection beyond ULS loading of 9 kN/m²,
just before collapse above each
supporting wall and the slab is with wide
cracks: simple support conditions

$$s_{SLS} := 6 \, \frac{kN}{m^2}$$

Maximum SLS loading

$$f_{SLS} := \frac{\left(p + s\right) \cdot l^4}{384 \cdot E \cdot \left(\dfrac{h^3}{12}\right)} = 1.444 \cdot mm$$

deflection with spans with fixed ends :

Table 6.1 SLS flexion stress in function of UDL

s in kN/m²	2	3	3.50	4	5	6
f in N/mm²	1.62	1.91	3.50	3.73	4.23	4.74

The test slab at TU Eindhoven (NL) and its maximum positive moment flexural stress are a function of the imposed Uniformly Distributed Loading intensity, as shown in Table 6.1:

Such a table is of interest since we'll see that the SLS flexion stress level at cold temperature has a large and very significant influence on the fire resistance.

Conclusion 3

The test slab shows a maximum SLS loading intensity of up to 5–6 kN/m² together with a very limited experimental deflection of 6 mm of span/δ= 900, much stiffer than the 500 of the Eurocode.

At 8 kN/m²–10 kN/m², indeed a stable ULS stage with significant cracking and large deflections beyond the L/500 limit, the structure is still stable.

Some more ductility was even available beyond the ULS stage as further, the loading was increased to record at 12.18 kN/m² just before the slab collapse occurred.

The TU Eindhoven test slab at design stage was not provided with the continuity APC set of rebars to prevent progressive collapse, solely for academic reasons, although it is mandatory according to the ACI 544-6R15 report.

The academicals refused the APC rebars against my opinion and experience because the effect of steel fiber reinforcement could not be identified clearly as being hidden behind the rebars effect.

At cold temperature test, it did not matter as the test clearly showed a reliable SFRC structure capable to undergo SLS loading and criteria, stable at ULS and to carry more loading beyond ULS. Clearly, the SFRC here is fit for use in the residential application-way of tunnel slabs.

We'll see now that these APC rebars become important in case of fire.

The fire test slab procedure has been decided because the ACI 544-6R15 report states that *"some of the needed areas of research in this subject includes fire resistance" (4.5. limitations and areas of needed research - 4.5.2 Research needs"* in p. 9/38.)

6.2.3 Full-scale test under standard fire (ISO 834) temperatures of a free-suspended elevated one-way slab

The entire fire testing process was completed on behalf of and financed by Primekss SIA, in collaboration with ArcelorMittal at the FIRES laboratory in Slovakia, and reported in Reference (5).

Primekss SIA, Riga (LV), formed and manufactured the slabs in Riga, using local ready mixed concrete under the usual Primekss tight quality control and specification, including 50 kg/m³ of ArcelorMittal steel fibers type HE+1/60 of 60 mm length, 1 mm diameter, provided with hooked ends and of 1500 N/mm² constituent steel wire strength. The slabs were trucked from Riga to FIRES s.r.o, Poprad (CK) and left there for ca. 3 months stored in open air to let them harden and dry out prior to testing.

The concrete matrix was typical Primekss concrete, including 350 kg CEM I with 35 kg/m³ of the proprietary Primekss DC ettringite-generating cement to provide 400µs net restrained expansion after hardening.

The mix aggregates were of limestone nature, which always has a positive effect on the fire resistance. The Ettringite, with its high content of embedded chemically bounded water, also has a positive effect on the fire resistance.

6.2.3.1 The fire test of SFRC slabs: Purpose of the tests

A multistory typical residential building in Riga-Latvia, whose scaled model can be seen in Figure 6.11, was taken as first case to design a slab and wall structure of twelve levels built in SFRC. The whole frame is built using tunnel forms where the walls and the slab of one level are cast in one single operation, as shown in Figure 6.5. After 18 hours, when the concrete reaches 14 N/mm² compressive strength, the tunnel forms are removed while shoring is placed, to install the forms at the next level up.

Each level of the building is designed of 180 mm thick walls to carry the cast in situ slabs spanning 5.40 m, a span to depth ratio of 30 and where both slabs and walls are only reinforced by 50 kg/m³ of ArcelorMittal HE+-1/60 hooked ends steel fibers of 1 mm diameter, 60 mm length, 1500 N/mm² steel wire strength. In the slab depth including 60 mm coverage from the bottom face, one B500 B rebar of 16 mm diameter is included at every 1.5m distance apart in order to meet the anti-progressive collapse (APC) condition following the ACI544.6R-15 provision J.1, p.36. in case a supporting wall should disappear due to an accident or explosion.

Following the success of the full-scale test at cold temperature and the subsequent validation of the design, a full-scale fire test was decided in order to gain the approval.

In buildings higher than 13 m, generally, a fire resistance of minimum of 120 minutes is required.

Figure 6.11 Overview of the project in Riga, Latvia.

6.2.3.2 The fire resistance test set-up

The test setup is made according to EN 1365-2 and EN 13381-3-2015 standards, "Test methods for determining the contribution of the fire resistance of structural members" where a statically determinate 180 mm thick slab of 4.40 m net span, 3 m width is subjected to two line loadings so that the flexural moment is constant and maximum over 2.20 m length, as shown in Figure 6.13 and in Figures 6.14 and 6.15, including as well the location of the thermocouples and deflection meters.

The slab is also provided with a set of APC rebars according to ACI 544.6R-15 (2015) and the Canadian CSA-A23.3-04, to prevent the slab from falling in the event of a supporting wall collapse due to an accident, explosion, or terrorist attack. The APC rebars, in the span direction only, are 16 mm diameter rebars at the bottom, spaced 1.50 m apart. These APC rebars have 60 mm of coverage, providing a very small proportion of 0.07% reinforcement, or only 3.6 kg of steel rebars per cubic meter of concrete. This is a very small proportion, as shown during the slab installation in Figure 6.12.

The principal reinforcing is obtained by adding 50 kg/m³ of ArcelorMittal steel fibers, with a 1 mm diameter and 60 mm length, made from 1500 N/mm² steel wire, to the ready-mixed concrete. The steel fibers (HE+1/60 type) are provided with hooked ends.

The concrete matrix, supplied as a C25/30 with 280 kg CEM I and W/C < 0.50, with a 16 mm maximum aggregate size was of 23.83kN/m³ density but, indeed, showed a 52.6 N/mm² average compressive strength at 28 days! As shown in Figure 6.12, the S4 flowability was such that no poker vibrating was needed. This is another important advantage of using SFRC slabs,

Figure 6.12 Installation of the test slab strips.

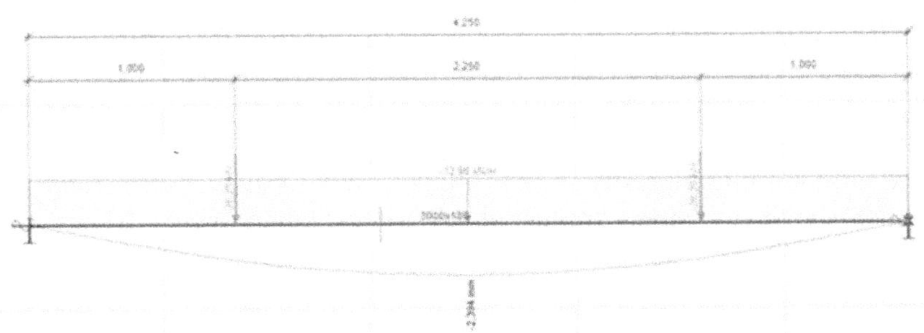

Figure 6.13 Scheme of the test slab: a statically determinate slab simply supported at edges.

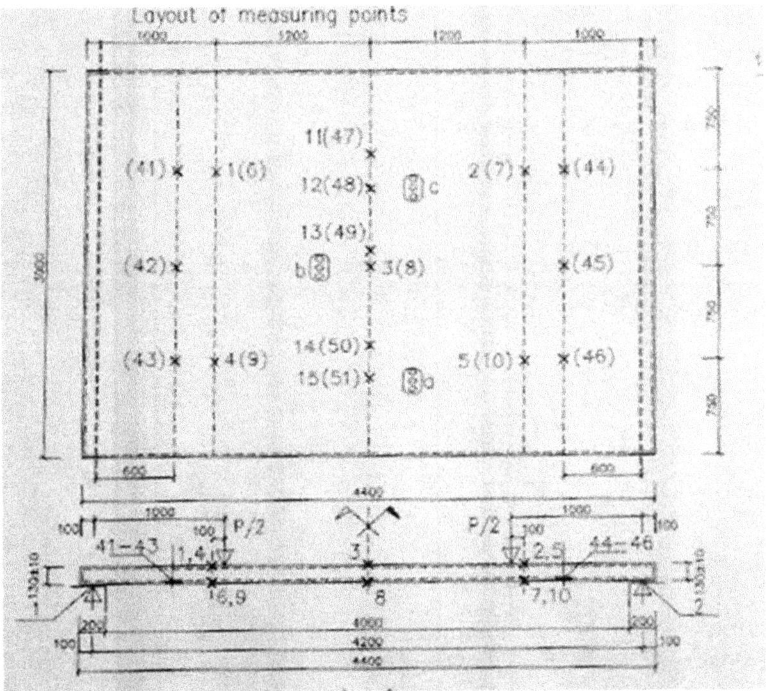

Figure 6.14 Test set-up with loading pattern.

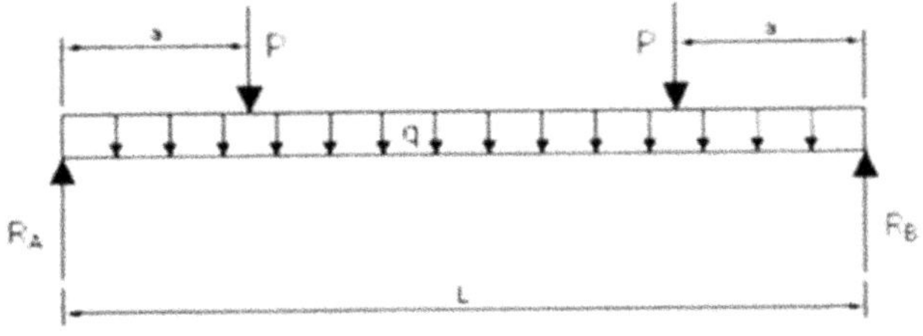

Figure 6.15 Test set up schematic.

as there is no need for mechanical vibrating, saving time and reducing environmental noise, nuisance, and noise of poker vibration in urban areas.

The test setup is shown in Figures 6.13 and 6.14, with the different recordings in temperature in Figure 6.14, as well as the loading schema in Figure 6.15.

The total loading intensity is 2P = 48.3kN over the 3 m width slab.

Hence, the flexion stress applied at the fire test beginning. is derived..

Further under fire, the section shall be a reduced section subjected to more stress.

$$H := 0.18m$$

$$P := 48.3kN$$

$$B := 6m \qquad u := 1m$$

$$p := \frac{P}{B} = 8.05\frac{kN}{m}$$

$$p \cdot u = 8.05 \ kN$$

$$a := 1m$$

$$M := p \cdot u \cdot a = 8.05 \ kN \cdot m$$

$$W := 24\frac{kN}{m^3}$$

$$w := W \cdot H = 4.32\frac{kN}{m^2}$$

$$w \cdot u = 4.32\frac{kN}{m}$$

$$L := 4.2m$$

$$M_{wu} := \frac{w \cdot u \cdot L^2}{8} = 9.526kN \cdot m$$

$$M_T := M + M_{wu} = 17.576kN \cdot m$$

$$f_T := \frac{6 \cdot M_T}{u \cdot H^2} = 3.255\frac{N}{mm^2}$$

Thus, under 48.3 kN total loading according to the loading schema, the slab undergoes a 3.255 N/mm² flexion stress.

6.2.3.3 Temperature regime of the fire resistance test

The test itself has been organized and carried-out by the FIRES lab in Poprad (Slovakia - SK) with a furnace set as per ISO 834, temperature vs. time diagram according to the following formula:

$$T = 345\log(8t + 1) + 20 \tag{1}$$

As shown in Figure 6.16, the ISO 834 and ASTM E119 standards are almost similar, the ISO being still of a slightly higher temperature from

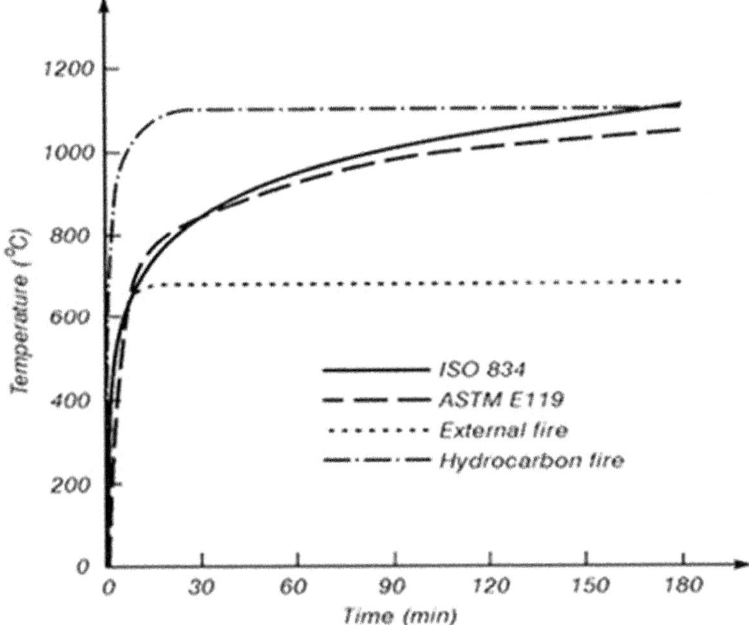

Figure 6.16 A comparison of time/temperature curves.

60 minutes to 180 minutes fire duration and is thus slightly more stringent than the ASTM E119.

Such an ISO/ASTM fire is also called cellulosic fire and is typical of houses and condos. The hydrocarbon fire is much more severe but only applicable where oil and gas are involved like in petrochemical plants.

As shown in Figure 6.16, the ISO 834 and ASTM E119 standards are almost similar, the ISO being still of a slightly higher temperature from 60 minutes to 180 minutes fire duration and is thus slightly more stringent than the ASTM E119.

6.2.3.4 Test results

The first statically determinate slab was tested under the initial flexion stress of 2.25 N/mm² (under 2 P = 15 kN instead of 48 kN) and passed the test after 4 hours (240 min) so that the test was stopped.

The three next statically determinate slabs have been tested under a 3.25 N/mm² flexion stress. All three passed the test brilliantly at 149, 192, and 204 minutes under fire, meeting the following five criteria:

- All four slabs resisted longer than 120 minutes.
- The deflection was smaller than 250 mm.
- The increase of deflection has been smaller than 11 mm per minute.
- The slabs remained flame and gas tight up to the end of test.

- The four slabs met the isolation conditions with a temperature above the slab between 75°C and 94°C, much colder than the 140°C average limit and 180°C local limit, as shown in Figure 6.17.

The tested slabs were all of the REI 120 class (resistance, integrity, and isolation, as shown in Figure 6.17), a figure found in reference by the Belgian building Research Institute, called now "Build Wise" as of year 2023.

Let's repeat here that the tested slabs were tested as statically determinate as specified in the EN 1365-2 fire rating testing standards. This differs from real world conditions where the slabs are statically indeterminate and can benefit of the plastic moment redistribution to provide even more ductility and safety resulting from the addition of both the material and the structural ductility.

Unexpectedly for SFRC statically determinate slabs, a large number of edge bottom cracks (approx. 16 cracks) are observed and have grown up from the bottom as it is visible in Figure 6.18: indeed resulting from the high SFRC ductility even at high temperature as shown earlier in Figure 6.2.

Also very positive is the fact that no spalling was observed during the 192-minute fire test, and after the completion of the slab test, there was still

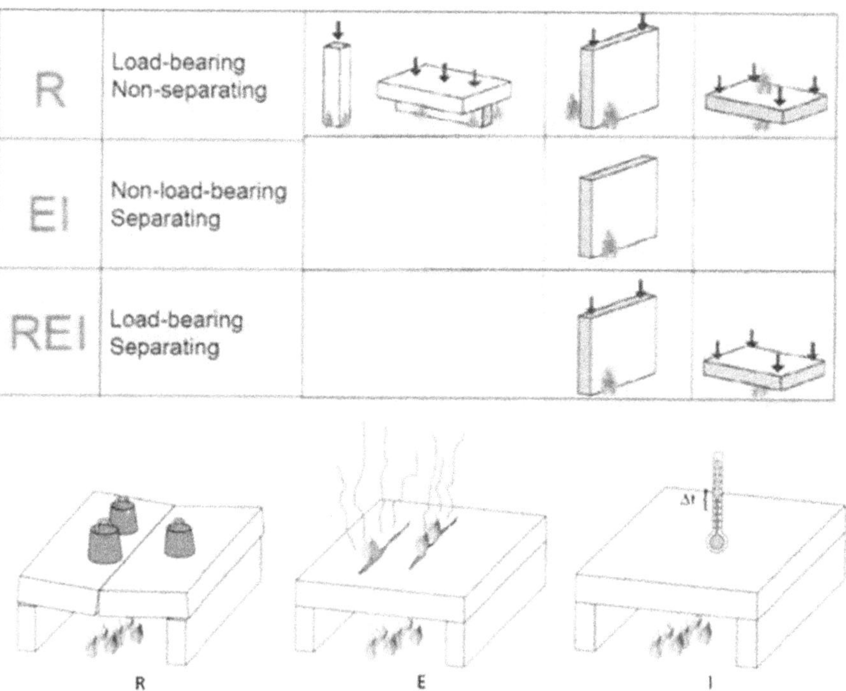

R	Load-bearing Non-separating			
EI	Non-load-bearing Separating			
REI	Load-bearing Separating			

Figure 6.17 Resistance, integrity, and insulation.

no collapse after 4 hours of fire exposure, starting at an initial flexural stress of 2.25 N/m², as shown in Figures 6.19 and 6.20.

This is an important finding. It is confirmed further by the fire tests of the SFRC walls in 6.3.

It is important to refer to References (7, 8) where P. Heek, J, Tkocz and P. Mark (T.U. Bochum, Germany) reports a similar EN 1365 -EN 13381 test of SFRC slab up to 40 kg/m³ dosage rate of high performance high strength hooked end steel fibers of the same family as the fibers used here for our chapter, in a C35-45 matrix resisted only 7 minutes before collapse under the standard fire. The report is indeed, without the right understanding, a fatal killer of the structural SFRC applications!

Why 7 minutes instead of more than 180 minutes in our case of tests? We have to understand.

Figure 6.18 The test slab at 192 minutes fire resistance under 3.25 N/mm²initial cold flexion stress.

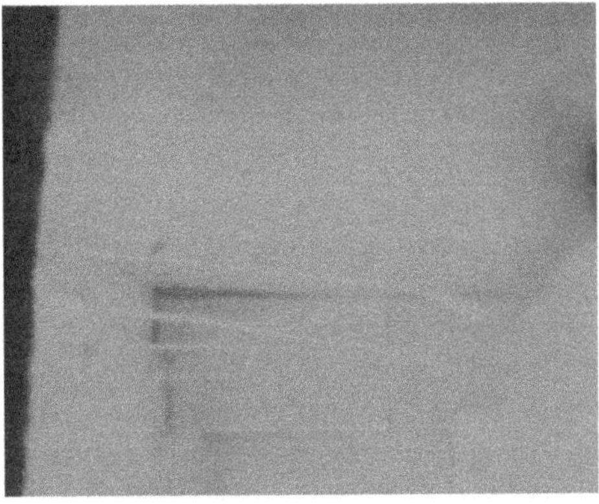

Figure 6.19 Inside the furnace at 181 minutes: no spalling at all observed.

Figure 6.20 After 242 minutes of test without collapse, no spalling at all observed.

According to the ISO 834 fire temperature law, after 7 minutes, the furnace temperature is not higher than 600°C so that at 30 mm depth above the bottom, the plain concrete temperature exceeds barely 120°C, a temperature without any effect on any structural plain concrete type what so ever. Something else happened in that 7 minutes collapse test. Wasn't the slab pre-cracked, and has not been detected prior to the tests. Or the small size of slab, or wrong heating, and anyhow slab size non-conform to EN specifications.

At 40 minutes, at 30 mm depth from the bottom, the plain concrete temperature reaches about 400 °C, a temperature that doesn't really damage the plain concrete.

It is not understandable why the collapse happened at such a short time under fire when the temperature inside the slab is still "cold" like.

We observe also that the slabs were only 1 m in width instead of 3 m by the EN fire standard, so the volume of tested material is very small with only 1 sample per type of slab.

As the test is of a statically determinate narrow slab, the slightest preexisting localized crack could not be controlled by an uncracked matrix further away in the slab.

The article also doesn't show in a control diagram that they followed the ISO furnace temperature.

It is also strange that the researchers mention limiting the total time of test to 2 hours due to the high cost of gas. Were they "cutting corners"?

At mid-span, we can imagine that immediately after the fire started, the preexisting single crack became wide open, a way inward of the heat that could quickly ruin the whole section.

In contrast to our tests presented in this chapter, the test failed massively, as the service capacity of the slab was greatly exceeded.

We must remind the reader that, under the German SFRC design standard, DAFSTB-DIN 1045, the steel fiber concrete (ULS)-design flexion strength is limited to ca: $(0.7 \, fr_{3m} /2)/1.5$ thus in this case with a f, = ca. 0.7 x 4.0 N/mm²/ (2 x1.5), thus $f._{ULS}$ = 0.93 N/mm² so that the maximum flexion stress in service could not be more than 0.93/ 1.5 = 0.62 N/mm².

In test case, the initial cold temperature moment should have been limited, by German standard, to 0.62*200²/6 = 4.13 kNm/m, which is equal to 37 % of the test moment M $_{Rd,\theta=20°C}$ = 11.1 kNm/m.

Why, in Germany, do academics test in fire at 1/0.37 =2.7 times more than twice the allowable loading intensity at cold?

Moreover, in real structure, there is a 0.7 coefficient applied to reduce the loading intensity under fire as it is very unlikely that the maximum service load is applied together with a fire event.

In the test case, the moment shouldn't thus have been taken more than 0.7 x 4.13 = 2.89 kNm/m thus 26% (one-fourth!) of the test load at cold!

As a result, usually in Germany by academics, the SFRC will be excluded in slab structures for ever unless the fibers are used as a marginal enhancement of heavy traditional rebar reinforcing.

In my opinion, the manufacturer of the supplied test fibers, is permanently resigned to that kind of low-profile backward uneconomical, and less sustainable solution.

Unlike in most advanced countries, such a severe limitation makes the SFRC structural use impossible in Germany if not used together with a larger quantity of rebars and wire meshes and this, at both ambient cold and fire temperatures.

It is also regrettable that the test didn't include the mandatory ACI 544 6R15 APC bottom rebars, which are essential to prevent the progressive collapse of real structures, where slabs could drop down to the level below and continue collapsing, ultimately leading to the failure of the entire structure.

In this case EN standard test of a 4 m x 3 m x 0.18 m slab subjected to 5 kN/m² variable loading, our test slab included an APC section of ((*4 m × 3 m × 0.18 m × 24000 N/m³*) + (*5000 N/m² × 3 m × 4 m*))/ (*0.85 × 435 N/ mm²*)) = 337 mm² section of steel rebars or typically here, 2 diameter 16 mm in the bottom (indeed 1 diameter 16 mm every 1.5 m distance apart) with 50 mm coverage.

These APC rebars are of 7 kg steel per cubic meter of concrete, thus quite economical with the 50 kg/m³ dosage rate of steel fibers.

Unlike the severe spalling of the traditional reinforced concrete, as shown in Figure 6.21 from Reference (6), the SFRC doesn't spall under fire.

Figures 6.21 and 6.22c have been reproduced here from Reference (6). of "The Structural Engineer", p.43, January 2018 (UK).

The significant advantage of steel fiber-reinforced concrete over the traditional reinforcing in preventing spalling is clearly evident and unquestionable in Figures 6.20, 6.21, and 6.23 which are self-explanatory. The wire mesh doesn't prevent the concrete matrix from spalling.

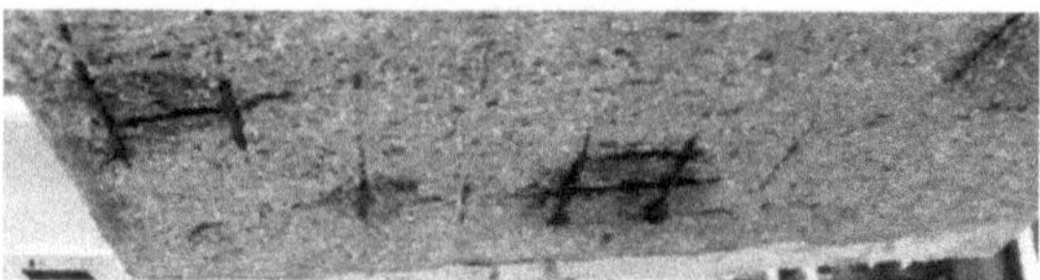

Figure 6.21 Spalling of traditional reinforced concrete slab after a standard fire test.

Figure 6.22 Spalled concrete samples after testing.

Figure 6.23 Real fire in car park building in the U.K.

In Figure 6.23, the fire of a single car on the third floor caused the collapse of the entire car park building, which housed 1000 cars, in 30 minutes, Despite being an open-air building without windows, the slab's concrete spalled, leaving the wire mesh reinforcement exposed after a while.

During our fire tests, the steel fibers, at a dosage rate of 50 kg/m³, were able to control the cracking so that the fiber concrete could contain the vapor pressure building-up that happens during the first hour. This was also a result of the high ductility that SFRC provides at 600°C, as shown earlier in Figure 6.2. (Kodur).

The insulation properties of the SFRC are also worth noting, as the temperature above the slab never exceeded 94°C by the end of the test, well below the 140°C insulating temperature limit, as shown in Figure 6.24.

The SFRC was able to prevent explosive spalling from the bottom, so that a layer of burnt but insulating concrete remained attached around the fibers, providing very effective protection for the concrete core. The beneficial effect is shown in Figures 6.25 and 6.26, where the temperatures at various depths from the top surface (30 mm depth) to the bottom (150 mm depth) are displayed. It can be observed that, at 150 mm depth from the top surface, after 4 hours of fire exposure, the temperature remains as low as 500°C, while the concrete still retains 60% of its initial strength.

The prevention of spalling thanks to the steel fibers explains why the Eurocode temperatures obtained with traditional reinforced concrete, are significantly higher than recorded in the first 60 mm bottom during the full-scale fire tests here, as shown in Figure 6.25. This is verified at 120 minutes, 180 minutes, and 240 minutes of fire. Figures 6.24–Figure 6.26 are rom Reference (7).

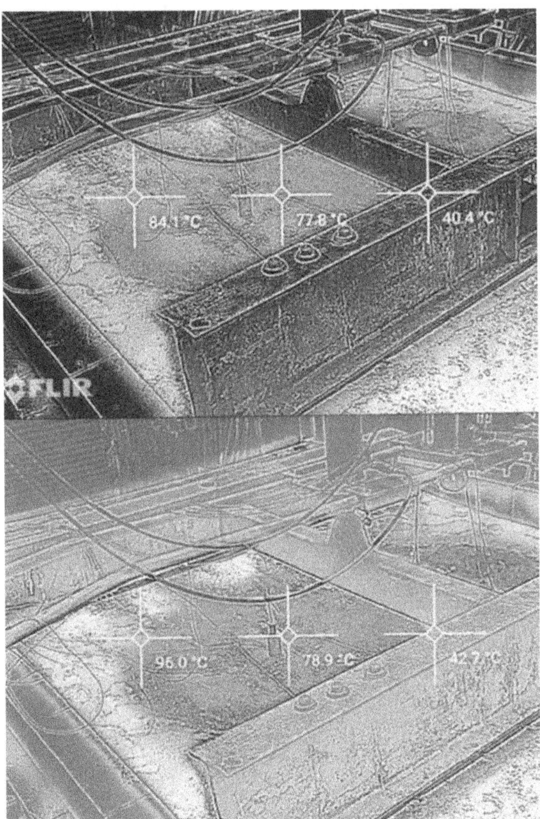

Figure 6.24 Temperature recorded above the slab near the end of test.

In Table 6.2, the superiority of our SFRC vs. Eurocode compliant reinforced concrete appears clearly: as here at 120 minutes duration of fire.

Table 6.2 is self-explanatory.

The same advantage is seen when experimental temperatures are compared to the ACI temperatures.

One can say that ACI temperatures in the depths are more accurate than EuroCode.

Figures 6.24, 6.25, and 6.26 are found in Reference 6.7 (M. Suta)

After 180 minutes fire, as shown in Figure 6.26, at 120 mm depth from the surface (60 mm above the ceiling of the fire room) the temperature is 350°C thus below the 400°C temperature when the concrete starts to loose rapidly his initial strength at ambient temperature and where the reinforcing steel enjoys still ca. 95% of its initial strength as shown in Figure 6.27.

In the top surface, the temperature did not exceed 75°C, which is a lot colder than the maximum 140°C average temperature EuroCode limit.

The SFRC tested slab here has shown a fire resistance significantly superior to extruded hollow core slabs, as outlined in Reference (11) released by the Technical University of Denmark. It is shown that extruded hollow cores slabs (which lack any transverse reinforcement, making it impossible to include in the extrusion manufacturing process) had only 25 minutes of fire resistance instead of the 60 minutes predicted by calculations , because of

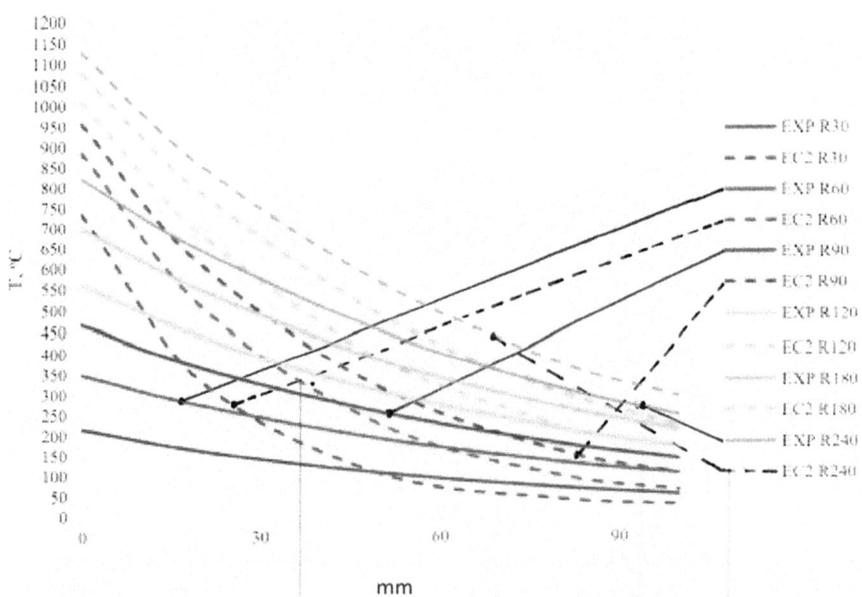

Figure 6.25 Experimental temperatures vs. Eurocode temperatures compared at various depth from the bottom up.

Table 6.2 Temperature in function of the depth from the bottom

Depth from bottom up/Temperature	SFRC fire test concrete	Eurocode compliant concrete
0 mm (bottom surface)	575°C	1000°C
30 mm	400°C	500°C
60 mm	275°C	350°C

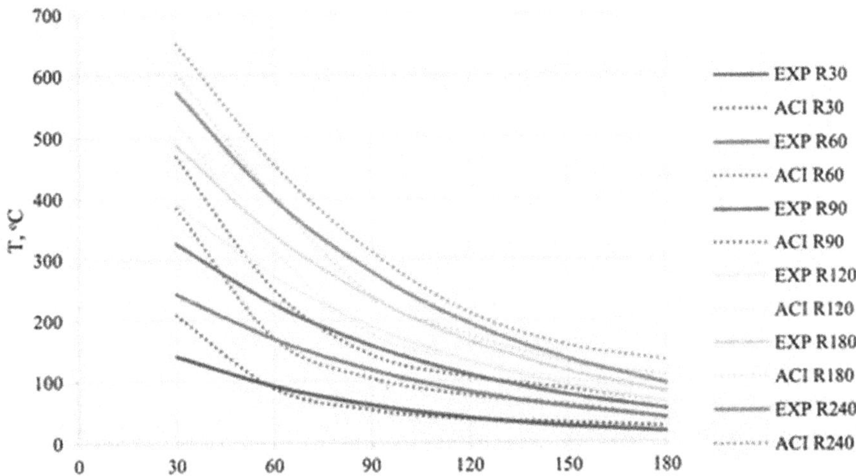

Figure 6.26 Experimental temperatures vs ACI temperatures compared at various depths from the bottom.

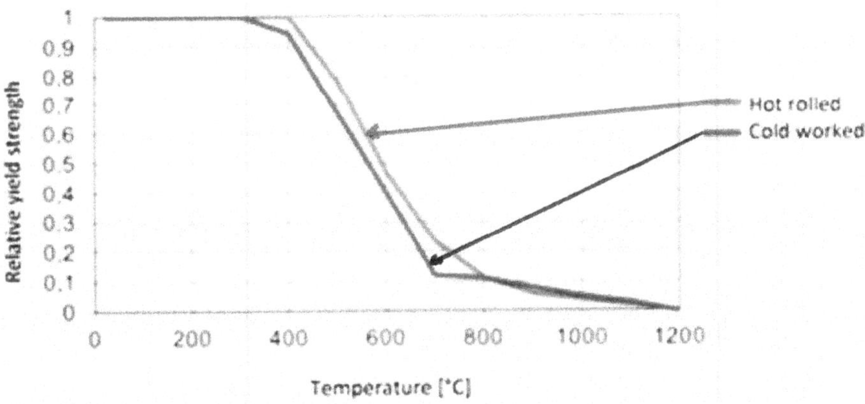

Figure 6.27 Decrease of the steel yield strength in function of temperatures.

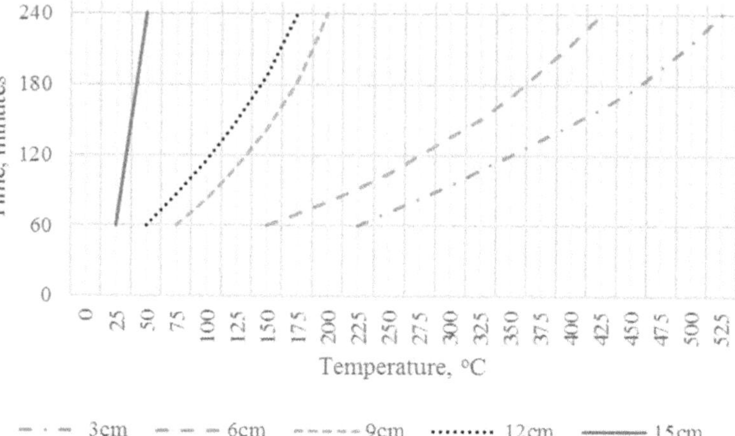

Figure 6.28 Time needed to reach a given temperature at various depths from the bottom face exposed to the fire [from Reference (7) by M. Suta] 3 cm: from the bottom face up; 15 cm at 3 cm below the top surface.

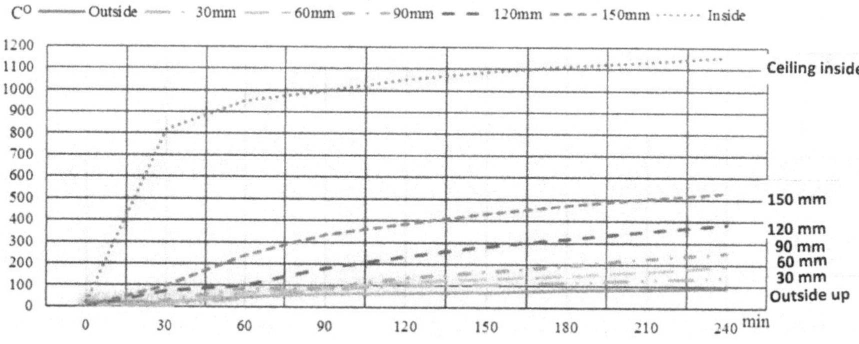

Figure 6.29 Temperature vs time at various depths (from M. Suta in ref. (6)).

"the bottom flange falling down" and "large cracks developing to other channels."

The Danish test procedure was decided following "a 2007 fire in a car park building in the Harbor Edge building in Rotterdam (Holland) as shown in Figure 6.30 in Reference (8) where a number of hollow core slabs delaminated and partially felt down", despite a thick overlay on top of the hollow cores slabs. "The thermal stresses may cause splitting and delamination of the webs, so that the bottom half of the slab falls down."

Figure 6.30 View of the hollow cores slab collapses after a fire (Park and Kang; from ref. (10)).

Similarly, In Reference (9) (FIRE), K. Cowie reports that a U.K. fire test at Aston University (published in The Structural Engineer of 7th October 2008) of a post-tensioned slab designed for 2-hour fire rating achieved only 66 minutes! It has been stated later by the Standing Committee on Structural Safety that *"Post-tensioned floors may be particularly vulnerable and specially so if not constructed correctly"*.

In Reference (10), S. Park and T. Kang report regarding unbonded tendons post-tensioning that *"Early concrete spalling occurred between 14 and 30 minutes after the fire test began."*

And further stated that

> The depth of spalling and potential induced damage during the fire test was restricted with the placement of bottom reinforcing bars. Otherwise, concrete spalling, which results in section area loss, would have weakened the fire-resistance performance due to heat penetration to the steel tendons.
>
> The tendon profile had an effect on concrete delamination, with the maximum damaged depth due to delamination being 260 mm, which equates to 93 % of the original thickness of the slab thickness.

There are so many cases reported regarding the collapses under fire of hollow core post-tensioned slabs. It is weird today that steel fibers are not included in the mix design of all of them to prevent spalling and collapse. It should have become a standard requirement.

6.2.4 Back-calculation of the 180 mm thick slab at 180 minutes fire

The data needed are: thickness, f_{r3} (EN14651), loss of thickness. 60 mm cover of the traditional reinforcing as APC preventing, at 0.075% geometrical percentage of the section.

The section of APC steel rebars is also calculated as a cable to carry at 85% of its yield strength, the slab's own weight, and the variable loadings.

At rupture, the unaffected concrete section starting at 60 mm depth from the bottom undergoes a flexion stress equal to f_{r3} (EN 14651, cold temperature).

$u := 1m$ — Width of slab

$H_c := 180mm$ — Thickness of slab

$f_{r3c} := 5\dfrac{N}{mm^2}$ — EN14651 flexion strength at CMOD = 2.5 mm

$f_{ys} := 435\dfrac{N}{mm^2}$ — Limit the strength of steel

$d := 16mm$ — Diameter rebar

$S := 2 \cdot \dfrac{\pi \cdot d^2}{4 \cdot 3} = 134.041\ mm^2$ — S: section of steel in 1 m width 2 diameter. 16 mm rebar in a 3m width slab

$\rho := 0.95$

$c := 60mm$ — Residual steel bars strength at 375°C: ρ coverage of rebar: c

$M_{rs} := \dfrac{0.9 \cdot \left(H_c - c - \dfrac{d}{2}\right) \cdot S \cdot \rho \cdot f_{ys}}{u} = 5.584\dfrac{1}{m}kN \cdot m$ — Resisting moment of rebars at 375°C

$M_{ref} := \dfrac{f_{r3c} \cdot H_c^2}{6} = 27kN \cdot \dfrac{m}{m}$ — Rupture moment of fiber concrete at initium of test

$f_i := 3.25\dfrac{N}{mm^2}$ — Flexion stress for concrete at initium of test

$M_i := \dfrac{f_i \cdot H_c^2}{6} = 17.55kN \cdot \dfrac{m}{m}$ — Initial test moment

$M_{rc} := M_{rs} + M_{ref} = 32.584kN \cdot \dfrac{m}{m}$ — Rupture moment of 1 m width slab at initium

$\Delta H_{180} := 60mm$ — Loss of thickness at 180 minutes fire test

$M_{R180} := \dfrac{f_{r3c} \cdot \left(H_c - \Delta H_{180}\right)^2}{6} + M_{rs} = 17.584kN \cdot \dfrac{m}{m}$ — Rupture moment at 180 min fire test

$f_{req} := \dfrac{6 \cdot M_{R180}}{H_c^2} = 3.256\dfrac{N}{mm^2}$ — Equal to flexion stress at initium

---

6.2.5 Back calculation of the 180 mm slab at 120 minutes fire

It shows here that at 120 minutes, the calculated resisting moment becomes larger than at cold, so that the cold temperature design prevails.

$u := 1m$ — Width of slab

$H_c := 180mm$ — Thickness of slab

$f_{r3c} := 5 \dfrac{N}{mm^2}$ — EN14651 flexion strength at CMOD = 2.5 mm

$f_{ys} := 435 \dfrac{N}{mm^2}$ — Limited strength of steel

$d := 16mm$ — Diameter rebar

$S := 2 \cdot \dfrac{\pi \cdot d^2}{4 \cdot 3} = 134.041 \cdot mm^2$ — S: section of steel in 1 m width 2 diameter. 16 mm rebar in a 3m width slab

$\rho := 0.95$ — Residual steel bars strength at 375°C: ρ coverage of rebar: c

$c := 30mm$

$M_{rs} := \dfrac{0.9 \cdot \left(H_c - c - \dfrac{d}{2}\right) \cdot S \cdot \rho \cdot f_{ys}}{u} = 7.079 \dfrac{1}{m} \cdot kN \cdot m$ — Resisting moment of rebars at 375°C

$M_{ref} := \dfrac{f_{r3c} \cdot H_c^2}{6} = 27 \cdot kN \cdot \dfrac{m}{m}$ — Rupture moment of fiber concrete at initium of test

$f_i := 3.25 \dfrac{N}{mm^2}$ — Flexion stress for concrete at initium of test

$M_i := \dfrac{f_i \cdot H_c^2}{6} = 17.55 \cdot kN \cdot \dfrac{m}{m}$ — Initial test moment

$M_{rc} := M_{rs} + M_{ref} = 34.079 \cdot kN \cdot \dfrac{m}{m}$ — Rupture moment of 1 m width slab at initium

$\Delta H_{120} := 30mm$ — Loss of thickness at 120 minutes fire test

$M_{RI20} := \dfrac{f_{r3c} \cdot (H_c - \Delta H_{120})^2}{6} + M_{rs} = 25.829 \cdot kN \cdot \dfrac{m}{m}$ — Rupture moment at 120 min fire test

$f_{req} := \dfrac{6 \cdot M_{RI20}}{H_c^2} = 4.783 \cdot \dfrac{N}{mm^2}$

6.2.6 A 140 mm thick slab at 120 minutes of fire

The same concrete is used but the span is reduced to 4 m in order to keep the same span-to-depth ratio (30).

Again here, the cold stage is more critical.

$u := Im$

Width of slab

$H_c := 140mm$

Thickness of slab

$f_{r3c} := 5\dfrac{N}{mm^2}$

EN14651 flexion strength at CMOD = 2.5 mm

$f_{ys} := 435\dfrac{N}{mm^2}$

limit strength of steel

$d := 14mm$

diameter rebar

$S := 2 \cdot \dfrac{\pi \cdot d^2}{4 \cdot 3} = 102.625 \cdot mm^2$

S: section of steel in 1 m width 2 diam. 16 mm rebar in a 3m width slab

$\rho := 0.95$

Residual steel bars strength at 375°C: ρ coverage of rebar: c

$c := 30mm$

$M_{rs} := \dfrac{0.9 \cdot \left(H_c - c - \dfrac{d}{2}\right) \cdot S \cdot \rho \cdot f_{ys}}{u} = 3.931\dfrac{1}{m} \cdot kN \cdot m$

Resisting moment of rebars at 375°C

$M_{ref} := \dfrac{f_{r3c} \cdot H_c^2}{6} = 16.333 \cdot kN \cdot \dfrac{m}{m}$

Rupture moment of fiber concrete at initium of test

$f_i := 3.25\dfrac{N}{mm^2}$

Flexion stress for concrete at initium of test

$M_i := \dfrac{f_i \cdot H_c^2}{6} = 10.617 \cdot kN \cdot \dfrac{m}{m}$

Initial test moment

$M_{rc} := M_{rs} + M_{ref} = 20.265 \cdot kN \cdot \dfrac{m}{m}$

Rupture moment of 1 m width slab at initium

$\Delta H_{120} := 30mm$

Loss of thickness at 180 minutes fire test

$M_{Rl20} := \dfrac{f_{r3c} \cdot \left(H_c - \Delta H_{l20}\right)^2}{6} + M_{rs} = 14.015 \cdot kN \cdot \dfrac{m}{m}$

Rupture moment at 180 min fire test

$f_{req} := \dfrac{6 \cdot M_{Rl20}}{H_c^2} = 4.29 \cdot \dfrac{N}{mm^2}$

Again, the cold stage calculation prevails, with an SLS flexion stress of 3.25 N/mm².

The 140 mm thickness is the limit for concrete thickness to meet the Insulation requirement.

As a conclusion. any SFRC slab, free suspended elevated, in at least C30-37 concrete matrix, with 50 kg/m³ of HE+1/60 ArcelorMittal steel fibers of f_{r3} = 5 N/mm²(EN 14651), provided with the APC rebars following the AC 544 6R15 provisions bottom reinforcing with 60 mm coverage, will show at least REI 120 minutes fire resistance (note that the average of 3 SFRC tests—149, 192 and 204—is 182 minutes).

6.2.7 Two-way slabs

The SFRC two-way elevated slabs, built so far, follow the provisions of the ACI 544 6R15 where the reinforcing includes the APC rebars in the bottom following the grid line of the columns.

As outlined in a previous chapter, the column strips remain stable so that the whole slab remains stable.

Regarding the fire resistance, the central part of the span, thus limited by the columns strips lateral limit, must also be stable under fire.

Deriving from the one-way statically determinate slab fire test, we know that there is a fire resistance of 120 minutes under flexion stress of 2.25 N/mm² in service conditions (SLS).

On the other hand, the maximum flexion stress at the center of span of a 2-ways slab is:

$$\left(p+q\right)\times L^2 / 8 - \left(p+q\right)\times L^2 / 10 = \left(p+q\right)\times L^2 / 40$$

where p is the own weight and q the variable loading

Thus, at mid-span, the flexion stress is ¼ of the stress above the columns at most.

The 2.50 N/mm² stress level at mid-span is not reached before 9 N/mm² flexion stress in SLS is attained above the columns supporting t the slab, a level far beyond the SLS state.

In conclusion, the 2.50 N/mm² flexion stress in the bottom should never be attained at SLS in most cases.

In the case of a minimum thickness of 180 mm, the slab field between the column strips, provided with the APC rebars following the ACI 544-6 R15 report, requires some additional reinforcing if the flexion stress under the service loading exceeds 2.50 N/m² in order to offer a REI 120.

6.3 SFRC ALL FIRE TEST

The wall test was organized under the same conditions as the case of slab tests. The walls were also 180 mm thick, like the slabs, with a height of 3 m and a length of 4 m. They were built using the same concrete as the slabs, including 50 kg/m³ HE+1/60 steel fibers, but without any APC rebars. As a result, the walls contained no traditional rebars, only steel fibers..

The walls are subjected under the test fire to a 323 kN per lineal meter of vertical loading intensity, typical of a nine or ten-floor high building outer walls.

The fire test program included three identical walls subjected to a 330 kN/m linear loading under fire. So far, one test has been completed and

Figure 6.31 The tested walls at 241 minutes fire under 330 kN/m vertical loading.

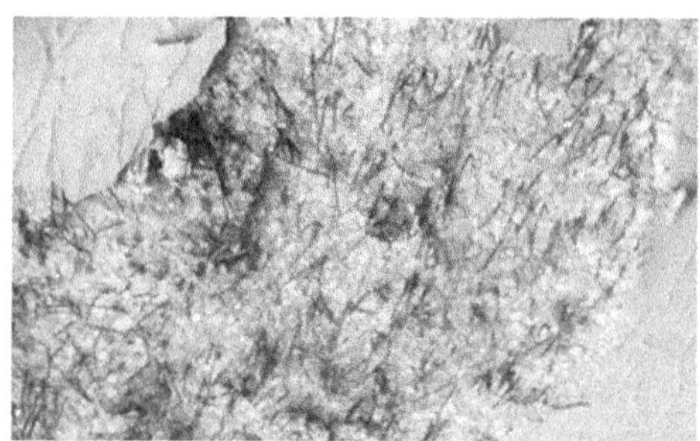

Figure 6.32 Wall spalling at the flame contact to concrete.

Performance criterion	Time till the performance criterion is achieved
load-bearing capacity – vertical contraction [mm]	242 minutes no failure
load-bearing capacity – rate of vertical contraction [mm/min]	242 minutes no failure
integrity – sustained flaming	242 minutes no failure
integrity – gap gauges ⌀ 6 mm and ⌀ 25 mm	242 minutes no failure
integrity – cotton pad	242 minutes no failure
insulation – average temperature (140 K)	242 minutes no failure
insulation – maximal temperature (180 K)	242 minutes no failure
radiation 15 kW.m⁻²	242 minutes no failure [1]

[1] Regarding to low temperatures on unexposed specimen surface below 300°C, the performance criteria of radiation is be complied as satisfied.

Figure 6.33 Failure criteria and wall performance for the tested wall.

Figure 6.34 SFRC wall bearing structure at South City, Brussels.

reported. All walls resisted for more than 4 hours because the test was stopped after 4 hours without collapse, as shown in Figure 6.31.

The wall has also been stable and flame-tight, with a cold surface temperature at 70°C well under the 140°C limit. The wall met and exceeded as well all standard requirements of resistance to fire. Indeed, the fire test has been stopped at 242 minutes without any observed failure to meet the EN1365-3 requirement. spalling has been observed in the wall surface with the exception of the very concentrated flame contact area where the concrete has spalled, as shown in Figure 6.32.

More detailed results of the test are given in the Figure 6.33 as well as that the standard criterions criteria of fire REI resistance are met

SFRC walls have been used for many wall-bearing structures such as in examples shown in Figure 6.34.

Figure 6.34 shows SFRC walls in C25-30 with 30 kg/m³ HE1/50 -1100 MPa reinforcing. Only one layer of starter rebars are used in wall top and bottom of each level.

REFERENCES

1. Kodur and Lie. *Fire Resistance of Fibre-Concrete by Kodur and Lie in Fiber Reinforced Concrete, Present and Future* pp. 189, The Canadian Society for Civil Engineering. (1998) Editors: N. Banthia, A. Bentur.
2. Y. Nurchasanan, M. Massoud, M. Solikin. "Steel fiber reinforced concrete to improve the characteristics of fire-resistant concrete." *Applied Mechanics and Materials*, 845, 220–225, 2016.

3. M. Herremans. "Staalvezelbeton toegepast in woningbouwcasco." SVB Proefproject TU/Eindhoven, 2010.

4. C. Kleinman, X. Destrée, A. Lambrechts. "Steel fibre as only reinforcing in free suspended elevated slabs: design conclusions of a tunnel formed slab and walls based upon full scale testing results." In *BEFIB 2012*, Eight RILEM International Symposium, pp.273, 2012.

5. "Test Report FIRES- FR-019-19-AUNE" The FIRES s.r.o laboratory (CK) - The Experts on Fire Safety. 2019.

6. "The Structural Engineer", p.43, January 2018 (UK).

7. M. Suta. Distribution of temperature a nd reduction of compressive strength in steel fibre reinforced concrete structures under the influence of high temperatures, Riga Technical University, Master Thesis, 2020.

8. K. D. Hertz, L.S. Sorensen, L. Giuliani. "Evidence of fire resistance of hollow-core slabs." *Fire Safety Day* 2015. p.1–11.

9. K. Cowie. "Fire rating questioned on post tensioned and prestressed concrete slabs." *FIRE 21th April 2009*. Steel Construction Institute New Zealand, Inc, 2009.

10. S. Park, T. Kang. "Behavior of unbonded post-tensioned concrete slabs exposed to fire." *ACI Structural Journal*, Vol. 120, Title 120-S58, p. 217–222, May 2023.

11. B. Fettah, *Fire Analysis of car park building structures*. Instituto Politecnico de Braganca, Portugal, June 2016.

OTHER RELEVANT REFERENCES

V.K.R. Kodur and T.T. Lie. (Fire risk management Program, Institute for Research in Construction, National Research Council of Canada) in their article Fire Resistance of Fibre Reinforced Concrete, pp189, in Fiber Reinforced Concrete, Present and Future, The Canadian Society for Civil Engineering, Montréal, 1998.

ACI/TMS Committee 216 "Standard method for determining fire resistance of concrete and masonry construction assemblies", Ch 2, ACI publication 216.1-97 published September 1997.

K. Lie. "Mechanical properties of fibre-reinforced concrete at elevated temperatures." *Internal Report N° 687*, National Research Council Canada, 1995.

Klingsch. Explosive spalling of concrete in fire" Doctoral thesis, ETH Zürich, 2014.

D. Widhianto. "Fire resistance of normal and high-strength concrete with contains of steel fibre." *Asian Journal of Civil Engineering*, 15(5), 655–669, 2014.

M. Nurchasanah. "Steel fiber reinforced concrete to improve the characteristics of fire-resistant concrete." *Applied Mechanics and Materials*, 845, 220–225, 2016.

ACI544.6R-15. Report on Design and construction of Steel Fiber-Reinforced Concrete Elevated Slabs, 2015.

X. Destrée, A. Krasnickovs. "Fire resistance of steel fibre R.C. elevated suspended slabs: ISO fire tests and conclusions for design." *FIB 2020*, Shanghai.

P. Heek, J. Tkocz, P. Mark, "A thermo-mechanical model for SFRC beams or slabs at elevated temperatures." *Materials and Structures*, 21, p 1–16. 51:87 June 2018.

J. Tkocz, P. Heek, C. Thiele, G. Vitt, M. Docevska, P. Mark. "Test and numerical simulation of SFRC slabs exposed to fire." *ALITinform International Analytical Review*, 41(6), 36–53, 2015.

Chapter 7

SFRC foundations

Sometimes, the raft foundation can be more economical than isolated footings due to the high costs of isolated footings and the time required to complete them.

The raft is often shallow but can also be installed at a deeper level.

Steel fiber-reinforced concrete (SFRC) rafts have been built, at least from our experience, since 1992. In this case, the steel fibers are used for structural, temperature, and shrinkage reinforcement. One of the great advantages of using steel fibers is the control and reduction of concrete cracking.

Since 1992, thousands of SFRC rafts have been built for housing projects, including detached houses, condominiums, factories, very high-racking warehouses (clad-rack systems), tank farms, silos, and more, all to the total satisfaction of owners and tenants.

The advantage of steel fiber reinforcement is also its simplicity, from the simpler drawings at the design stage to the much quicker installation on-site in just a few days, compared to weeks with traditional methods.

Typical thicknesses of SFRC rafts range from 200 mm for detached houses to 300, 400, 500, and 600 mm for condominiums, clad-rack warehouses, or office towers, up to 1 m for water treatment plants, and up to 1.8–2 m and 3.0 m for windmill foundation pads or atomic bomb detonation protection (2.7 m thick at SHAPE* base, Casteaux, Belgium, 1982). (*Supreme Headquarters Allied Powers in Europe).

The suitable steel fibers for such applications are generally 0.8 to 1 mm in diameter and 50 to 60 mm in length, with a dosage rate of 30 to 50 kg/m^3 in the mix design to allow easy concrete pumping on-site and installation without mechanical vibration. Compaction is achieved naturally by the own weight of the relatively flowing concrete mixture.

7.1 EXAMPLE 1

The first reference, shown in Figure 7.1, was completed by Silidur S.A. in 1993 using Twincone fiber-reinforced concrete. It was a 1 m thick slab for a

DOI: 10.1201/9781003188315-7

Figure 7.1 The AIDE agency SFRC raft that was built in 1993.

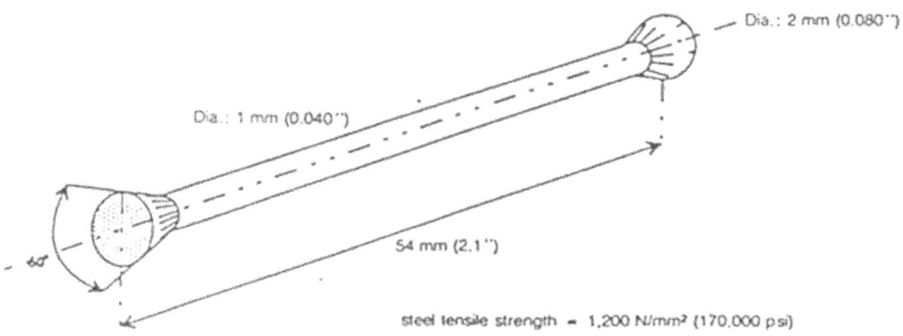

Figure 7.2 Twincone fibers provided with conical ends for a total anchorage.

water treatment plant, covering a rectangular-shaped aeration basin of 1500 m² at the AIDE agency (Embourg/Liège, Belgium).

The entire slab, reinforced with 40 kg/m³ of Twincone fibers, as shown in Figure 7.2, was installed in a single day, with 10 hours of pumping 1500 m³ of concrete using two concrete pumps. This resulted in the integration of 3 tons of steel fiber per hour per machine.

The mix design consisted of 350 kg of cement, including 125 kg of GGBS (ground granulated blast furnace slag), a 0-20 mm continuous aggregate grading with 800 kg/m³ of 0–5 mm river sand, and 1.0% superplasticizer to boost the slump test from 50–75 mm to 180–200 mm.

The design moment, caused by underground water pressure against the empty reservoir, was 750 kNm/m, resulting in the design flexural stress $f_d = 6 \times 750 \cdot 10^3/1000^2 = 4.5 \text{N/mm}^2$. The design and installation were approved by SECO, the official government technical controller in charge for the AIDE Agency.

In comparison to the traditional, where the reinforcing schema includes 80 kg of rebars per cubic meter of concrete, the savings in hours were

Figure 7.3 Typical steel bower for 15 tons capacity of SF integration in 8 hours on site.

tremendous: approximately 1500 m³ x 80 kg x 10 hours/ton = 120 tons of steel on-site, requiring 1200 workman-hours. This resulted in a saving of 1200 hours of labor.

The traditional raft would have been installed in four days, with two very expensive full-depth construction joints, approximately 80 m in total length, to divide the rectangular raft into four equal parts.

Two layers of starter rebars projected out of the SFRC raft into the perimeter walls installed on top of it.

Since 1993, a design flexural stress of 4.5 N/mm² for rafts has been accepted and widely used for cases involving 40 kg/m³ of 1 mm diameter, 50-60 mm length fibers with a 1200 N/mm² tensile strength.

7.2 STEEL FIBER BLOWING MACHINES

To avoid dry fiber balling, both the Twincone fibers, which we invented (shown in Figure 7.1) and the undulated 1 mm x 60 mm steel fibers must be blown separately at high speed into the truck mixer using a specific steel

Figure 7.4 Two steel fiber blowers in action on site.

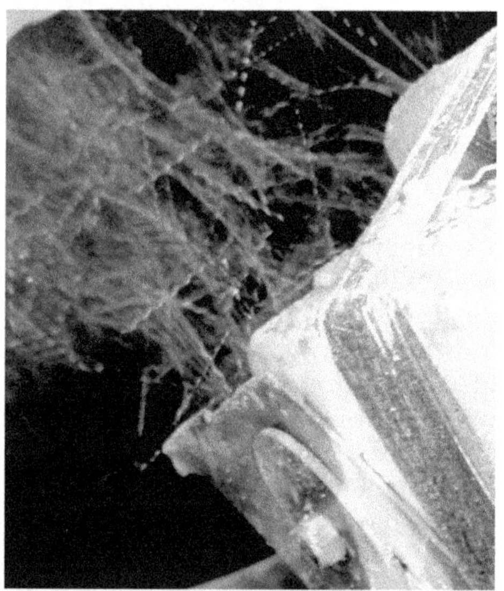

Figure 7.5 Air borne separated loose fibers from a steel fiber blower into the truck mixer.

fiber blowing machine (also called a steel blower or fiber blast machine), as shown in Figures 7.3. and 7.4.

Such a stainless steel-built machine blows the steel fibers at a velocity of 30 m/s, separately, as shown in Figure 7.5, over the entire length of the mixer at high rotation speed in the fresh concrete. In just 8 minutes, an 8 cubic meter capacity truck mixer is ready.

Figure 7.6 Typical steel fiber conveyor belts.

The machine is equipped with hydraulic jacks at its four corners to position the fiber ejection pipe deep inside the mixer, ensuring that the fibers are blown into the concrete and not onto the dry blades at the entrance.

The machine also has a loading bucket at its rear, sometimes referred to on-site as "the coffin," in which a 20, 25, or 30 kg box of steel fibers is emptied. The operator can add a precise quantity of superplasticizing admixture via pumps and pipes to the truck mixer. The machine is typically diesel-hydraulic powered.

I designed and built the first steel-blowing machines in 1984 (1). This method is very convenient for concrete contractors who prefer to prepare ready-mixed reinforced concrete on-site, as opposed to others who prefer purchasing fiber concrete from a concrete plant.

For more user-friendly fibers, such as those with hooked ends and an aspect ratio smaller than 60, conveyor belts, as shown in Figure 7.6, are used to discharge the fibers into the truck mixers on-site.

7.2.1 Practical standard deviation on the nominal fiber dosage rate

It has been demonstrated by Khudair (2), that the precision of fiber integration in this case is not as accurate when using conveyor belts as shown in Figure 7.6, compared to when suitable blowing machines are used. Khudair reports that on five different sites with the same fiber type, when using Primekss steel fiber blowers on-site, the standard deviation at a 40 kg/m^3 dosage rate was 2.4 and 2.0 kg/m^3, unlike with the conveyor belt, where the standard deviations were 10.6, 5.5, and 11.6 kg/m^3.

The same measurements were taken at the bottom of the slabs, and the results were even more severe. With Primekss steel fiber blowers, the standard deviation was 1.6 and 3.9 kg/m³, while with the conveyor belt integration, it was 14, 10.1, and 8.1 kg/m³.

One can conclude that the steel fiber blowing with the Primekss machine results in an integration of 10 to 15 kg/m³ of steel fibers. In other words, when using 40 kg/m³ of blown fibers, you would need to use 50 to 55 kg/m³ when conveyor belts are used.

In Reference [2] in Chapter 7, Khudair reports that when using the Primekss fiber-blowing machine, the process is precise, ensuring optimum performance.

The machine is mounted on hydraulic jacks, so the fiber outlet pipe is positioned exactly along the truck mixer's longitudinal axis. It extends down into the truck mixer, ensuring that the fibers are blown over the fresh concrete surface rather than hitting the steel blades of the truck mixer.

The introduction and mixing time is 1 minute per cubic meter. The mix is fluid but not at the collapse of slump. Excessive workability, as well as a stiff mix, can harm the precision of the result.

7.2.2 Steel fiber types used

In general, for floors, footings, and rafts, we mostly used hooked-end or undulated steel fibers with a 1 mm diameter x 50 mm length, as well as 1 mm x 60 mm fibers with a higher strength of 1500 N/mm², to maximize the post-cracking ductility of the SFRC.

The production of Twincone fibers was stopped in 2011 due to their high cost compared to HE+1/60 fibers and their lower residual flexural strength (f_{r3}) according to EN 14651, as the Twincone fibers tend to break in tension from the bottom up, resulting in a higher level of f_{r1}.

While small prismatic specimens in flexure showed that the Twincone fiber concrete exhibited very safe multiple cracking, ending up with more yield lines, the disadvantage is that Twincone fibers were more difficult and slower to introduce and mix into the concrete truck with the fiber blast machines.

7.3 EXAMPLE 2

Another example, shown in Figure 7.7, is from downtown Brussels in 2007, where an SFRC raft of 20 m x 80 m was pumped into a deep basement between foundation walls made of secant piles. The project involved a multistory condominium building with 5 floors above ground and two basement levels.

The raft is subjected to a 6 m height of underground water pressure and has a thickness of 450 mm, with a thicker central area of 750 mm where more counterweight was needed to offset the water pressure.

Figure 7.7 SFRC raft installation in Brussels downtown.

The mix was C30/37 with 50 kg/m³ of hooked-end steel fibers (1 mm diameter x 60 mm length, with 1500 MPa tensile strength, generally referred to as HE+1/60).

The mix included 350 kg of cement type I, with 35% GGBS, in a 0-20 mm continuous aggregate grading; the water content was adjusted to ensure a W/C ratio < 0.53, with 1% superplasticizer by cement weight to achieve a flowing concrete mix suitable for pumping. No poker vibration was needed, as usual.

A construction joint is visible in the photo, at the mid-length of the building.

Later, the perimeter wall secant piles were covered by an SFRC wall cast in situ, with 20 kg/m³ of hooked-end fibers (1 mm x 50 mm and 1200 MPa tensile strength).

7.4 SFRC RAFT DESIGN

The specialist geo-engineers soil report is an important document that the designer must know and read carefully. In general, the type of foundations to use is clearly defined in the report.

The report, in case of raft foundation, provides a ground bearing capacity of q (kN/m²), expecting the typical long-term settlements and possible eventual differential settlements.

The ratio q/s (in kN/m³ or N/mm³) is a global sort of k-value that can be used in the elastic design method.

A 150 kN/m² ground bearing capacity with a 10 mm settlement means:

$$k = 0.15\left(N / mm^2\right) / 15\left(mm\right) = 0.015N / mm^3$$

Quite often, this value is increased by importing and compacting a thick layer of suitable granular materials, raising it to around 0.03 or 0.04 N/mm³.

Hence, the structural engineer prepares an elastic FEM analysis to produce a moment diagram, where peak values are found underneath the columns. These peak values result from the elastic calculation, in which cracking and micro-cracking are not considered, so in reality, the peak stresses are somehow spread over a larger area.

Peak values are then reduced by 25%.

7.4.1 Example of design: a raft to support a grid of columns and walls

$$\text{Peak } M_d = 150 \, \text{kNm} / \text{m under columns}$$
$$\text{Assume } H = 400 \, \text{mm as raft thickness.}$$

The design flexion stress: $f_d = 6 \times 150,000 / 400^2 = 5.625$ N/mm² and note that the unfactored flexion stress $f_{SLS} = 5.625/1.5 = 3.75$ N/mm². This is smaller than the flexural cracking strength at the limit of proportionality (LOP), which is 4.5 N/mm² for f_{r1} and f_{r3} stresses.

The raft does not even crack in flexure under the most critical moment when the steel fiber-reinforced concrete is considered as a purely linear elastic material.

At ULS, however, the post-cracking deformability of the fiber-reinforced concrete must be taken into account. This refers to the f_{r3} flexural strength at a 2.5 mm crack (CMOD, or crack mouth opening displacement) as per the EN 14651 standard small notched beam test in flexure.

At this stage, the SFRC is considered a plastic material in flexure, so cracking develops into multiple cracks at the bottom and top, generating yield lines of rupture.

In addition to the material ductility, structural ductility must also be considered to distinguish between a statically determinate structure (without any redistribution of moments) and a statically indeterminate structure, which benefits from significant moment redistribution due to the nature and geometry of the structure.

The Swedish standard EN-SS 812310, "design of fiber reinforced structures" defines structural indeterminacy factor η_{det}, for the SFRC design of structures., For statically determinate structures, $\eta_{det} = 1$ as there is no moment redistribution possible.

$\eta_{det} = 1.4$ for one-way continuous slabs and beams (parallel yield lines at top and bottom)

$\eta_{det} = 2$ for two ways slabs with OX and OY direction yield lines at top and bottom.

In the earlier chapter regarding the full-scale testing of SFRC structures, both material and structural ductilities are demonstrated and used for back-calculating the test.

Hence, we define here the M_{rd}, the resisting moment of the SFRC, as $M_{rd} = f_{rd} \times H^2/6$ where

$f_{rd} = \eta_{det} \times f_{r3} / (\gamma_c \times H^2)$ where $\gamma_c = 1.5$ the concrete material factor. In our example, where $\eta_{det} = 2$, what value of f_{r3} should we need? Indeed $f_{r3} > \gamma_c \times f_{rd} / \eta_{det} = 1.5 \times 5.625 / 2 = 4.21$ N/mm². This value is achievable with a C30-37 plain concrete mix, likely containing 40 kg/m³ of fibers with 1 mm diameter, 60 mm length, 1200 N/mm² tensile strength.

All f_{r3} values are derived from 12 small notched beams—500 mm x 150 mm x 150 mm—tested using the EN 14651 method with a 25 mm notch depth at mid-span. The average f_{r3m} value from these 12 specimens is used.

As the EN 14651 test method is subject to a huge scatter in the results, the characteristic values are irrelevant for designing statically indeterminate structure, as this scatter is inherent to the test method itself, as explained in an earlier chapter.

7.4.2 Example of equivalent calculations of a raft

$f_y := 435 \ \dfrac{N}{mm^2}$

Thickness of 220 mm
10 mm diameter
150 mm spacing
wire mesh in top and bottom rebar diameter

$h := 220 \ mm$

$D := 10 \ mm$

$p := \dfrac{\pi \cdot D^2}{4} = 78.54 \ mm^2$

$d := 150 \ mm$ $u := 1000 \ mm$ Spacing center to center of rebar

Section of rebar per meter run

$S_x := p \cdot \dfrac{u}{d} = 523.599 \ mm^2$

$c := 40 \ mm$

$M_R := 0.9 \cdot (h - c) \cdot f_y \cdot S_x = 36.898 \ kN \cdot m$ Resisting moment of the wire meshes

$f_d := \dfrac{6 \cdot M_R}{u \cdot h^2} = 4.574 \ \dfrac{N}{mm^2}$ EN 14651 equivalent flexion stress.

$\eta_{dct} := 2.0$
$\gamma_c := 1.5$

Statical indeterminacy factor for a two-way slab bny EN - SS 812310

$f_{R3} := \dfrac{\gamma_c \cdot f_d}{\eta_{dct}} = 3.431 \ \dfrac{N}{mm^2}$ SFRC f_{R3} equivalent flexion stress needed

$$f_{t3} := 0.37 \cdot f_{R3} = 1.269 \; \frac{N}{mm^2}$$

and tensile strength

$$\rho_s := 78.5 \; \frac{kN}{m^3} \quad V := 1m^2$$

$$w_s := S_x \cdot \rho_s \cdot u = 41.103 \, N$$

$$c_s := \frac{4 \cdot w_s}{\dfrac{h}{u}} = 747.318 \, N$$

Weight of wire meshes needed per m³ of raft

$$C_s := \frac{c_s}{V} = 0.747 \; \frac{kN}{m^3}$$

$$C_{total} := 1.10 \cdot C_s = 0.822 \; \frac{kN}{\left(m^3\right)}$$

83 kg/m³ rebars without chairs. Equivalent SFRC with 35 kg/m³ of HE 90/60 steel fibers

7.4.3 SFRC raft design

Table 7.1 shows typical design examples of SFRC rafts.

This table shows the raft thickness and the dosage of fiber reinforcement needed based on the unfactored column load and the global K-value of the ground for a C30/37 concrete mix design.

The ground report prepared by the specialist geo-engineers is a key element to study in depth before any attempt to design a raft foundation. The raft concrete contractor should always refer to the ground report in his bid.

The mix design typically includes up to 350 kg of Portland cement, with up to 30% replacement by materials such as GGBS (Ground Granulated Blast Furnace Slag) and Fly Ash (ASTM) Type F. In Europe, this would include CEM I, CEM II, and CEM III.

Experience has shown that CEM V, with a high rate of cement replacement, is prone to early-age cracking, particularly when the ambient temperature is low. The reason is likely due to high fresh water loss by evaporation, and the concrete mortar's early strength is not sufficient to resist shrinkage strain.

As mentioned in an earlier chapter, a 20 mm aggregate size (no smaller sizes, but larger 25 mm sizes are possible depending on the mix design and the concrete pump line diameter of 150 mm) is more than recommended. Never use a 100 mm diameter pumpline

Placement with a pump is possible if the pump line has a 125 mm diameter, and the mix includes 450 kg of fines smaller than 200 microns, including cement. Smaller contents could cause excessive friction in the pump line, leading to blockages and delays.

The freshly mixed concrete should have a minimum density of 2375 kg/m³ with a water content of W/C < 0.53 (on water-saturated, surface-dry aggregates), typically 185 kg of water.

Table 7.1 Raft design ed thickness in SFRC in function of columns loads intensity

SLS column reaction (kN)	Number of levels (residential)	Section of column	$K_w = 0,01 N/mm^3$ Raft thickness Dosage rate	$K_w = 0,02 N/mm^3$	$K_w = 0,03 N/mm^3$
400 kN (40T)		250×250 mm²	250mm 40 kg HE1/60	250 mm 35 kg HE 1/60	250 mm 35 kg HE1/60
600 kN (60T)		250×250 mm²	300mm 40 HE1/60	300 mm 35 kg HE1/60	300 mm 35 kg HE1/60
800 kN (80 T)	3 levels	300×300mm²	300mm 45 kg HE+1/60	300 mm 45 kg HE+1/60	300 mm 40 kg HE1/60
1000 kN (100T)	GF + 3	300×300mm²	360 mm 40 kg HE+1/60	350 mm 40 kg HE+1/60	340 mm 40 kg HE+1/60
2000 kN (200T)	GF + 5	370×370mm	500 mm 40 kg HE+1/60	490 mm 40 kg HE+1/60	480 mm 40 kg HE+1/60
3000 kN (300T)	GF + 7	450×450mm²	620 mm 40 kg HE+1/60	600 mm 40 kg HE+1/60	590 mm 40 kg HE+1/60
4000kN(400T)	GF + 9	525×525mm²	700mm 40kg HE+1/60	670mm 40kg HE+1/60	660mm 40kg HE+1/60
6000kN(600T)	GF+ 14	650×650 mm²	800mm 45kg HE+1/60		

GF: Ground Floor.

HE1/60: hooked ends steel fibres of 1 mm diameter – 60 mm long – f_y =1200 MPa tensile strength.

HE+1/60: as HE 1/60 but f_y = 1500 MPa.

Indeed AS TM type I steel fibres with steel wire at 4 % or more tensile strain at rupture.

K_w: global reaction coefficient as defined in this chapter.

The plain concrete slump before any additions should be a maximum of 75 mm, so that the final slump at the supply from the pump is 150 mm, thanks to the addition of a superplasticizer.

Note that the addition of steel fibers will cause a slump loss of 30–50 mm, and the pump line length, along with the number of 90° elbows, could also cause a loss of 30 mm or more.

All in all, the slump of the steel fiber concrete when pouring into the concrete pump hopper should be around 200 mm.

It is important to ensure that the concrete temperature does not exceed 25°C during installation, as higher temperatures could cause more slump loss and increased shrinkage.

For installations with a thickness of more than 400 mm, a walk-on-top wire mesh is recommended, placed at a minimum depth of 100 mm (with a minimum A252 size of 8 mm x 200 mm, supported by 1 chair per m²) to prevent the crew from being quickly exhausted by the fresh concrete weight over their boots.

Since 1993, we have seen a significant number of SFRC raft designs. As a rule of thumb for preliminary budgeting, the dosage rate for HE 1/60 1500 MPa steel fibers can be taken as 45% of the standard traditional rebar quantity per cubic meter of concrete.

For example, a 110 kg/m³ rebar quantity in the raft can be replaced by 110/2.2 = 50 kg/m³ of HE 1/60 – 1500 MPa steel fibers.

As with piled slabs and elevated suspended SFRC slabs, shear reinforcement is generally not needed in SFRC rafts, as flexure and shrinkage are more critical and prevail, especially when the span-to-depth ratio (column spacing/thickness of raft) is smaller than 20 for rafts.

In a pure punching-out test of a 200 mm thickness C30/37 SFRC slab, a point load of 1500 to 2000 kN is required to punch it out. Unlike a suspended test slab, the raft is supported by the ground over a large area with a diameter D,

$$D = 2 \times L_e + H + d,$$

where

$$L_e \text{ is the elastic radius of rigidity} = \left(E \times H^3 / 12 \times \left(1 - v^2\right) \times K \right)^{1/4}$$

As an example of an 800 mm thick raft:

E = 25000 MPa; H = 800 mm; v = 0.2 coefficient of Poisson; K = 0.02N/mm³ thus L_e = 2730 mm and, d = 1000 mm being the column diameter.

Hence,

$$D = 2 \times 2730 + 800 + 1000 = 7260 \, mm$$

The maximum possible column load is determinate by the concrete strength and is

1000^2 x $\Pi/4$ x 14 N/mm² = 10994 kN, but too high for the ground bearing capacity of the site.

The average pressure onto the ground is

9400 kN/7260^2 x $\Pi/4$ = 9400000 N/785397mm² = 140 kPa as the ground bearing capacity doesn't exceed 140 kN/m²

The settlement to be expected is:

$$0.140 \, N/mm^2 / 0.02 \, N/mm^3 = 7 \, mm$$

Table 7.1 shows that the same 800 mm raft carries 6000 kN only for a column size of 735 mm diameter. Indeed, the design of a SFRC raft with Table 7.1 is reliable and conservative.

7.4.4 SFRC clad rack building

A quite common application of SFRC is the clad-rack slab, which serves as the foundation slab of a very high racking warehouse, where the racking system is also the building structure or skeleton. The storage height can reach up to 35 m and higher.

When the warehouse is empty, a critical design consideration is the high winds. In general, a raft thickening is needed underneath the two long sides, regardless of the type of concrete reinforcement used. The raft thickening typically spans a width of 4 m to 8 m, depending on the case and the most onerous loading combination.

In my experience, these rafts are typical 300 mm to 350 mm thick, with a 600 mm width along the long side, and are provided with 40 kg/m³ to 50 kg/m³ of steel fiber reinforcement, with HE +1 mm diameter x 60 mm length fibers and 1500 MPa tensile strength.

I believe I have seen about 30 of these SFRC raft slabs built over the past 20 years. The first one I was involved in was the Tesco Unity Freezer in Crick (UK) in 1999, covering 40,000 m² of surface and operating at −34°C, as shown in the background of Figure 7.8.

The only rebar used was to tie the slab together at each construction joint, ensuring that it would not open and suffer traffic damage. The slab was installed in pours of approximately 600 m³ each.

Years later, the GC told me that they had to inject 200 m of cracks, which was much easier, cheaper, and more effective than repairing joints.

7.4.5 Economical advantage

We can now estimate the savings generated by using SFRC for a 1000 m² raft with 800 mm thickness.

Figure 7.8 A typical SFRC raft under a clad rack building at Crick (UK).

In 2023, installing rebars in a raft still requires 10 hours labor per ton on-site. For 88 tons the installation of rebars(indeed 110 kg traditional reinforcing per cubic meter), 880 man-hours are spent. An on-site crew of ten would need 880 hours to install the rebars, about 10 calendar days, compared to completing the SFRC in a single day on the day of the pour.

Hence, the rebar placement crew wage cost is estimated to be 55 €/m³, with the rebars material cost at 88.000 kg x 0.8€/kg / 800 m³ = 88 €/m³. Thus, the total cost for traditional rebars is 88 + 55 = 143 €/m³ (average typical market in 2023).

For SFRC, the steel fibers cost 1.20€/kg x 45 kg/m³ = 54 €/m³, and the mixing cost of the steel fibers in concrete at 0.15 €/kg is at most 0.15 x 45 = 6.75 €/m³.

The mix will likely require a bit more super plasticizer at a cost of ca. 4 €/m³ cost so that we obtain a Steel fiber material cost of ca. 54 + 6,75 + 4 = 64,75 €/m³ instead of 143 €/m³ with traditional rebars thus 78 €/m³ savings! (in 2023).

Moreover, the installation of the raft is on the critical path of the jobsite planning, so that an additional 9 days in traditional reinforcing are needed at full fixed costs of the jobsite: power supply, water, fences, cranes, safety, insurances, taxes, and administration.

We can estimate for a small site as in the example a 2000 €/day fixed cost to run the site thus 18,000 € in total to add to the cost of rebars, thus 18,000 €/ 800 = 22.5 €/m³

Total cost of steel fibers in the concrete on site: 64.75 €/m³.

Total cost of the traditional rebars: 143 +22.5 = 166.75 €/m³.

Despite the significant saving of 102 €/m³, most traditional engineers today are not familiar with steel fibers, and if they have some knowledge, they are very reluctant to design SFRC structures due to the fear of taking on responsibility. Furthermore, most design standards currently do not even mention the existence of steel fibers.

Moreover, based on existing normative documents like the FIB Model Code 2012, the Eurocode draft for fiber reinforcement (for 2027 at the earliest), and DIN-1045 DAFstB, the structural design of SFRC appears incredibly complicated for a non-specialist engineer, making these documents more of a hindrance than a help.

In the case of underground water pressure, its maximum intensity is also given. The ground water pressure causes negative moment up to cracking which can affect the water tightness of the raft.

7.4.6 Case of underground water pressure p_w

As a preliminary opinion derived from experience, if a structure needs to be watertight, the application of specific waterproofing techniques is essential. The concrete material itself should not be guaranteed waterproof, and even more so, a concrete structure cannot be assumed to be watertight. Reinforced concrete cracks and leaks. The concrete on-site is not always perfectly compacted, nor is it crack-free. Leaks will occur, and there will be a need for specific treatments to stop the flow of water. The structure is often limited by joints, but these joints frequently leak, such as between a slab and a wall, or between two adjacent walls.

Cold joints can also develop when concrete supply is not continuous, causing the concrete to set too early and harden prematurely along the cold joint.

The definition of water tightness is often unclear: sometimes a humid spot on the concrete surface is considered unacceptable, while in other cases, limited humidity and minor leaks are tolerated.

Always define water tightness clearly, as it can be a cause of litigation. Be explicit about what you are supplying and what you are not supplying.

Even 2 m thick walls heavily reinforced with traditional concrete won't be watertight (as expected), as shown in the Figure 7.9 of locks in Central America.

Suppose our 800 mm thick raft under 5 m height of underground water pressure or 50 kN/m², , is subjected to a net pressure deducted by 80% of the raft **own weight,** *that is, 24 kN/m³ x 0.8 m x 0.8 = 15.36 kN/m²)* of 50 kN/m²- 15.36 kN/m²= 34.64 kN/m² affected by a 1.20 load factor thus, corresponding to 42.56 kN/m² pressure or a 4.20 m net height of water.. The column spacing is 8 m, so the uplift force per column is calculated as 42 kN/m² × 8 m × 8 m = 2688 kN. The own weight of the building is approximately 4500 kN.

Flexion moment of underground water pressure is calculated on the net span of L_N:

$$L_N = L - d - H = 8\,m - 0.73\,m - 0.8\,m = 6.47\,m$$

To evaluate the flexural moments, we can use a typical summary of moment graph, derived from Timoshenko (Theory of plates and shells), as shown in Figure 7.10 and Table 7.2

Hence,

$M < 0$: - pw x L_N^2 / 15 = 42 x 6.47² / 15 = -117.21 kNm/m, the negative intensity on top of raft halfway between the columns.

Table 7.2 Table of n-values to obtain the moment M = qL²/n

q: U.D.L L: span	M_0	M_1	M_2	M_3	M_4
q L²/n	n = −10	n = 22	n = −50	n = 40	n = −15

The resulting factored (1.2) flexion stress is 6 x 117,210 / 800² = 1.10 N/mm².

Underneath the columns, there is a positive moment:

M > 0: P_W X LN² /-10 = 175 kNm/m, and a bottom stress 6 x 175 000/800² = 1.64 N/mm²

The effect of the 9400 kN column loading intensity (unfactored) is a flexion moment:

M+ = p x (L − b)²/8 for an isolated footing.

However, we affected the moment by a 10/8 =1.25 reduction factor in case of a continuous raft:

$$M_r+ = 1/1.25 \times p \times \left(2L_e + d + H - d\right)^2 / 8 = 1/1.25 \times 140\left(kN/m^2\right)$$
$$\times \left(2 \times 2.73\,m + 0.800\,m\right)^2 / 8 = 548\,kNm/m$$

The resulting raft unfactored flexion stress is 6 x 548,000/800² = 5.14 N/mm².

The total bottom positive design moment stress is 1.4 x 5.14 + 1.64 = 8.84 N/mm² where 1.4 is the average loading factor between 1.35 and 1.5 with a predominance of the own weight over the variable loading.

The flexion strength f_{rd} is derived from $f_{r3\,m}$ = 5 N/mm² (for 45 kg/m³ HE 1/60 – 1500 MPa) following EN 14651. Taking into account the structural indeterminacy factor (EN-SS812310) η_{det} = 2 for a two-way slab, together with γ_c = 1.5 concrete material factor, we have:

$$f_{rd} = 2 \times 5 / 1.5 = 6.67\,N/mm^2$$

Because 6.67 N/mm² < 8.84 N/mm², the example raft is insufficient to carry 9400 kN unfactored column load and 5 m of unfactored underground water height, as there is resisting strength deficit of 2.17 N/mm² or a resisting moment deficit of 2.17 x 800²/6 = 231 kNm/m.

Hence, M_{r+} should be reduced to 548 kNm/m − 231 kNm/m = 317 kNm/m by 42%.

The maximum unfactored column load of the 800 mm raft is then 9400 X 0.58 = 5452 kN.

Table 7.1 gives a conservative figure of 6000 kN in dry ground conditions thus without any groundwater pressure.

The maximum stress on top of the raft is obtained at halfway between columns are as follows:

From the water pressure we had already 1.10 N/mm²

From the column loads, we have at halfway between columns in the top surface:

317 kNm/m x 10/15 = 211 kNm/m, and a flexion stress of 6 x 211,000
Nm /800² = 1.98 N/mm²,

Hence the total factored stress:

$$1.4 \times 1.98 \, \text{N} / \text{mm}^2 + 1.10 \, \text{N} / \text{mm}^2 = 3.87 \, \text{N} / \text{mm}^2 < f_{rd} = 6.67 \, \text{N} / \text{mm}^2$$

A 3.87 N/mm² factored stress, although still leaving a significant gap to resist restrained drying and thermal concrete stresses, doesn't mean the raft will be crack-free.

Indeed, all concretes crack:- the traditional reinforced concrete cracks.

- the post-tensioned concrete cracks.
- SFRC cracks without any adverse effect up to 1 mm opening in dry conditions.

In wet and dry areas, SFRC can crack up to 0.5 mm opening without any adverse effects, even when deicing salts are used. The durability of SFRC has been reviewed in the Chapter 2 based upon the experiences and Reference [7] in Chapter 2 regarding the crack opening limit, including based on other references such as [8–11] in Chapter 2.

The crack opening calculation using standard methods for traditional reinforced concrete does not necessarily reflect the actual cracks that occur. The crack opening calculations by standard methods are used to demonstrate that the calculations have been conducted according to approved procedures, thereby indemnifying all designers who follow these methods. In reality, cracks are often caused by factors more closely related to the installation of the slab.

We commonly hear the following reasons for issues with concrete slabs:

- The ground was softer than expected, leading to higher moment cracks.
- The external temperature was too high, or the concrete temperature was too high.
- The night temperature was too low, causing an unexpected temperature gradient.
- The pump was too slow, the pump line too narrow, or the pressure too high, resulting in more friction and heat.
- The placing crew arrived late, causing too many concrete trucks to wait before discharging, or there were cold joints hidden by power troweling on top of them.
- The cement was too fresh and had too high a temperature.

- The raft surface was not adequately cured.
- Water was added to the truck mixer due to long waiting times on-site.
- Concrete transit time was too long due to unforeseeable traffic jams.

None of these valid reasons or explanations are typically accounted for in the engineer's calculations, but they will be analyzed by experts in the event of litigation.

Only the supply and installation of the raft could be blamed, as, based on experience, ready-mixed concrete is always compliant with all standards.

What is the solution to overcome this type of frequent issue?

I would recommend that the owner select a Design-Build-Maintain solution with a guarantee of functionality for 10 or 15 years by contract, so that the concrete contractor becomes the sole source of responsibility for the entire project.

In this case, the concrete contractor's job is to supply a slab or raft that meets the service conditions as defined in the specification—essentially limits regarding cracking, deflections, and durability at SLS.

At ULS, the structure can still meet requirements when sufficient material and structural ductility are available, although there will be more cracking and deformations beyond the SLS limit. At ULS, the structure remains stable, and beyond ULS, additional loading is required to reach collapse.

The Design-Build guarantee is a much better solution than relying on dozens or hundreds of pages of calculations that only guarantee compliance with standards.

Labor-only finishers are not equipped for this and cannot supply and install design-built rafts, as the suitable concrete contractor must have in-house materials, design engineers, lab technicians, and a fully implemented quality management system in real time.

At the very surface (y = 0 mm) of the raft the free shrinkage strain is ca. $\varepsilon_R = 4 \times 10^{-4}$ and is

$$\varepsilon_R = 0 \text{ at } 400 \text{ mm depth} \left(y = 400 \text{ mm} \right).$$

Between 0 mm and 400 mm depth , The shrinkage strain decreases as a parabola of second order, as an approximation of the exponential function indeed, of variables being $X = \varepsilon_R$ and $Y = depth$

The parabola equation is of the type $Y = A x^2$

The integral over 400 mm depth and $4 \times (10^{-4})$ maximum shrinkage amplitude is $400/4^2 \times X^3/3 = 25 X^3/3 = 25 \times 64/3 = 1600/3 = 533$ to be compared to the rectangle area of $400 \times 4 = 1600$. Thus the 800 mm thickness raft will be stressed by shrinkage as if it was only of 400 mm thickness. Taking an elasticity modulus of rather long term (5 year) or 20 000 N/mm², the stress is

$$4 \times 10^{-4} \times 533 / 1600 \times 15.000 \text{N} / \text{mm}^2 = 2\,\text{N} / \text{mm}^2$$

The fiber concrete at 45 kg/m³ offers such a resistance if we refer to the f_{r1} (EN 14651) value = 4.5 N/mm², at 0.5 mm crack opening, equivalent in uniaxial tension of 0.45 x 4.5 N/mm² = 2.02 N/mm². We can conclude that it is important for the concrete contractor to organize the installation smoothly, managing all related operations from material supply and transport to mixing, discharging or pumping without delays, temperature control, crew management on-site, and curing.

Pre-pour meetings are necessary to plan, organize, and coordinate the installation.

It is the Concrete Contractor's responsibility to complete the installation successfully, without cold joints. If a cold joint begins to form, traditional rebars or wire meshes should be installed at the top and bottom in the still fresh concrete.

In the case of underground water pressure, all construction joints within the slab or between the slab and walls should be provided with a water-stop strip or similar details to prevent leakage. Resin injection where needed is a part of the Design-Build- Maintain contract.

If water puddles are later found on the slab, don't immediately blame the slab itself. First, check the joints in the walls, under the ceiling edges, or between the slab and walls, as these are often the source of leakage.

In traditional construction, no one is responsible beyond the design standards. The purpose of the standard design is to indemnify the engineer, provided the standard design methods are followed. The materials supplied are standard materials from approved suppliers. The material supplier's responsibility ends at the door of their labs.

Following the design and material standards does not guarantee that the structure will be fit for service. Even though the materials comply with standards, the ready-mixed concrete could still suffer from excessive shrinkage, leading to undesirable cracking.

Note that fr1 of EN 14651 is recorded at 0.5 mm crack mouth opening displacement and is used at SLS, that is, the service limit state.

The SFRC raft is not supposed to be crack-free but well serviceable with controlled crack opening up to 0.5 mm opening.

The SFRC doesn't corrode at all up to 0.5 mm crack opening under the most severe chloride environment (3, 4).

Table 7.3 clearly summarizes that, even with a 100-year life expectancy for the SFRC, all crack openings up to 0.5 mm are considered acceptable.

The steel fibers visible at the surface may, over time, produce rusty spots; however, this does not have any negative consequences on the structure itself, unlike in traditional reinforced concrete, as shown in Figure 7.11.

The Table 7.3 shows the recommended limits of the crack opening $w_{\max}$ (mm) for fibre concrete only as far as durability is concerned

Figure 7.9 Important leaks in a traditional Reinforced Concrete wall of 30 m high wall of a lock.

Table 7.3 Reprinted from (4), p.16

Exposure class	$L50\ w_{max}{}^a$	$L100\ w_{max}{}^a$
X0, XC1	_b	_b
XC2, XC3	0.5	04
XC4	0.4	0.3
XS1, XS2, XD1, XD2	0.3	0.2
XS3, XD3c	0.2	0.1

[a] For structure members with a combination of fibers and conventional or pre-stressed reinforcement, see SS ^{137010} concerning allowable crack widths with respect to exposure conditions.

[b] For X^0, XC^1 exposure classes, crack width has no influence on durability. Therefore, its limit should be set to guarantee acceptable appearance and deformations.

[c] In exposure class XS^3 or XD^3 corroding steel fibers must be combined with conventional reinforcement for suspended decks slabs or beams.

In case maintenance is needed, SFRC cracks are simply filled with fluid resins or sometimes injected.

Practically during the installation, it is always beneficial to allow the concrete to bleed slightly, as this keeps the surface moist. Never let the surface dry out in the fresh state, as this can cause fresh cracking, which is not covered by any calculations or prevented by any type of reinforcement. It can be however mitigated by the fiber reinforcing.

To avoid thermal cracking in thick slabs due to hydration heat at an early age, which no reinforcement can prevent or eliminate, it is recommended to use slow-setting Portland cement combined with up to 35% supplemental cementitious materials, such as GGBS and Class F (ASTM) fly ash. It is also essential to ensure that the concrete temperature remains under complete control, ideally not exceeding 25°C, with a temperature drop of no more than 15°C, and a minimum of 10°C.

Hot days during concrete pouring, followed by cool nights, are a common cause of thermal cracking.

The heat of hydration from Portland cement becomes more critical in slabs thicker than 500 mm. The use of supplemental cementitious materials, in limited concentrations of up to 35%, can help significantly, along with slow-setting cement.

However, too high a concentration of SCMs can cause early-age cracking in cold weather or in high-temperature conditions if rapid moisture loss occurs at the surface.

7.5 SFRC CAST IN SITU "FORMED IN THE GROUND SOLFIBRES CFA PILES"

Steel fibers are often a very adequate solution, equivalent to up to 0.5% of traditional reinforcement, as shown in Figure 7.12, compared to the cumbersome and time-consuming process of introducing a rebar cage into the fresh concrete pile.

In the case of continuous fly auger machines (CFA), where fiber concrete is pumped through the hollow core of the auger, it is advisable to limit the fiber length to 50 mm.

The mix should be flowing and very stable, with no risk of segregation that could block the pumping and CFA machine.

Savings are significant, as the fiber dosage rate for steel fibers ranges from 0.35% to 0.62% (25 to 60 kg/m^3), depending on the fiber type. The

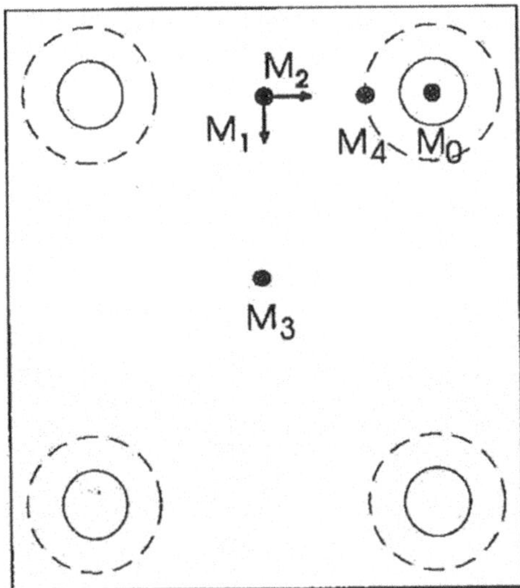

Figure 7.10 Location of critical flexural moments of ground bearing rigid raft to carry column loads.

cementitious content is high (up to 450 kg/m³ together with the use of pea gravels for the mix design) to ensure smooth pumping along the flexible line.

Superplasticizers are used to make the mix flow up to the point of slump collapse.

An example mix design includes 30 to 40 kg/m³ of HE 1/50 – 1500 steel fibers, 375-400 kg of cement with 35% SCM, and a 0-12 mm continuous aggregate grading, resulting in a fine content (< 250 μm) of 500 kg/m³. The mix design includes a relatively high content of fines smaller than 250 μm to prevent segregation and the formation of fiber or cement balls in the concrete pump or along the line, which could delay execution by up to 1 hour per event.

The mix in place is almost at slump collapse and is therefore never vibrated. Generally, a set retarder is also added to maintain a stable slump for 2 to 3 hours, which is often needed during drilling operations, as the execution can face many unknowns on site.

The omission of a rebar cage results in significant savings in time and cost. The SFRC eliminates the cumbersome placement of the rebar cage in the fresh concrete pile, whether done by hand, as shown in Figure 7.7, or by using the piling rig, which significantly reduces the daily number of piles produced. The daily rate of such a machine can be quite high, ranging from 2000 to 5000 €/day (2023), so the time spent placing the rebar cage (15 minutes minimum to 60 minutes in worse cases) could cost anywhere between 50 € and 500 € (2023).

The mix described here replaces 0.5% of rebars in the section. Starter rebars of 0.5% proportion are placed in the fresh pile by hand to a depth of 6 times the pile diameter and are left sticking out of the pile to connect to the foundation reinforcement.

In Figure 7.13, the steel fiber reinforced concrete shows a number of steel fibers sticking out between the starter rebars.

The equivalent section of a 600 mm diameter pile with 6 rebars of 16 mm diameter is equivalent to a 35 kg/m³ dosage rate of HE 1/50 steel fibers.

Figure 7.11 Typical corroded rebars and damage to concrete in a Mediterranean environment.

It is not uncommon to speed up the installation of 400 mm diameter SFRC auger piles, with a 12 m length, to as many as 12 piles per shift (approximately 18 m³ of SFRC per 8-hour shift).

This system was developed in France in 1987 and rapidly gained popularity, with approximately 300,000 CFA piles completed between 1987 and 2003.

The Solfibres system was used for a wide range of buildings, including residential, commercial, and industrial plants, power plants (including nuclear), bridges for highways and railroads, and in seismic regions of the Alps, where the fibers used were of the Twincone type, provided with conical ends for total anchoring to the concrete matrix.

The SOLFIBRES system has been approved by official certification bodies such as Bureau VERITAS and SOCOTEC, based on a complete specification of materials, processes, design, and quality control. Several French special foundation contractors have used the SOLFIBRES system.

A number of full-scale loading tests have been completed, along with numerous real-scale laboratory tests at the CEBTP central labs.(Centre d'Etudes du Bâtiment et des Travaux Publics)at the St Rémy-Les-Chevreuses research station

In the diagram shown in Figure 7.14, a Solfibres SFRC CFA pile, with a 30 kg/m³ dosage of undulating steel fibers (1 mm diameter x 60 mm length, 1200 MPa constituent steel wire tensile strength), is compared in flexure to a standard reinforced concrete pile with 0.6% traditional reinforcement.

The horizontal force is represented on the abscissa in kN, while the angle of rotation is shown on the vertical axis.

The SFRC Solfibres pile demonstrates greater ductility under the same or higher horizontal force.

I recall that the rebars in the traditional reinforced pile fractured in a brittle manner, in contrast to the highly ductile behavior of the Solfibres pile.

Figure 7.12 Cumbersome introduction of the traditional rebar cage in a cast in-situ a CFA pile.

A decision by the French bureaucracy abruptly ended the Solfibres activity. It was decided that control bodies like Socotec and Bureau Veritas could no longer certify the system and liability could no longer be insured.

The only solution would have been to apply for an "Agreement Technique" from the CSTB, a Ministry of Equipment agency. This required redoing all tests in laboratories and real-scale, as the CSTB could not accept third-party results, even from universities or business federation labs.

The alternative was to include the process in the DTU, the French reinforced concrete design standard, a process that could take 10 years or more.

The costs, time implications, and overall burden were so great that M.S. Lamotte, the owner and founder of Solfibres, decided to stop, sell the equipment, and lay off the staff.

For the same reasons, unlike in other countries where such business is still done, steel fibers are no longer used in structural concrete slabs in France. Despite early agreements with SOCOTEC, Bureau Veritas, and others, a number of SFRC piled slabs were successfully built from 1994 to 2003.

This is the result of the power of central government bureaucracy, which can halt successful innovations and ruin businesses. It is clearly a hindrance to progress.

It is amazing that such an agency, which brands itself as the "future of construction," has contributed to such stagnation. To me, this whole French validation system seems far more "Jurassic" than futuristic.

No report—whether statistical or technical—has ever shown that concrete construction in France is safer or better than anywhere else.

A contrasting example is the collapse of the 2E Terminal building, an arch concrete structure at Charles de Gaulle (CDG) Airport in Paris, during the night of a May 2004 day.

Figure 7.13 Si bars of 16 mm diameter as starter rebars of a 600 mm pile diameter.

Figure 7.15 shows the disaster that killed four people, which occurred in the very early hours of the day.

However, in theory, it was a terminal building project that followed the bureaucratic system, requiring a number of approval stamps.

I don't know if anyone ended up in jail, but what I do know is that, in general, design standards and bureaucratic approval processes exist simply to relieve or limit the involved parties of responsibility.

7.6 READY MIXED SFRC FOR FOUNDATION CONTRACTORS

Some examples of equivalent calculations for piles are provided below, in the case of SFRC supplied by a concrete plant to a foundation contractor.

The first example demonstrates how to convert a flexural moment into the appropriate dosage rate of steel fibers.

The second example involves a tensile pile provided with 8 rebars of 16 mm diameter, which is then modified into an SFRC tensile pile, as shown in Figure 7.16.

$D := 1180\,mm$ — Pile diameter

$I_d := \dfrac{\pi \cdot D^4}{64} = 0.095\,m^4$ — Moment of inertia of the pile

$M := 450\,kN \cdot m$ — Design moment (factored loadings) estimated value based on 14 rebars of 16 mm dia (14 mm × 202 = 2828 mm² - 0,3% re-steel percentage)

$v := \dfrac{D}{2} = 590 \cdot mm$ — neutral axis

$f_d := \dfrac{M}{\frac{I_d}{v}} = 2.79 \cdot \dfrac{N}{mm^2}$ — design stress in flexion

$\eta_{det} := 1.4$ — SS812310 structural indeterminacy factor for a continuous beam

$\gamma_m := 1.5$ — material factor

$f_{r3} := \dfrac{f_d \cdot \gamma_m}{\eta_{det}} = 2.989 \cdot \dfrac{N}{mm^2}$ — EN14651 required average flexion strength at CMOD = 2.5 mm

$f_{r3} := \dfrac{f_d \cdot \gamma_m}{\eta_{det}} = 2.989 \cdot \dfrac{N}{mm^2}$ — 30 kg/m³ HE 1/60–1500 MPa

---

Conclusion: use 30 kg/m³ HE1/60-1500 MPa steel fibers—connection/starter rebars needed on top of the piles.

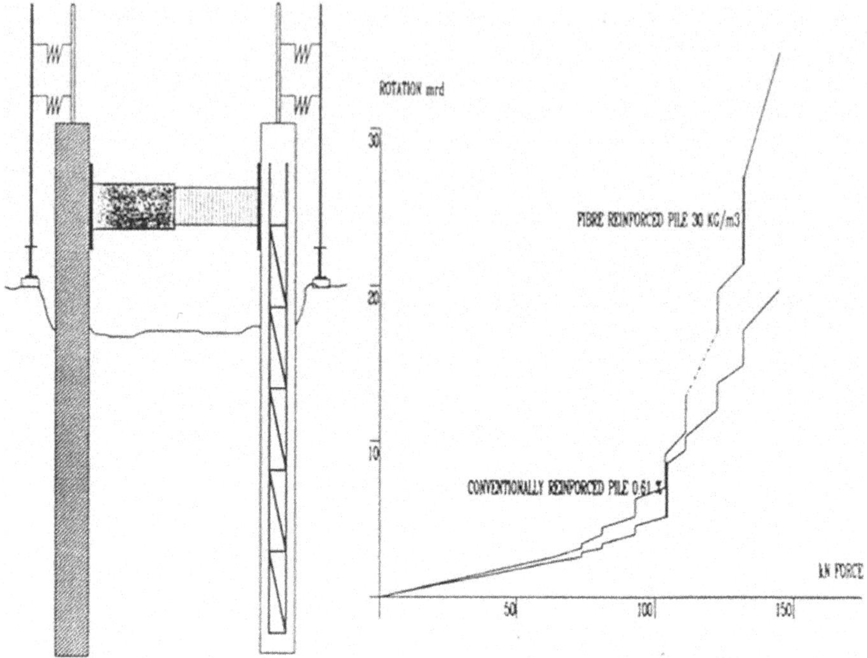

Figure 7.14 Comparison between a traditional Reinforced pile (right) and a Solfibres piles (left).

In Figure 7.14, a 1.10 m diameter tension pile is provided with 8 rebars of 16 mm diameter.

To calculate the SFRC section, the concrete section is assigned a tensile strength due to the inclusion of steel fibers.

Initially, the conversion from rebar stresses to concrete stresses is made by considering only the core concrete between the rebars, as the outer concrete could be cracked, deteriorated, or contaminated by the ground.

However, with fiber concrete, there is good reason to consider the entire section diameter, as the fibers will control cracking up to the skin of the pile. This improves durability, as cracks serve as pathways for ion ingress in concrete, whereas the fibers do not corrode within the matrix.

In the first calculation, only the core section is considered, which leads to a relatively high fiber concentration.

In the second step, we'll analyze the second option of tensile pile with the whole diameter section to be taken into account. Note that steel fiber reinforcing in this case of uniaxial tensile stresses, lacks the redundancy: a centrical continuous steel bar must remain. It is still an advantageous solution as a centrical rebar is much easier and quicker to install that a rebar cage.

$w_x := 0.65m \quad w_y := 0.65m$	sizes of rebar cage
$d := 16mm$	diameter of rebars
$n := 8$	number of rebars in the cage
$f_y := 500\dfrac{N}{mm^2} \quad \gamma_s := 1.15 \quad \gamma_c := 1.5$	steel strength and material factors
$S := \dfrac{\pi \cdot d^2}{4} \cdot n = 1.608 \times 10^3\, mm^2$	section of n rebars
$f_{td} := \dfrac{f_y \cdot S}{\gamma_s \cdot w_x \cdot w_y} = 1.655\dfrac{N}{mm^2}$	equivalent tensile strength of concrete
$f_{r3} := 6\dfrac{N}{mm^2}$	EN 14651 flexion strength of the fiber concrete—C40-50
$\eta_f := 0.75$	fiber orientation factor
$f_{ftr} := \dfrac{0.37 \cdot \eta_f - f_{f3}}{\gamma_c} = 1.11\dfrac{N}{mm^2}$	Tensile strength of the fiber concrete
$\Delta f_{rd} := f_{td} - f_{ftr} = 0.545\dfrac{N}{mm^2}$	Strength deficit
$s := \dfrac{\Delta f_{rd} \cdot w_x \cdot w_y}{\dfrac{f_y}{\gamma_s}} = 529.853\, mm^2$	section of steel rebar to fill the deficit 1 diam 28 mm = 708 mm²

The 28 mm centered rebar is also used a connecting rebar the external structure. Therefore its design strength must be higher than : 435×1608/708 = 987 N/mm² thus a 28 mm dia 1200 MPa threadbar Conclusion: replace the 8 dia 16 mm rebars from top to bottom by 1 dia 28 mm 1200 MPa strength centered plus 50 kg/m³ of HE 1/60 - 1500MPa. - No stirrups needed.

The second option shows the advantage of the steel fibers, as the high durability of the steel fiber concrete allows the entire section without a dead outer surface layer.

As shown below, the dosage rate of steel fibers is significantly reduced from 50 kg/m³ to 30 kg/m³.

$D := 1.10m$	Diameter of pile
$d := 16mm$	diameter of rebars
$n := 8$	number of rebars in the cage
$f_y := 500\dfrac{N}{mm^2} \quad \gamma_s := 1.15 \quad \gamma_c := 1.5$	steel strength and material factors
$S := \dfrac{\pi \cdot d^2}{4} \cdot n = 1.608 \times 10^3 \cdot mm^2$	section of n rebars

$$f_{td} := \frac{f_y \cdot S}{\gamma_s \cdot \pi \cdot \dfrac{D^2}{4}} = 0.736 \cdot \frac{N}{mm^2}$$

equivalent tensile strength of concrete

$$f_{r3} := 2.5 \frac{N}{mm^2}$$

EN 14651 flexion strength of the fiber concrete—C40-50 25 kg HE 1/60-1500

$$\eta_f := 0.75$$

fiber orientation factor

$$f_{ftr} := \frac{0.37 \cdot \eta_f \cdot f_{f3}}{\gamma_c} = 0.463 \cdot \frac{N}{mm^2}$$

Tensile strength of the fiber concrete

$$\Delta f_{rd} := f_{td} - f_{ftr} = 0.273 \cdot \frac{N}{mm^2}$$

Strength deficit

$$s := \frac{\Delta f_{rd} \cdot \pi \cdot \dfrac{D^2}{4}}{\dfrac{f_y}{\gamma_s}} = 597.58 \cdot mm^2$$

section of steel rebar to fill the deficit 1 diam 28 mm

The 28 mm threadbar, along the full depth of pile is of regular 435 N/mm² design strength, and is used as a connecting bar.

Conclusion: replace the 8 dia 16 mm rebars from top to bottom by 1 dia 28 mm centered plus 30 kg/m³ of HE 1/60 - 1500MPa. - No stirrups needed.

The 28 mm centrical threadbar of 435 N/mm² design strength is over the full depth of pile.

7.7 SFRC GROUND BEAMS

Traditional reinforcing ground beams are often substituted by SFRC ground beams, cast together with the ground bearing or piled ground slab.

In the example here below, it is about a ground beam cast together with the ground slab, so that the ground slab is a kind of table of compression as it is for a T beam or in this case a Γ beam.

$$h_{rib} := 450mm$$

rib depth

$$e_{rib} := 400mm$$

rib thickness

$$S_{rib} := h_{rib} \cdot e_{rib} = 0.18 \ m^2$$

rib section

$$h_s := 250mm$$

top slab thickness

$$b_s := 1000mm$$

width of slab cooperating (4 × slab depth) total thickness

$$H := h_{rib} + h_s = 0.7 \ m$$

$$S_s := h_s \cdot b_s = 0.25 \ m^2$$

top slab section

$$\mu := \frac{S_s}{S_{rib}} = 1.389$$

ratio Slab/rib

$$I_{rib} := e_{rib} \cdot \frac{h_{rib}^{3}}{12} = 3.038 \times 10^{-3} \ m^{4}$$

rib moment of inertia

$$I_{s} := b_{s} \cdot \frac{h_{s}^{3}}{12} = 1.302 \times 10^{-3} \ m^{4}$$

slab moment of inertia

$$E_{s} := 30000 \frac{N}{mm^{2}}$$

modulus of elasticity of concrete of the slab

$$E_{rib} := 30000 \frac{N}{mm^{2}}$$

and same for the rib

$$n := \frac{E_{s}}{E_{rib}} = 1$$

same material rib and table

$$d := \frac{H}{2} = 0.35 \, m$$

distance between centers of gravity of rib and table

$$I_{t} := I_{s} + \frac{I_{rib}}{n} + S_{rib} \cdot \left(\frac{\mu}{1 + n \cdot \mu} \right) \cdot d^{2} = 1.716 \times 10^{10} \cdot mm^{4}$$

total section moment of inertia

$$H_{eff} := \left(12 \cdot \frac{I_{t}}{b_{s}} \right)^{\frac{1}{3}} = 0.591 \, m$$

Effective equivalent slab thickness

$$v_{bot} := H - \frac{\mu}{1 + \mu} d = 0.497 \, m \quad v_{top}$$
$$\phantom{v_{bot}} := H - v_{bot} = 0.203 \, m$$

neutral axis position

$$M_{d} := -350 \, kN \cdot m$$

Maximum Given Design Moment from the FEM diagrams regarding the ground beams

Design flexion strength of SFRC

$$f_{d} := \frac{M_{d}}{\dfrac{I_{t}}{v_{top}}} = -4.151 \cdot \frac{N}{mm^{2}}$$

$$\gamma_{c} := 1.5$$

concrete material factor

$$\eta_{det} := 1.4$$

EN-SS812310 statical indeterminacy factor 1.4 in case of a continuous beam

Minimum EN 14651 SFRC flexion strength required

$$f_{R3} := \frac{\gamma_{c} \cdot f_{d}}{\eta_{det}} = -4.447 \cdot \frac{N}{mm^{2}}$$

Conclusion: 45 kg/m³ of HE1/60 - 1500 MPa in a C30-37 instead of all traditional reinforcing.

The following Figure 7.17 shows a perimeter ground beam with 40 kg/m³ of HE 1/50 – 1200 MPa steel fibers, placed using a "Tele-Belt," a truck-mounted conveyor belt with extenders that replaces the concrete pump. It is much easier and cheaper to operate. As shown in Figure 7.15, the concrete slump is smaller than with a pump. The Tele-Belt is much cheaper to operate

Figure 7.15 2E concrete terminal building collapse in 2004.

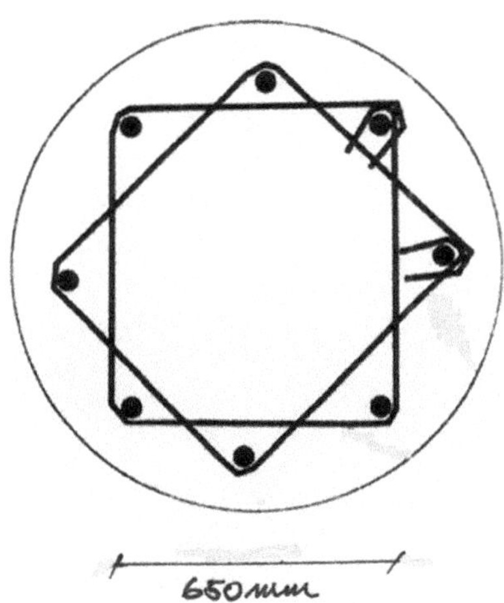

Figure 7.16 A traditional section of pile with eight rebars of 16 mm diameter.

than a concrete pump, and it doesn't affect the mix design in the same way that a concrete pump does.

Here, we can see a chute-pipe to control the free fall of the fresh concrete.

7.8 SFRC SHALLOW FOUNDATIONS FORMED IN THE GROUND

These windmills pads, as shown in Figure 7.16 a and b, are quite massive, even for small units of 10-15 MW production, in order to prevent tipping

Figure 7.17 SFRC ground beam being installed by a tele-belt.

over due to the weight of the foundation pad. Thousands of foundation pads for such units have been built using SFRC.

The example in Figure 7.18 shows a 10 MW unit, 25 m in height, resting on a 3.3 m x 3.3 m foundation pad, 1.2 m thick, made with C30 flowing concrete and 20 kg/m³ of HE 1/60 – 1500 MPa steel fibers.

The steel fiber reinforcing speeds up so much the process that the 13 m³ concrete pad is installed in half a day!

In Spain, as shown in Figure 7.19, the SFRC strip footings and slab are formed in the ground and poured together to complete a structural raft with 40 kg/m³ of HE 1/60 steel fiber reinforcement. The entire slab was cast and finished "joint-free" without sawn cuts.

Significant savings in time and materials provide a convincing advantage.

By digging out the ground, without the need for any formwork, the three anchor legs, 700 mm deep, are placed and immediately followed by the supply of fiber concrete from a ready-mixed truck with a built-in tele-belt, placed without the need for any poker vibrating.

7.9 SFRC FREEZER RAFT SLAB ON PILES

In Figure 7.20, we see the form-bearing structure of an SFRC free-suspended raft for a freezer warehouse. Round pile heads of 800 mm diameter in a 2.5 m x 2.5 m grid support a 500 mm thick raft with 50 kg/m³ of HE 1/60 – 1500 MPa steel fibers. No other reinforcement is used. The bay size between construction joints is 500 m³ in volume or 1000 m² in area. It goes without

Figure 7.18 a and b: installation of a windmill foundation pad (by courtesy of THY Moellen, DK.)

Figure 7.19 SFRC foundations formed in the ground.

Figure 7.20 Form bearing structure of an elevated suspended SFRC raft.

Figure 7.21 The pumping of an elevated suspended SFRC raft.

saying that each bay is jointless. The freezer warehouse slab covers approximately 10,000 m².

The loading intensity is 100 kN/m².

Figures 7.20 and 7.21 show the forming and two concrete pumps in action to cast the slab. It is important to note that SFRC with high dosage rates is often pumped in many countries. In this case, 50 kg/m³ of the same fiber type (HE 1/60 – 1500 MPa) is used, but 70 kg/m³ has also been applied, and up to 100 kg/m³ dosage rates are used with thicker fibers, such as those with an L/d ratio of 38 (e.g., 1.3 mm x 50 mm length). Also note that there are no visible rebars, as all traditional rebars have been completely omitted.

Figure 7.22 shows the traditional power troweling in action to finish the surface.

Figure 7.22 The trowel finishing of the raft surface.

7.10 RESIDENTIAL APPLICATION OF SFRC RAFT

Figure 7.23 shows an SFRC raft being finished between secant pile walls. This raft serves as the foundation for a six-level residence. The total raft area is 4000 m², with a thickness of 400 mm, and local thickening up to 600 mm beneath some heavy columns or walls with a 5000 kN unfactored loading.

The SFRC is made with 50 kg/m³ of HE1-60 – 1500 MPa steel fibers.

The base under the raft is a natural local silty sand, capable of offering a 200 kN/m² ground bearing capacity.

Note that in all these rafts, like almost all those we have been involved with over the last 30 years, no top or bottom mesh reinforcement is included.

In the case of rafts thicker than 500 mm, however, it may be necessary to install a walk-on wire mesh, such as 8 mm x 200 mm, supported by one chair per m².

Figure 7.24 shows a ground-bearing raft under a 12-story wall-bearing residence in Riga, Latvia. The raft is approximately 80 m long and 14 m wide, supporting several transverse bearing walls spaced 4.50 m apart.

This is a 500 mm thick raft with 80 kg/m³ of undulated steel fibers, 1.3 mm in diameter and 50 mm in length, with 850 MPa wire tensile strength. The SFRC is installed using a concrete pump. The concrete is of a flowing type, so no poker vibrating is required.

The linear unfactored loading intensity of the walls is approximately 5000 kN/m.

7.11 SFRC PILED FOUNDATION RAFT

The Friends Arena, a 65,000 seats multipurpose arena in Solna (Stockholm, Sweden), is an example of suspended SFRC raft foundation.

Figure 7.23 Part of a 4000 m² raft at Triumph Gardens, Brussels.

Figure 7.24 A 12 floors high residence Raft (courtesy of Primekss SIA, Riga-LV).

Figure 7.25 Friends Arena SFRC raft installation by pumping.

Figure 7.26 Overall view of the inside of the Friends Arena building raft in Stockholm/Solna.

Approximately 20,000m³ of SFRC was installed at a thickness of 300 mm on top of a grid of piles spaced 3 m x 3 m for the main arena slab and up to 5 m x 5 m in the service areas. The structure was designed to handle point loadings up to 600 kN unfactored and 40 kN/m² uniformly distributed loading.

The SFRC used was of 45 and 50 kg/m³ dosage rate of HE 1-60 1500 MPa steel fibers in a C35-45 mix design

The savings in time and costs achieved were phenomenal since in traditional reinforcing, 175 kg/m³ were given in the original specification.

We were told afterward that the SFRC solution had been able to shorten the planning of the works by 6 months!

The raft installation happened mainly in the open sky during the winter under wintery conditions of snow or icy rain and wind as shown in Figure 7.25 were the pumping of the SFRC is shown.

Think of the 3000 Tons of reinforcing bars that were omitted and could save about 30000 men x hours on site or ca. 4000 men for x days! It is also much safer as traditional rebars cutting, bending, placing and handling are today still causes of injuries on site.

On the same figure, we can observe a construction joint provided by rebars across in order to provide the slab continuity.

Figure 7.27 Friends Arena raft completed and 600 metric tons cranes on it.

Figure 7.26 reveals the interior of the Friends Arena, featuring the main central field and seating for 65,000 people. The technical areas are located beneath the seating and are not visible here.

Figure 7.27 shows an overview of the completed Friends Arena piled raft foundation, with 600-metric-ton cranes operating to install the roof!

Figure 7.28 shows the pile grid along with the different areas: the central field in green, the services area under the seating in red, and the external pavement in orange.

7.12 GROUND RETAINING WALLS IN SFRC

SFRC ground retaining walls with up to 3 m spans between slabs are widely used for basement applications, typically up to 2 underground levels in residential and commercial projects.

Starter rebars at the mid-depth of the walls are kept out of the raft and slabs.

In the case of underground water pressure, watertightness is achieved using regular specialty waterproofing techniques.

The walls are then calculated as a continuous beam supported by 2 or 3 points with fixed ends.

Ground retaining walls with a free top and a fixed bottom in the footing are also possible.

Figure 7.29 shows a typical example of a 1.5 km long retaining wall built in Beringen, Belgium, in 2004.

The SFRC wall height-to-thickness ratio is 6000/300 = 20, which is considered "thin" compared to the 10 often recommended by standard guidelines.

Note that, in terms of standard performance, 45 kg/m³ of 1 mm x 60 mm hooked-end fibers with 1200 MPa wire strength are equivalent to the 80 kg of 1.3 mm x 50 mm steel fibers used here.

One layer of 20 mm diameter rebars with 200 mm spacing was still needed to provide the appropriate negative moment capacity and serve as starter rebars. At ULS, the resisting moments from the fiber reinforcement and the additional rebars are additive.

The wall was built in 30 m long sections between construction joints, each provided with water stops.

The SFRC was pumped, as a flowing mix was used and only slightly poker vibrated during installation in the smooth form. Very little or no fiber was visible on the finished surface. It is important to ensure that the form is tightly sealed at the joints and angles to prevent leakage of cement laitance, which could cause steel fibers to protrude from the joint or angle.

Cracking due to hydraulic or thermal causes (cracks from the bottom up to the mid-height of the wall) has never been reported here, unlike in traditional concrete, where such cracking is common.

Regarding the calculations, the rebar resisting moment and the SFRC resisting moment are simply added at ULS. The elimination of double or single wire meshes offers a significant advantage in terms of cost and time, as it does for rafts. Double meshes, in particular, are difficult to install in wall forms.

7.12.1 Water retaining wall with free top

$D := 300 \cdot mm$	$b := 1 \cdot m$	wall thickness
$h_w := 2.5 \cdot m$		height of water table maximum height of
$\lambda_s := 1.20$	$H_{max} := \lambda_s \cdot h_w = 3\ m$	watertable
$\gamma_Q := 1.5$		loading factor
$\gamma_w := 10 \cdot \dfrac{kN}{m^3}$		volume weight of water
$M_d := \dfrac{\gamma_w \cdot \lambda_s \cdot \gamma_Q \cdot h_w}{2} \cdot \dfrac{h_w \cdot h_w}{3} = 46.875\ kN \cdot \dfrac{m}{m}$		design moment
$f_d := 6 \cdot \dfrac{M_d}{D^2} = 3.125\ \dfrac{N}{mm^2}$		design stress
$f_{r3} := 4.7 \cdot \dfrac{N}{mm^2}$		flexion strength EN14651 45 kg/m³ HE + 1/60
$\eta_{det} := 1$		No structural indeterminacy material
$\gamma_m := 1.5$		factor

$$f_{rd} := \eta_{det} \cdot \frac{f_{r3}}{\gamma_m} = 3.133 \ \frac{N}{mm^2}$$

design flexion strength

$$\frac{f_d}{f_{rd}} = 0.997$$

<1, OK

$$f_y := 435 \cdot \frac{N}{mm^2}$$

$$c := 30 \ mm$$

coverage

$$S := \frac{M_d}{0.9 \cdot (D - c) \cdot f_y} = 443.451 \ \frac{1}{m} \cdot mm^2$$

section of starter rebars 10mm diam - 150mm up to 800mm height above the joint with the slab (524 mm²/m)

It is important to remind the reader that steel fiber reinforcement, like rebars, is not a waterproofing technique. Therefore, the wall must be completed with the application of all suitable waterproofing techniques.

An example of calculations for an SFRC retaining wall is provided in Reference [23] in (Chapter 5) on page 334, Figure 17.11.

7.13 FOUNDATIONS PADS

Isolated footings are very rigid pads with a size-to-depth ratio ranging from 4 to 8 at most. Steel fibers are a very suitable reinforcement, so all traditional rebars can be omitted. The starter rebars in the column above are bent at the bottom of the footing, as in traditional designs.

In cases of extreme loading, a light bottom mesh and steel fibers are required together.

Typical solutions are given in Table 7.4.

The column on the left represents the footing thickness, and the line on top indicates the traditional reinforcement to be replaced.

The solution is given in kg/m³ dosage rate of HE 1/60 steel fibers (hooked ends, 1 mm diameter, 60 mm length, 1200 MPa tensile strength) in C30-37 concrete. In some cases, an 8 mm x 150 mm bottom wire mesh is still needed.

Such a table of solution replaces all side rebars, top and bottom and stirrups.

7.14 PRIMEKSS SFRC RAFT

In Chapter 9, regular SFRC slabs are compared to Self-Stressing SFRC, which is obtained through a restrained chemical expansion reaction, typical of the Primekss proprietary process. This results in zero shrinkage and permanent compressive stress in the slab, effectively creating a self-stressing

Table 7.4 SFRC foundation pad design

	8/150 335mm²	10/150 524mm²	12/200 565mm²	12/150 Or 10/100 754mm²	16/200 1010mm²	12/100 1130mm²	20/200 1570mm²
300mm	25kg/m³	30	35	45	45+8/150	45+8/150	/
350mm	25	30	35	45	45+8/200	45+8/150	/
400mm	25	30	35	40	40+6/150	40+6/150	/
450mm	20	25	30	35	40+6/150	40+6/150	50+8/150
500mm	20	20	30	30	45	45	45+8/150
600mm	20	20	25	25	45	45	45+6/150
800mm	20	20	20	20	30	30	40
1000mm	20	20	20	20	20	25	35

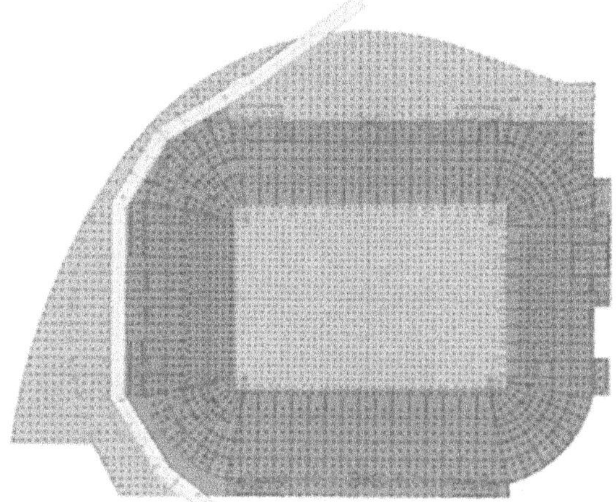

Figure 7.28 Pile grid underneath the Friends Arena raft foundation.

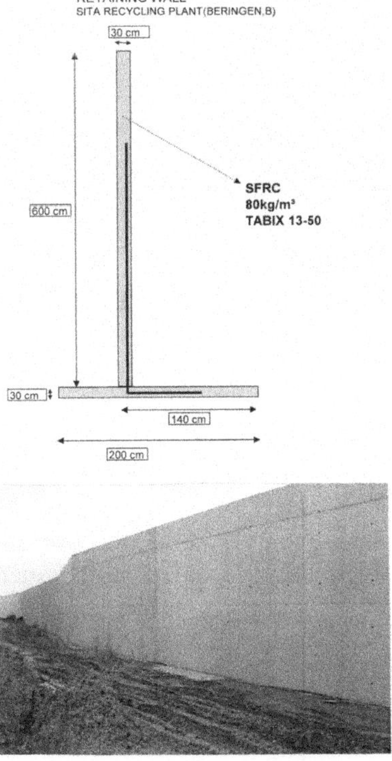

Figure 7.29 Shows a 300 mm thick footing and wall made from C30-37 concrete with 80 kg/m³ of TABIX 1.3 mm diameter by 50 mm length, undulating steel fibers with an 850 MPa tensile strength.

Figure 7.30 Overall view of the Primekss Alpha-Con SS-SFRC in Alberta, Canada.

Figure 7.31 Detailed view of the ALPHA-CON/Primekss Self-Stressing SFRC under the 146 ft high legs (courtesy of Alpha-CON Inc.)

structure. These two advantages, beyond SFRC, contribute to a greater stiffness, allowing the slab to be designed much thinner than traditional reinforced concrete.

An ultrahigh clad rack building, 146 ft in height (44 m), has been built on such rafts by Alpha-Con Inc. in Calgary, Alberta, Canada, as shown in Figures 7.30 and 7.31.

The self-stressing SFRC Alpha-Con raft was built over an XPS-type insulation mattress under license from the Primekss system, with the following characteristics: a footprint of 2000 m² and a thickness of 535 mm (21 ¼ in). It operates as a freezer at –20°C. Closely spaced leg loads of 400 kN intensity are shown in Figure 7.31.

Figure 7.32 Handling of the traditional reinforcing on a site.

There are a number of references existing today like herebelow:

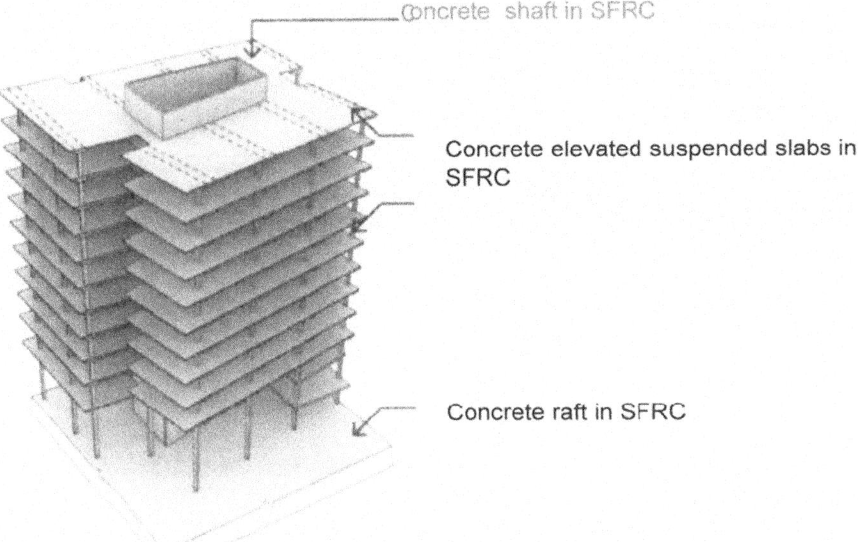

Figure 7.33 Example of actual applications of SFRC in buildings.

7.15 CONCLUSION

In most foundation rafts, there is no reason to continue using traditional reinforcement. Steel fibers, as explained here and in other chapters of this book, are the most suitable reinforcement to use and are recommended as advantageous for all stakeholders in a project.

Figure 7.32, from my own collection, taken on a site in the UK, is typical of the traditional handling of concrete reinforcement. The right composite

SFRC solution is the ideal choice, with 30 years of successful track records across 4 continents. Traditional reinforcement is becoming outdated when it comes to meeting the challenges of the 21st century.

Figure 7.33 summarizes the structural SFRC applications in a schematic building: foundation rafts (either load-bearing or pile-supported), ground beams, footings, bearing walls, and all suspended slabs cast in situ above forms or concrete planks. These applications are not new, as they began in the early 1990s, and thousands of cases have been completed successfully. SFRC elevated suspended slabs have been in use since the early 2000s.

REFERENCES

1. "System evenly mixes fibers", Engineering News Record, the McGraw-Hill construction weekly, p. 16, February 20,1986.
2. A. Khudair. "Fiber distribution of steel fibers in industrial floors-Field Measurements" MS Thesis, KTH Royal Institute of Technology, Sweden, 2016.
3. M. Marcos, F. Solgaard, S. Edvardsen. "Corrosion resistance of steel fibre reinforced concrete- a literature review." *Fib Symposium* 2016, Downloaded from orbit.dtu.dk
4. Swedish standard EN-SS 812310: p.16, 7.3.1 Table 4.

Chapter 8

Shrinkage of slabs

8.1 BACKGROUND

The shrinkage of concrete has been, and is still considered inevitable; it is almost part of the very definition of concrete. Shrinkage begins to concern everyone involved in a project when cracking and joint movements exceed the usual, standard, or agreed-upon limits. G. Garber, in his book "Design and construction of concrete floors" (1), offers a simple and very clear summary of the issues related to slabs.

In floors, shrinkage is always a significant concern, as the external exchange surface area to volume ratio is high. The higher the surface area to volume ratio (in m^{-1} units), the quicker the shrinkage develops; for example, a 200 mm thick slab has a ratio of 5, while a 100 mm thick slab has a ratio of 10. An internal slab of 100 mm thickness can show excessive shrinkage within 6 months, while a 200 mm slab may only exhibit noticeable shrinkage after 18–24 months.

Shrinkage will induce various movements or displacements in the slab, depending on whether the movements are free or restrained, or somewhere in between.

8.2 CURLING

Under free shrinkage contraction, joints open significantly and the slab curls upward, depriving the slab of ground support along the joints, which leads to severe cracking as the slab rocks up and down at the joint.

When subjected to vehicle traffic, such a slab will always experience significant cracking caused by constant rocking. Surface cracking develops under repetitive dynamic negative moments, including a fatigue effect.

The curling of slabs is a result of high free shrinkage at the top of the slab, in contrast to the bottom, where shrinkage is much smaller due to the zero evaporation rate down on the bottom. More shrinkage leads to more curling,

DOI: 10.1201/9781003188315-8

especially in higher-strength concrete. High-strength concrete, particularly when including micro-silica and superplasticizers, can curl much more than regular concrete.

The curling increases as the square of the distance between adjacent joints, and the joint opening increases in the same manner as curling.

In Figure 8.1, the influence of air relative humidity and maximum aggregate size is shown, as presented by Soum.

Note that smaller aggregates with a diameter (D) require more paste in the mix design of concrete to coat a larger surface area. For example, when the maximum aggregate size decreases from D = 20 mm to D = 14 mm, the increase in curling is $\sqrt{(20/14)}$ = 20%. Additionally, when the relative humidity decreases from 0.70 to 0.4, the curling increases by $\sqrt{((1-0.7)^3/(1-0.4)^3)}$ = 35%, according to Soum, as shown in Figure 8.1.

Slab cracking between relatively close joints can also act as a relief for curling stress, so that after cracking, the amplitude of curling may decrease.

The effects of slab shrinkage, cracking, curling, and joint opening on slabs on the ground have been clearly addressed in the literature, as shown in Figure 8.1 (Bissonnette, Attiogbe, etc).

8.3 FREE AND RESTRAINED SHRINKAGE

Under free shrinkage, slab displacements occur both horizontally, resulting from shortening, and vertically, resulting from the gradient of shrinkage between the top and bottom of the slab.

From 1992 to 1994, we conducted an important test on free shrinkage slabs at the Université d'Artois (IUT de Béthune, France). Slabs of 4 m × 4 m with a thickness of 150 mm were cast on a smooth plywood base isolated by a double poly sheet. Vertical and horizontal shrinkage were measured over several weeks. The slabs were wet-cured under plastic sheeting. The tests included slabs with W/C ratios of 0.52 to 0.60, with or without superplasticizer, without steel fibers, and with 40 kg/m³ of 1 mm × 60 mm undulated fibers.

The main conclusions were:

- A 40 kg/m³ dosage rate of steel fibers reduces horizontal shrinkage by 25% and curling by 40%.
- Curling begins the day after the plastic sheet is removed from the top.
- With a W/C ratio of 0.60 and superplasticizer, there is a 35% increase in shrinkage.

These results highlight the importance of reducing water content by using optimal concrete constituent materials. The proper use of superplasticizers helps reduce water content while maintaining or increasing workability in the final mix.

JOINT CURLING

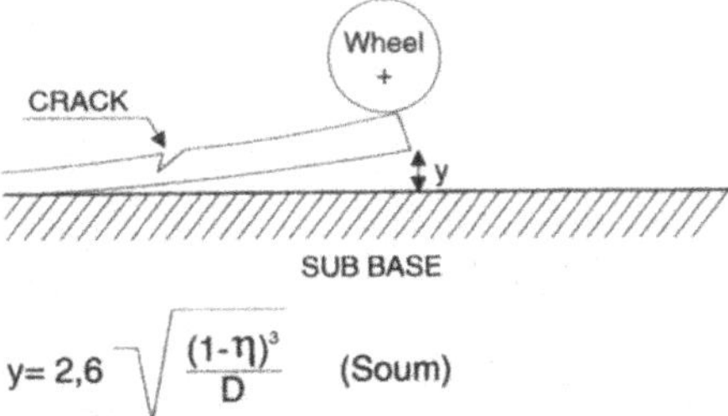

$$y = 2{,}6 \sqrt{\frac{(1-\eta)^3}{D}} \quad \text{(Soum)}$$

η = percentage water content in air with reference to saturation

D = size of aggregate

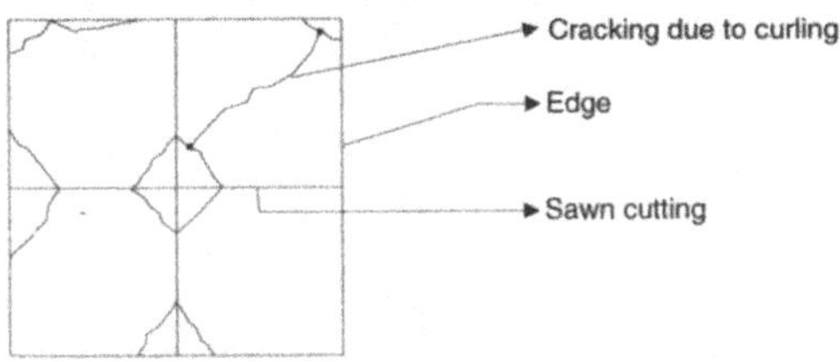

Figure 8.1 Soum theory of curling and detrimental effects of shrinkage.

When shrinkage contraction movement is restrained, the floor will crack, but the cracks are likely to be more evenly distributed, with a higher number of smaller, random cracks. Excessive detrimental opening can be prevented with suitable concrete reinforcement, an optimal mix design for a "low shrinkage content," and careful detailing in crack-prone areas such as loading docks, manholes, reentrant corners, and joint crossings.

Slab restraint is caused by external friction against the base and any edge-blocking conditions. Restraint increases under permanent loadings.

8.4 JOINTS AS CRACK PREVENTING: REALLY?

Construction joints and saw-cut joints, also called shrinkage joints, are designed to relieve the tensile stresses caused by restrained shrinkage contraction.

Traditionally, closely spaced joints are used in thick, unreinforced concrete slabs to manage shrinkage and the resulting cracking. The difficulty and cost of adequately reinforcing these slabs often lead to the use of minimal reinforcement. I don't use the term "plain concrete" as today, most slabs contain a small percentage (0.15% geometric percentage of the section) of reinforcement, often poorly placed. This reinforcement is often more like a "chicken" mesh in the mud, installed by hand during the pouring process.

This type of concrete does not qualify as reinforced concrete, but rather as psychological reinforcement. It does not eliminate or reduce curling, nor does it control any cracking, whether top, bottom, hairline, or crazing cracks. There are many options available to designers to manage slab cracking, and they should consider the following:

Minimizing total shrinkage: An optimal mix design to ensure a high unit weight of the concrete is achieved through a suitable grading of the best available constituent materials.

The base as a cause of friction restraint

- o A hard base with a high Westergaard K-value (e.g., 0.08 N/mm³ - 300 pci or more) provides higher restraint. This scenario helps minimize horizontal slab contraction movements and increases the number of random cracking of finer cracks. However, it can also increase curling at the joints.
- o A hard base is suitable for long, wide, jointfree slabs (2000–3000 m² and larger), provided the slab has a suitable percentage of reinforcement.
- o A soft base (K-value of 0.025 N/mm³ – 100 pci) is the opposite, promoting larger horizontal displacements and reducing curling. A low K-value is more suitable for traditional jointed slabs. However, the slab will need to be significantly thicker to account for the high curling-induced moments. Sawn cuts and construction joints must be designed to ensure minimum load transfer, as curling is inevitable over time and increases.
- o The reinforcement: The type, nature, and percentage of reinforcement used to limit cracking.

Joints, specification, and design: Although all joints will curl over time, the more shrinkage, the more curling.

8.5 THE INSTALLATION FACTOR

Most engineers specify concrete based on its standard strength to suit durability and environmental conditions. This typically results in a water-to-cement ratio between 0.45 and 0.55, or even 0.60. Many engineers expect shrinkage to have been adequately addressed within these specifications.

In reality, however, the same concrete "in theory"—at least as far as the compressive strength target is concerned—can experience up to 400% more shrinkage, as summarized in an ACI presentation slide:

- Higher concrete temperature: 8% more shrinkage
- Too high a slump: 10% more shrinkage
- Too long a haul in the transit mixer: 10% more shrinkage
- Too small aggregate: 25% more shrinkage
- High-shrinkage cement: 25% more shrinkage
- Dirty aggregates: 25% more shrinkage
- High-shrinkage aggregates: 50% more shrinkage
- Admixture: Up to 30% more shrinkage

The cumulative effect can result in up to 400% more shrinkage, increasing the factor from 1 to 5.

High shrinkage variations between different mixes are caused by a variety of factors and parameters typical of any concrete, including:

- Concrete supply: The concrete supply itself plays a role if we consider the nature of the constituent materials of the ready-mix concrete, including cement and admixtures, at the time of arrival on the job site.
- Mix design: The strength and workability parameters.
- Haul in the transit mixer: The waiting time before discharging and the conditions of the haul.
- Installation on site: This includes whether the concrete is discharged directly or needs to be pumped. Conditions like pipe diameter, total length, and the number of elbows in the line affect the shrinkage.
- Weather conditions: Temperature and wind velocity on the site.
- Placement: The method of curing and the time of application.

8.6 THE MIX DESIGN

The goal is to obtain well-compacted and dense concrete with a high fresh density of around 24 kN/m³ at supply. Lower densities, such as 23.50 kN/m³ or even as low as 22.80 kN/m³, are sometimes supplied and will meet the required strength but will exhibit much higher shrinkage.

Low concrete densities are often caused by excessive mortar content, too many fines smaller than 200 μm size, and even mineral dust used as a cement replacement. Too much water content can lead to an enormous amount of free water. Note that the cheapest constituents of concrete are the fines, including dust, and poor-quality sands.

The problem with these poor quality mixes is that they can still meet, and even exceed, the standard required strength to satisfy the engineer's needs. However, the slab's serviceability and durability may become very unsatisfactory, primarily due to overall shrinkage.

The human factor and the organization on the jobsite also critically influence slab shrinkage, factors that are not accounted for in the standards.

The evolution of cement from the 1940s to today has focused on meeting minimum strength requirements at early ages (1 day, 7 days, 14 days, and 28 days) but without clear upper limits. Modern cements have become more reactive to deliver higher strengths earlier, thanks to different chemical constituents and proportions in C3A, C2S, C3S, C4AF, and alkalis, along with finer grinding (less than 3000 Blaine in the past to 5000 Blaine and higher today).

As a result, it is now possible to supply a 25 MPa slab mix with as little as 200 kg of cement per cubic meter. However, to obtain a stable and workable slab mix that won't segregate during transit or on-site, other fines are used as cement replacements.

The inert fine-to-cement ratio then becomes too high, requiring more water and leading to a considerable increase in the free water content of the concrete. This free water is the type that evaporates easily and quickly, both from the plastic and hardened concrete, leading to very high shrinkage.

The free water content can increase significantly due to the use of excessive amounts of very fine non-hydraulic or "semi-hydraulic" by-products, which are often sourced at little or no cost.

A mix with 330 kg/m³ of cement (denoted as A-mix) with a wet unit weight of 24 kN/m³ at W/C = 0.55 contains 182 liters of total water, of which approximately 66 liters is free water, making it easy to evaporate.

A mix with 220 kg/m³ of cement (B-mix) with high Blaine (around 5500), likely high C3A content with high alkalis (at the high side, a rapid-setting cement), of a wet concrete unit weight of 23.40 kN/m³, with 110 kg or more of replacement fines (33% hydraulicity), will have 0.66 x 110 + 66 = 139 liters of free water. This is twice as much free water as A-mix, resulting in significantly higher shrinkage, which causes early cracking, joint opening, curling, and rocking at the joints.

The A-mix concrete has a dry concrete volumetric weight after shrinkage of 23.40 kN/m³, representing a 2.5% mass loss from the fresh state. This is replaced by the second mix, which has a dry volumetric weight of approximately. 22.50 kN/m³ after shrinkage, corresponding to a 6.25 % mass loss from the fresh state and 4.3% mass loss in the dry state after curing. The difference between the two different mixes is significant, with a ratio of 4.3/2.5 = 1.72, indicating a 72% increase in shrinkage attributable solely to the type and quantity of cement, as well as the nature of the replacement fines.

At the same water content, B-mix does not flow as expected. Therefore, more superplasticizers and water are added to increase the flow, but this also leads to a significant increase in shrinkage.

Bissonnette and Attiogbe in Reference (1) observed that "Slabs made with a flowable high-strength concrete produced the greatest curl heights and largest joint openings even though the standard drying shrinkage values were lower than those obtained for the plain concrete"

8.7 REINFORCING

The reinforcing rate and type play a significant role, but even with a standard design and certified concrete quality, bad cracking can still occur. One of the main advantages of steel fiber reinforcement, which is widely accepted today, is its ability to reduce crack opening when the right type and dosage of steel fibers are used.

Further in this chapter, the effect of friction and the steel fiber effect will be considered in more detail, including parameters like fiber shape, size, and dosage in concrete.

Regarding joints, experience shows that all types of joints create local weaknesses in the slab. While joints do not eliminate cracking, they tend to induce more cracks due to joint movement after shrinkage. Even the best-designed joints, which are meant to allow for full load transfer, free contraction movement, and a smooth ride on the slab surface, degrade over time after excessive overall shrinkage.

A well-known phrase used in ACI seminars summarizes this issue: "All floors are good between joints and cracks". What I have frequently observed is that when remedial work is necessary and Garber's conditions as outlined in Reference (1) are exceeded, cracks are easy to fix by filling or injecting them with liquid resin. However, joints are much more difficult, expensive, and time-consuming to repair, when repair is even possible.

It is important to note that, like joints, cracks will curl once their opening becomes too wide. The greatest attention when designing slabs should focus on the following:

– Minimizing the shrinkage content of the concrete
– Providing the slab with a sufficient percentage of reinforcing, both in terms of type and proportions
– Minimizing the number of joints to control the cracking pattern.

A controlled cracking pattern ensures that these cracks and joints will always need repair over time. The controlled cracking pattern means that it is not detrimental to the end-users so these cracks don't need any attention, maintenance, and remedial works.

Excessive shrinkage can affect the serviceability of the slab. In such cases, after lengthy discussions and potential litigation, the slab might even need to be replaced, which is an expensive process.

Sometimes, sales agents or owners may offer or wish for a crack-free floor. In reality, however, all concretes will crack. Saw cuts are still used to control random cracking in concrete slabs, but experience shows that joints do not control or prevent cracking. In fact, joints can even be a cause of spalling and cracking. Figure 8.1 shows typical saw-cut joints (also called shrinkage joints) with slight spalling.

With a low shrinkage, sawn cuts will degrade slightly as shown in Figure 8.1.

Non-detrimental cracks, not crack-free floors, represent the ideal solution for the user. A floor with non-detrimental cracks is fully serviceable under its maximum capacity, as these cracks and joints are practically unnoticeable: they do not cause chipping, spalling, or other types of degradation, and the surface remains smooth without rocking joints or visible defects. A "crack-free floor" is, therefore, a floor without any detrimental cracks.

From my 45 years of experience, I've seen many floors, both traditional and steel fiber reinforced, in excellent condition for even after 10 years or more. However, I've also seen many floors where joints and cracks became a serious issue, detracting from the user experience.

My 45 years of experience have taught me that industrial slabs should be designed with a primary focus on shrinkage rather than flexion and shear. Constituent materials of concrete, the organization of the job site, and the personnel involved in the installation process should all receive full attention. While a design that complies with standards can be technically correct, the finished slab may still fall short in terms of durability and serviceability.

As discussed in the earlier chapter, full-scale tests on various types of slabs have shown that traditional design methods for flexion and shear at the ultimate state grossly overdesign slabs.

While these designs are safe and ultra-conservative, they may not be optimal in real-world applications. Many slabs, even when designed according to standard provisions, exhibit distress such as uncontrolled cracking, excessive joint openings, curling at joints, at edges, or in doorways—despite the inclusion of additional reinforcing. Joint-induced cracking, caused by slab rocking, remains a common issue.

The root cause of these issues is excessive hydraulic shrinkage. If shrinkage could be eliminated, slab thickness could be reduced by half, as we'll explore further in this book.

Unfortunately, codes of practice, reports, and standards pay little attention to shrinkage, beyond specifying jointing and the water-to-cement ratio.

The practice of jointing concrete dates back over a century. Originally, concrete slabs were poured dry with little workability, and compaction was achieved manually by stamping the concrete.

Very small square areas were placed at a time. By one crew of 10 finishers in a single day, two dozens of square meters were placed between pieces of timber. The surface was finished by a granolithic topping of high cement content and finished by hand. Afterwards, the timber were not even removed so they were built-in forever. The main square streets in Bellefontaine Ohio, have been built accordingly in 1903. From that time, people believe that concrete needs joints.

Before concrete was widely used, roads and pavements were made with bricks, wood, or macadam, which consisted of well-compacted layers of crushed aggregates. The first continuous concrete slab was laid in the 1920s by a paving machine, an effort to speed up placement and eliminate construction joints and built-in timbers. Hence, later slabs were provided with saw cutting at 10 feet distance apart to localize the shrinkage cracks. Sawn cuts were man made cracks in a regular pattern.

Today, suppliers and engineers debate the ideal depth for saw cuts, the correct distance between cuts, when to use meshes or dowels across joints, and how soon after hardening the slab should be saw cut. The term "dominant joints" has even emerged in recent years to describe joints that are intended to control cracking.

Construction joints are widely discussed, and various types are available in the market, all designed to accommodate free shrinkage contraction movement while ensuring full shear transfer to prevent slab rocking. These joints aim to manage shrinkage and maintain the integrity of the slab under movement.

However, the reality is that no matter how advanced these joint solutions are, or the type and percentage of reinforcement used in the concrete, they are ineffective if shrinkage is not properly controlled and becomes excessive. Despite the high costs of supplying and installing these joints, they fail to prevent cracking when shrinkage is allowed to run unchecked. Figure 8.2 shows the chipping and spalling that occurred at the surface of concrete, even though a high-quality, sophisticated, and heavy construction joint with a sinusoidal shape was used. This type of joint, designed to prevent such issues, was unable to stop the deterioration. In addition, spalling and chipping-out are clearly not prevented.

The photo in Figure 8.2 was taken at a logistics center in South Africa, where the concrete shrinkage content appears to have been excessively high. Figure 8.3 shows the same type of joint after repairs were made, but the underlying issue of excessive shrinkage still persists.

Figure 8.4 shows an armored joint and its typical spalling in a bad case.

8.8 CRACKING AND DURABILITY

Plastic shrinkage cracks occur very early before the concrete has fully hardened.

Figure 8.2 Example of heavy sinusoidal shape joints failure.

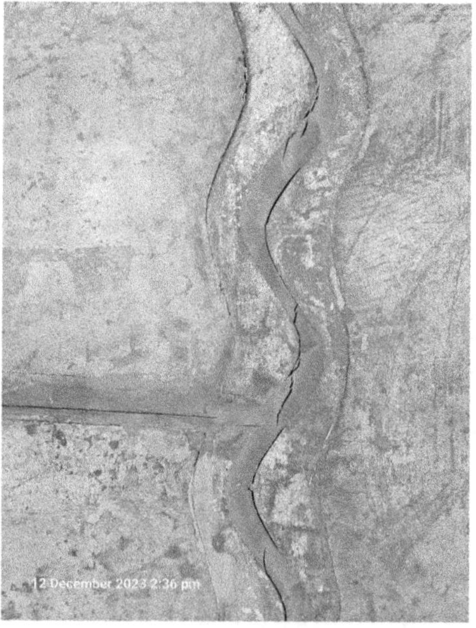

Figure 8.3 Same sinusoidal shape joint after repairation (in a Distribution Center in the Middle East region).

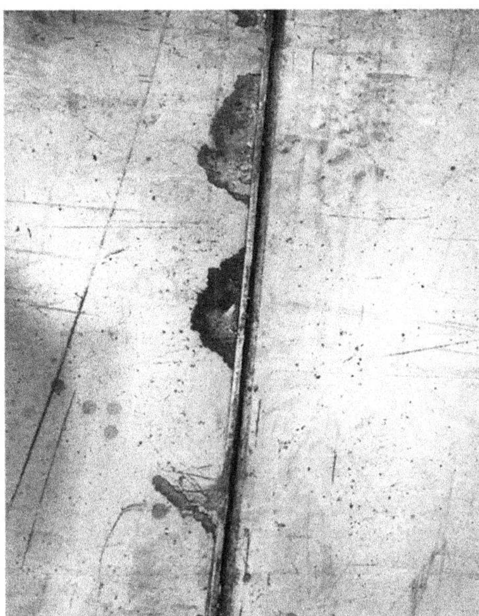

Figure 8.4 Spalled armored joint.

Hardened concrete cracks, on the other hand, are typically caused by drying shrinkage, thermal contraction, or structural factors such as bending, shear, or torsion. These types of cracks are less common in slabs on-grade.

Crazing cracks are shallow, fine cracks that are closely spaced, typically only 5-10 mm deep and with openings between 0.05 mm to 0.1 mm. These cracks result from surface shrinkage, which is restrained by the concrete beneath. Crazing is rarely a critical issue but can become problematic in areas with high traffic, where it may affect the surface appearance and functionality, as shown in Figure 8.5, resulting from a wrong mix design with excessive shrink of the hardener layer.

Such a crazing can show distress in the aisles after some usage.

Figure 8.6 is a sign of excessive shrinkage where the edge is subjected to local curling.

8.9 EXCESSIVE SHRINKAGE CONTRACTION AROUND A COLUMN

Figures 8.2, 8.3, 8.4, and 8.5 depict photos of SFRC slabs that comply with the TR34 4th Edition guidelines for "jointfree" construction, with approximately 35 meters between sinusoidal construction joints. Despite this compliance, the poor results seen in these photos were not prevented by adhering to the TR34 provisions. This indicates that the shrinkage issue was not adequately addressed during both the design and installation phases.

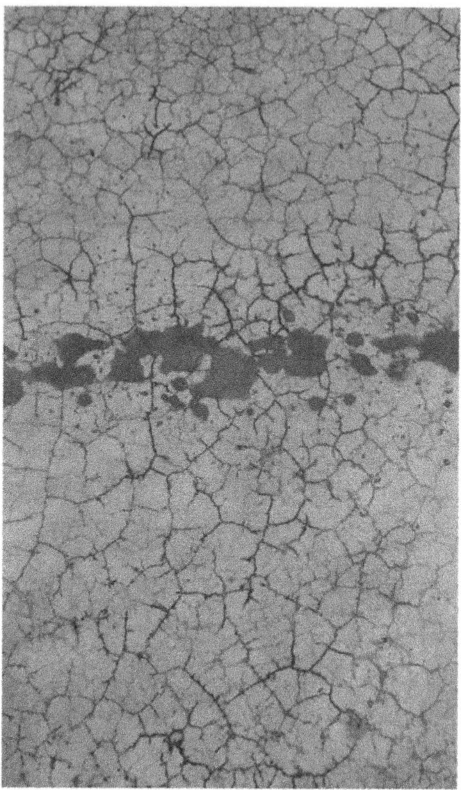

Figure 8.5 A case of critical crazing (more than 0.1 mm crack opening in top).

While cracking is inevitable in concrete, it is essential to differentiate between what is considered acceptable and what is not. Shrinkage remains the most common cause of cracking in both plain and reinforced concrete.

According to standard design practices, traditional reinforced concrete typically accepts cracking up to 0.3 mm to 0.5 mm in opening. However, in reality, cracks exceeding 1 mm in width are not uncommon.

As shown in Figure 8.7, typical cracks are quite higher than in slabs, which is depicted in a crash barrier wall and the bridge slab of a Singapore road next to the Marina Bay tower.

The closely spaced cracks suggest a high percentage of reinforcement, despite crack openings being placed well over 1 mm.

Cracking and durability are primarily serviceability concerns. Many slabs can become nonfunctional, and thus no longer serviceable, well before reaching the ultimate limit state (ULS) or collapse.

In industrial applications, cracks are rarely an aesthetic concern. Functionality takes precedence over appearance. A serviceable slab is one that remains functional despite the presence of cracks. All types of industrial

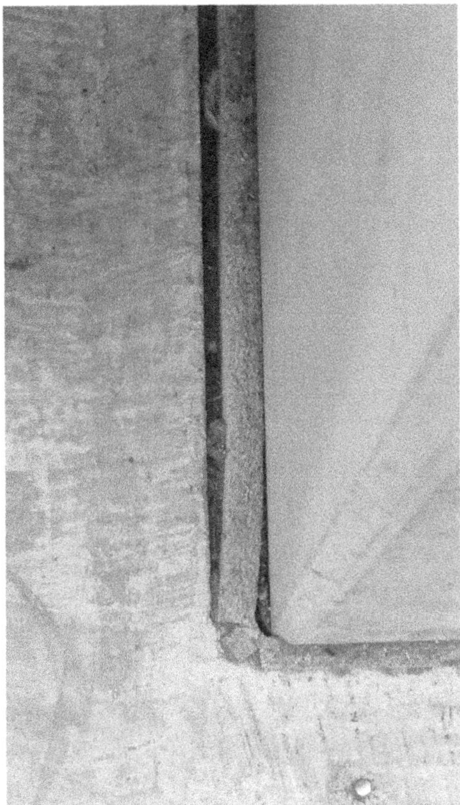

Figure 8.6 Large shrinkage contraction movement around a column.

concrete slabs will exhibit some cracking, regardless of the type and amount of reinforcement used. Even post-tensioned slabs, which are often assumed to be crack-free, can show cracks. This does not indicate failure of the post-tensioning technique, as noted by K. Bondy in Reference (2). According to Bondy Reference (3), restraint to shortening (RTS) is a major cause of cracking in post-tensioned structures.

The acceptable crack opening limit in industrial slabs is well-defined by G. Garber in "Design and Construction of Concrete Floors" Reference (4; p. 214):

"Many cracks - perhaps most - do not harm. Cracks rarely need repair if hidden under floorcoverings or exposed only to foot traffic. Even in an exposed industrial floor, a crack generally needs not to be repaired unless one or more of these conditions:

Its width exceeds 1 mm and it is exposed to hard-wheeled traffic.

It spalls under traffic, regardless of width.

It shows differential movement under traffic."

Figure 8.7 Shrinkage cracks of 1 mm in a structural crash barrier.

From our experience, a useful rule of thumb is that the depth of a slab shrinkage crack does not exceed 150 times its surface width. For example, a 0.5 mm wide crack in a 200 mm thick steel fiber reinforced concrete (SFRC) slab will not extend to the bottom of the slab.

We should also consider the EN 14651 flexural strength, where the post-cracking flexural strength is defined by either fr_3 or fr_4, corresponding to a 2.5 mm or 3 mm Crack Mouth Opening Displacement (CMOD). This value is used to determine the ultimate limit state (ULS) resistance in the FIB 2010 Model Code, the TR 34 from the U.K. Concrete Society, and the upcoming Eurocode, which is expected to be released in 2027 at the earliest.

Thus, cracking does not necessarily compromise the ULS of a fiber-reinforced slab design as long as Garber's limits are not significantly exceeded. Cracking is mainly a serviceability concern, which Garber addresses effectively. If any of his three conditions are exceeded, the slab's serviceability is hampered, and remedial work may be required.

In the Eurocode, the general upper limit of cracking in traditional reinforced concrete is 0.3 mm. However, slabs on grade, unlike slabs on piles, are not considered as reinforced concrete applications.

The Euro Code defines a maximum crack width of 0.3 mm for traditional reinforced concrete (0.41 mm in ACI 224) in dry air, accounting for detrimental effect of carbonation and the rebars corrosion. However, this doesn't apply to SFRC. It has been demonstrated that in a very aggressive chloride ions environment—conditions never found in industrial applications—the steel fibers do not corrode for long with a 0.5 mm crack opening. This is confirmed by research such as Lambrechts and Nemegeer Reference (5), which shows that steel fibers remain unaffected even under severe conditions with a chloride concentration of 3 g Cl⁻ per kg of concrete.

It is well known that steel fiber reinforcing reduces the crack width, thereby reducing the durability by preventing the ingress of aggressive elements.

Over my 40 years of experience, I have not encountered a single instance of SFRC slabs exhibiting distress due to excessive cracking or corrosion.

During mixing and compaction of concrete, steel fibers float freely in the mix as opposed to fixed rebars where voids and channels develop on top: steel fibers enjoy a denser interface hydration products as in Reference (6) of Raupach and Dauberschmidt.

In aggressive environments, steel fiber reinforcement has a clear advantage: only the fibers exposed at the surface may corrode, as the fibers are discontinuous within the matrix. Corrosion does not spread from one fiber to another. Moreover, the fiber-matrix interface does not offer a preferential path for chloride ingress. The uniform steel surface of cold-drawn wire, the material used for type I steel fibers, significantly delays the initiation of pitting corrosion compared to traditional rebars.

Short fibers have a smaller potential difference between the anode and cathode at their extremities than rebars, meaning corrosion is unlikely to develop unless the chloride concentration is 5 to 10 times higher than usual, which is highly improbable. Even if corrosion does occur at the single fiber level, it is minimal and does not cause the surrounding concrete mortar to break apart. The corrosion remains superficial and is typically an aesthetic concern, which is negligible in industrial applications.

Regardless of the case or the use, the durability of the SFR Concrete is limited by the durability of the matrix which should be of a W/C, water-to-cement ratio, anyhow smaller than 0.50 or less, and the compaction.

The 0.5 mm crack limit is also given as it is not clear that beyond that limit, the concrete matrix could take advantage of any self-healing.

In conclusion, the SFRC is more durable than plain and traditional reinforced concrete due to its superior ability to control cracking.

The Swedish standard (SS812310 – 2014) details are shown in Table 8.1, typical acceptable SFRC crack opening in function of the environments regarding structural and semi-structural members for 50 and 100 years life expectancy:

For the understanding of everybody, the Eurocode categories of exposition classes are shown below:

Exposure classes according to Eurocode 2 (EN 1992)

X0: No risk of corrosion or attack.

XC: Corrosion by carbonation
XC1: Dry or constantly wet
XC2: Wet, rarely dry
XC3: Moderate humidity (open halls), damp rooms.
XC4: Alternate wet and dry: structures in water change zones
XD: Corrosion caused by chlorides.
XD1: Moderate humidity, like in spray mist area of traffic areas
XD2: Wet, rarely dry, like swimming pools or industrial wastewater with chloride.
XD3: Alternating wet and dry: road surfaces, parking decks.
XS: Corrosion by seawater
XS1: Salty air, no contact with seawater
XS2: Underwater
XS3: Tidal area, splash water, and spray areas.

For industrial slabs inside, X0 is the prevailing exposure class. In most cases, if a cracking limit is specified, it should be clearly stated in the contract, for example, a 1 mm limit (Garber) or less for other reasons. Experience shows that in distribution center warehouses with heavy forklift truck traffic, SFRC cracks with openings under 0.6 mm do not degrade.

Obtaining such a limit in slabs requires focusing on the mix design and the slab installation process. The fiber type and dosage rate in the end result contributes to only 42%, while all the concrete aspects matter for 43%, with the rest being an indeterminate noise, according to a very recent study by E. Garcia-Taengua (University of Leeds, UK), who performed data mining and artificial intelligence analysis on several thousand types of SFRC in Reference (7).

The final satisfactory result for the slab end-user can't be met by just adding the "miracle" fiber in a mix and installation process that doesn't minimize the total shrinkage content.

The Swedish standard is already very stringent as it is applicable to all types of structures, not only for slabs. The L50 (50-year-long exposure) goal doesn't however apply to most industrial slabs, where 10 or 20 years at most matters practically, sometimes even less.

There are a number of SFRC slabs under XD3 conditions that perform perfectly after 20 years and longer usage, like for open-air gas stations and sea ports' intermodal pavements. Cracking in these applications has not been an issue in all cases where the shrinkage content of the final concrete has been taken care of from the beginning to the end of the building process.

Under XS3 conditions, there are numerous examples of SFRC dolosses that perform ideally after 30 or 40 years of usage.

8.10 REBAR CONCRETE COVERAGE

It is recommended to cover the rebars under a minimum thickness of concrete of 25–35 mm up to much deeper in case of a more aggressive environment, as shown in Table 8.2:

For example, a marine structural concrete reinforced by rebars of 40 mm diameter should have a 135 mm coverage thickness in marine environment application.

Despite the traditional reinforcing, the cover concrete can shrink and crack.

The coverage concrete should, however, be reinforced as well to prevent the wild cracking of plain concrete, as can be seen in massive pillar application where horizontal cracks are so severe that they are visible from 30 m away and more.

Steel fiber reinforcing is the best option, even at moderate dosage rate of 25–30 kg/m³ (41–50 lbs/cu.yd), to control such a cracking with less coverage.

8.11 CURLING OF SLABS

It is not the intention here to rewrite the general knowledge about curling, which is very well explained by at least two authors, R. Ytterberg in "How to deal with warping and curling" and G. Garber in Chapter 14 of his handbook "Design and Construction of Concrete Floors."

However, we can remind the reader of the essential facts derived from experience:

— Curling results from larger shrinkage at the surface of the slab compared to the bottom, as the surface alone is subjected to free water evaporation, unlike the bottom. The larger contraction at the surface tends to curl up the slab near joints.

Table 8.1 Recommended values of w_{max} for fiber concrete only with respect to durability

Exposure class	L50 $wmax^a$ mm	L100 $w_{max}{}^a$ mm
X0, XC1	$-^b$	$-^b$
XC2, XC3	0,5	0,4
XC4	0,4	0,3
XS1, XS2, XD1, XD2	0,3	0,2
XS3, XD3[c]	0,2	0,1

a   For members with a combination of fibers and conventional bar or prestressed reinforcement, refer to SS-EN 1992-1-1 for allowable crack widths based on exposure conditions.

b   For X0 and XC1 exposure classes, crack width does not influence durability. Therefore, the limit should be set to ensure acceptable appearance and control of deformations.

c   In exposure class XS3 or XD3, corroding steel fibers must be combined with conventional bar reinforcement for suspended decks, slabs or beams.

- Curling is counteracted by the own weight of the slab, so it stops at some distance away from the joints.
- Load-transfer devices within joints can limit curling but rarely eliminate it.
- More shrinkage content in the concrete leads to more curling development.
- Curling affects flatness and levelness of the slab with time, which is why flatness and levelness checks are generally performed the day after the slab is installed, and later when needed.
- Curling is a cause of severe cracking at some distance from the joints. Cracks caused by curling are difficult to repair.
- Curling-induced cracks will also curl later.
- The final curling displacement is not dependent on the slab's thickness: thicker slabs curl as much as thinner ones but take more time to achieve it, as the shrinkage gradient decreases with increased thickness.
- Higher reinforcing percentages reduce or even eliminate curling.
- Steel fiber (Type I) reinforcing reduces the amount of curling.
- Two-way post-tensioning can eliminate curling.
- Best curing methods only delay curling.

The addition of expansive cements, like some of the calcium sulfo-aluminate (CSA) types, in combination with different powders, can eliminate or reduce curling to a non-detrimental.

8.12 SFRC JOINTFREE FLOORS HISTORY AND EXPERIENCE GAINED

Since 1983, based on an international patent in reference (8) (Destrée, Eurosteel SA, a very large number of jointless (indeed without any sawn cuttings) SFRC ground-bearing slabs have been completed. From the mid-eighties onwards, 2500 m² to 3000 m² SFRC bay-sized slabs (by Silidur SA, Brussels) have been built with a 330 kg/m³ CEM I content, a water-to-cement ratio below 0.55, and 1 mm diameter undulated, 1200 N/mm² steel fibers (Eurosteel type) in a 40 kg/m³ dosage rate.

The first ever SFRC jointfree slab was constructed by SILIDUR SA for "Keystone Valves" in Holland at Breda-Noord in 1983. It was an 1800 m² slab, 150 mm thick, made of C 25-30 with a W/C ratio of 0.53 and 30 kg/m³ of 0.8 mm diameter x 60 mm length steel fibers made from 1100 MPa tensile strength steel wire. The base had a k-value of 0.03 N/mm³ with a compacted sand bed, designed to handle a 30 kN/m² loading capacity.

Over time, these floors exhibited some cracking, including minute cracking in large numbers, but without the need for concern, and mostly without any complaint from the warehouse tenant or owner. It was a practical demonstration that steel fibers are effective in reinforcing concrete, controlling shrinkage over distances as long as 50 meters or more. It was also discovered that the construction joint openings between adjacent bays do not increase in proportion to the length of the bay.

By the early nineties, some significant players in SFRC tried to infringe the patent but were stopped by a Paris court of appeal. By the end of the nineties, the provisions of the SFRC jointfree slab patent technique had become commonplace, particularly in Europe.

The next step came in 1993 with the completion of the first-ever built ground-suspended SFRC slab for the Belgian Post. This slab covered an area of 20,000 m² with a 200 mm thickness, constructed in 3000 m² bay-sized phases. The mix used was 350 kg/m³ concrete with a W/C ratio of 0.55, and it incorporated 40 kg/m³ of Twincone-type steel fiber reinforcing. The slab was supported over a 4 m × 4 m pile grid.

Based on the experience of the ground-bearing jointfree slabs, SFRC jointfree suspended slabs on piles have also been used since 1993, as shown in References (9, 10) by Destrée.

By 1995, approximately half a million square meters of SFRC slabs had been successfully completed in Belgium and Holland. This marked the first massive structural application of SFRC as the sole reinforcement, something I'm still proud to have invented.

This slab type spread globally, starting in Holland and Belgium, and reaching the UK, and later expanding further west and east after the 2000s. The goal was not to fully eliminate cracking but to control it. These SFRC "jointfree" floors have remained serviceable throughout their life cycle, with the crack width being carefully controlled to maintain desirable serviceability under heavy traffic.

Since then, jointfree floors without saw cuts have been built for nearly 40 years, with bay sizes reaching up to 4000 m². Today, the TR34 4th Edition limits the joint distance to typically 35 meters to keep the construction joint opening under 15 mm. However, this doesn't always hold true as it depends on various physical and chemical parameters not fully covered by the document.

Although SFRC jointfree floor systems are widely used, they have mostly been adopted by heavy-duty users.

Such a jointfree SFRC floor requires the integration of optimum constituent materials and processes to limit overall slab shrinkage and make cracking predictable.

Such a challenge was also summarized in 1983 by Destrée in the US patents (**ref 19**) and later by Alexandre and Bouhon in 2010 Reference (11), former colleagues of mine, based on their American experience.

However, existing methods of calculating crack opening in SFRC apply only when traditional reinforcement, such as rebar or wire mesh, is included alongside steel fiber reinforcement in the slab.

In this chapter, the reader will find a comprehensive, evidence-based method to predict the crack opening in SFRC slabs that are solely reinforced with steel fibers, for both ground-bearing and piled applications.

The main cause of cracking is the restraint associated with shrinkage. The method outlined here addresses the role of drying shrinkage in cracking, specifically the vapor water loss from fresh concrete. Plastic shrinkage cracking, or thermal shrinkage resulting from high-temperature differences, is not considered here. Ground settlement or swelling also contributes to cracking, but they are outside the scope of this chapter.

Several factors adversely affect the drying shrinkage of a slab, including excessive water content, high mortar content in concrete, use of high-shrinkage cements, and the use of inappropriate admixtures. Aggregate-related issues often arise from using dirty, flaky aggregates or those with high intrinsic shrinkage. Such issues are easily avoidable with proper attention to mixed design and constituent materials.

The amount of drying shrinkage is also influenced by the slab construction process, from the concrete plant to the final finishing and curing. Excessive concrete temperatures, long hauls in the transit mixer, adding water on-site, and failure to protect the slab from wind and air currents that cause high evaporation rates all contribute to excessive shrinkage.

A successful jointfree floor requires low-shrinkage concrete and no movement restraint along the edges, around columns, or at re-entrant corners. The length of the slab between construction joints also increases friction stresses, which can raise the probability of further cracking.

Steel fibers influence drying shrinkage based on parameters such as diameter, length, bond type, dosage rate, and strength. These factors affect fiber-matrix anchorage, crack opening, and load transfer across matrix cracks. The steel fiber type should be precisely specified in terms of shape and size, with an aspect ratio of no less than 50 and a dosage rate of at least 30 kg/m³.

Today, hybrid design types are used for slabs on grade, which include a top mesh (8 mm diameter by 100 mm or 150 mm spacing) in an attempt to better control cracking, given that modern cements have significantly changed since 1983.

8.12.1 The drying shrinkage

The present approach is based on key parameters identified through a historical perspective of numerous publications spanning the last 50 years. These documents detail the factors influencing drying shrinkage and the magnitude of their correlation with shrinkage, as outlined in References (12, 13). It is well-known that parameters such as water content, aggregate type and volume, cement quantity, and mix workability are interrelated and can influence each other. A simplified summary of these parameters and their impact on concrete drying shrinkage, based on experience, reported values, and conventional wisdom, is listed below:

- Excessive water content in the concrete can cause a 50-100% increase in shrinkage. The excess water also increases the porosity of the concrete, leading to faster and more pronounced vapor water loss.
- Low aggregate content in the mix design can lead to up to a 100% increase in shrinkage. When the aggregate ratio is low, the sand and fine content of the mix becomes higher, thus increasing water demand.
- Aggregates with low stiffness result in higher shrinkage, as they contract more easily than higher-stiffness aggregates. The difference in shrinkage can be as high as 30–50%.
- Excessive dirt in aggregate can cause up to a 25% increase in shrinkage. This impacts the aggregate-paste bond and the load transfer between the two phases. The use of smaller aggregate sizes can also increase shrinkage by up to 25%.
- Different cement types (I, II, and III) do not seem to significantly influence drying shrinkage. However, plastic shrinkage is more influenced by cement type, though it is not discussed here as the concrete mix design for jointfree floors typically uses CEM I or II with limited supplementary cementitious additions.
- Cement content is not discussed in detail, as it does not vary significantly from floor to floor. Cement contents generally range between 300 and 350 kg/m³ of CEM-type I, depending on the floor type and installation, while the concrete strength typically falls in the range of C25-30 or C30-37.
- Air content does not appear to affect shrinkage and is generally not recommended for power-troweled concrete floors, as it can lead to delamination. Therefore, it is not discussed here.
- Superplasticizers or high-range water-reducing admixtures can increase shrinkage up to a maximum of 35%, especially when the water content is not effectively reduced. This is a critical issue when the water content remains high despite the use of High Range Water Reducing Admixture.

The influence on the crack opening, of each of the above parameters as well as the friction from the grade, has been investigated in depth by Destree, Yao, and Mobasher in References (14, 15).

I believe that the unit weight of fresh concrete upon arrival at the job site is the simplest parameter to control the supply quality and understand instantly the potential to shrink. More fines and water further reduce the unit weight of the concrete, dropping it to 23.4 or 22.7 kN/m³.

However, there is no literature available that specifically addresses the influence of the unit weight of fresh concrete on drying shrinkage.

Jointfree floors should be made from optimally compacted concrete prepared according to state-of-the-art methods, with a concrete density of up to 24 kN/m³. When the density decreases to 23.5 kN/m³ or lower, it may indicate several factors—discussed earlier—that are out of control and can adversely increase shrinkage.

8.12.2 Slab installation

Slab installation parameters are not taken into account here, as it is assumed that jointfree floors are installed by experienced professionals using the best practices available. However, slab installation parameters that can adversely affect shrinkage include:

- Excessive air and concrete temperature
- Excessive haul time in the transit mixer
- Slab installation under windy or sunny conditions without direct protection

The combined effects of these adverse influences, along with the parameters related to the mix design and installation, can result in a possible 400% increase in shrinkage, as noted in Reference (16) of Bissonnette, Attiogbe, E.

It is well known that steel fibers in concrete can reduce free shrinkage contraction movement by up to 15% and delay associated cracking. When randomly distributed steel fibers are sufficiently close to each other, they form a bridge between aggregates, preventing the mortar phase from debonding or micro-cracking. As a result, the micro-cracking stage of concrete can develop much further along the shrinkage time history before a crack becomes visible at the surface.

Long fibers are not advantageous because they complicate the mixing process. Shorter fibers with an anchoring deformed shape have been the preferred type for almost 40 years, as they combine workability, stress transfer, and better post-cracking performance. Experimental results for the maximum crack width show a reasonable correlation with the product of fiber concentration and aspect ratio, as noted in References (17, 18).

There fiber anchoring shapes are available today, including hooked-ends, undulations, and conical ends. The anchoring is enhanced by increasing the rigidity of the fiber section, especially when the steel wire strength is increased. The shape and anchoring of the steel fibers also influence shrinkage.

The literature in Reference (17), dating back 25 years and beyond, shows that the free contraction movement of a slab can be reduced by 20% using steel fibers. Restrained shrinkage cracking is improved significantly compared to plain concrete, as the crack width can be reduced by a factor of 5, while the number of cracks increases. This is because steel fiber reinforcement provides effective crack distribution properties.

Steel fibers offer internal restraint, resulting in a reduction of free shrinkage. Mangat and Azari, in Reference (17), define a μ_f parameter that is typical of each fiber geometry. The coefficient μ_f varies from 0.04 for straight fibers to 0.08 for hooked fibers, and to 0.12 for undulated fibers. However, this variation does not imply that undulated fibers perform three times better than hooked fibers at controlling cracking. The influence of μ_f must be considered in context, as very successful jointfree floors have been completed with both types of fibers.

8.12.3 Slab Restraint by friction

Shrinkage stresses develop in a slab that is restrained and cannot accommodate free contraction movement. The degree of restraint can vary significantly due to static friction caused by the weight of the slab, the friction coefficient value, and, to a significant degree, the extent of the interlock mechanism, which results in a relatively high friction coefficient. These parameters are functions of the base preparation conditions.

Concrete floor cracks are most often the result of shrinkage restraint and much less frequently due to flexural moments caused by point loads. Most concrete floors are exposed to a constant indoor climate, so cracks due to temperature variations are infrequent.

When the slab is allowed to move freely along its edges, the degree of restraint depends on the slab length, which is defined as the distance between joints. This indicates that smaller joint spacing will inherently reduce shrinkage cracking. The effect of slab thickness is considered in terms of a thinner slab, which undergoes shrinkage more rapidly than a thicker slab, producing more friction force between the slab and the grade per unit thickness.

The friction variations caused by the degree of restraint result in localized stresses, as discussed by Silfwerbrand and Malmberg in References (19, 20).

The degree of restraint for various slabs on grade can vary from 0.25 to 0.60, representing a change as large as 240%.

Estimating and/or measuring frictional forces is difficult, and some designers recommend using a smooth separating sheet installed under the slab on top of the grade. However, the grade is not a perfectly horizontal surface with the same mechanical characteristics everywhere. Therefore, it is highly unlikely that simply installing smooth sheeting between the slab and the grade could sufficiently isolate the slab and the ground support.

The service load applied to the slab will prevent it from gliding freely onto the base, so the slab and its base form a two-elastic-layer system, where each layer resists the stresses depending on its modulus of elasticity. The higher the modulus of elasticity of the grade, the more restraint it will provide to resist the slab's movement. Field trials have supported the observation that the joint opening of a jointfree floor tends to be smaller when the support is stiffer, as indicated by a higher modulus (also known as the Westergaard k-value) in Reference (21).

The modulus of elasticity of the grade is not easily measured, but it can be empirically related to the Westergaard modulus k when the recorded settlement is back-calculated according to both the Westergaard elastic liquid base theory and Boussinesq's elastic E modulus theory. For example, using the DIN 18134 test method, the E modulus of the grade can be tested using a disk with radius R, and the Poisson coefficient υ, from which k is calculated as follows:

$$\beta := \frac{\pi \cdot R}{2 \cdot \left(1 - \upsilon_G^2\right)} = 0.314 \, \text{m} \quad k := \frac{E_G}{\beta} = 0.032 \frac{N}{mm^3} \tag{8.1}$$

In case of R = 150 mm, ν_G = 0.5, and E_G = 10 N/mm2, the equivalent k = 0.03 N/mm³ is obtained. A correlation of the slab restraint parameter μ_s as a function of the modulus k has been proposed as in 14 where u are the units of k:

$$\mu_S := 4.60 + \ln\left(\frac{k}{u}\right) \tag{8.2}$$

So that in the case here,

$$\mu_S := 4.60 + \ln\left(\frac{k}{u}\right) = 1.153 \tag{8.3}$$

The factor μ_s is still positive for k = 0.03 N/mm³ as a minimum value. Indeed, the logarithm function expresses the range of friction values as directly related to the elastic modulus of the support. The restraint increases by a factor of two between k = 0.03 N/mm³ and k = 0.1 N/mm³, which is similar in nature to the experimental work of Silfwerbrand in Reference (20).

8.12.4 A crack-opening model

We still need to review the influence of water content combined with superplasticizer (HRWRA). Too low W/C ratios are impractical in jointfree concrete flooring, as a large volume of concrete needs to be installed, compacted, and finished within 16 hours or less, all inside a closed building where the use of paving machines is not feasible.

A minimum fluidity of the concrete is necessary, along with a limitation of the water content to reduce overall shrinkage. Thus, the use of HRWRA is highly recommended. Sometimes, pumping concrete is required, which further increases the need for the concrete to flow.

The potential for cracking in concrete is expressed by its hydraulic shrinkage, $\varepsilon r = 2.10^{-4}$ (or β, 10^{-4}) for a reference mix made from ideal constituent materials and the same cement type and quantity. This coefficient β applies when 0.45 < W/C < 0.65. In reality, the W/C ratio is often closer to 0.55 or 0.65, or even higher values due to the factors mentioned above, causing β to increase from 2 at W/C = 0.55 to 3, and up to 5 with W/C > 0.65. High W/C ratios lead to a dramatic increase in free water content, which is the main driver of drying shrinkage in slabs.

The following expression has been proposed to obtain ε_r:

$$\varepsilon_R := \left(20 \cdot \frac{W}{C} - 8\right) \cdot 10^{-4} \tag{8.4}$$

The effect of the superplasticizer (HRWRA) when the water content is not reduced is translated by a coefficient λ that is increased from 1.05 to 1.15 and to 1.35 when W/C > 0.65. The following expression is then proposed to obtain λ:

$$\lambda := 1.75 \cdot \frac{W}{C} + 0.2125 \tag{8.5}$$

The distance S between two joints is the first important parameter in the crack opening expression O and the units are in mm.

The following expression enables to predict what likely shall be the crack opening of a SFRC jointfree slab:

$$O := S \lambda \varepsilon_R \frac{(4.60 + \ln(k))}{V_f \frac{1}{d}(1+\mu_f)^{2.5}} \tag{8.6}$$

where S is the distance between two construction joints in mm and V_f the volume percentage of fiber dosage so that fiber dosage rates of 25, 40, or 50 kg/m^3 correspond to

$$V_f = 0.32\%, \ 0.51\% \text{ or } 0.64\%.$$

As $\lambda \times \varepsilon_R$ is, as already shown, a function of W/C, the crack opening expression is modified so that only S (distance between the construction joints), W/C (the water to cement ratio), k (the Westergaard plate bearing coefficient), V_f (the volume percentage of steel fibers), l/d (the aspect ratio of the used fiber), and μ_f (the friction coefficient of the fiber type) are needed:

$$O := S \cdot \left[35 \cdot \left(\frac{W}{C}\right)^2 - 9.75 \cdot \left(\frac{W}{C}\right) - 1.7\right] \cdot$$

$$10^{-6} \cdot \frac{(4.60 + \ln(k))}{V_f \cdot \frac{1}{d} \cdot (1+\mu_f)^{2.5}} \tag{8.7}$$

8.12.5 Examples

A large number of floors were inspected on several occasions. To summarize the results in Table 8.3, three separate slabs were inspected, and the crack openings throughout the area were measured. The results were compared to the predictive model presented here.

Case A was a slab in Poznan, Poland, constructed for a superstore. The slab covered an area of 7000 m², consisting of jointless bay sizes of 36 m x 36 m, with a thickness of 200 mm and using a C25-30 mix design. The specified steel fibers were 25 kg/m³, with a length of 35 mm and a diameter of 0.75 mm, hooked end. The slab was installed onto a well-compacted sand base. The construction joint opening between the jointless bays was 15 mm. Shrinkage crack width measurements were taken after the slab had been in service for 2 years. The observed crack openings ranged from 0.5 mm to 0.70 mm. The total length of the shrinkage cracks was less than 30 m across the entire slab. The following calculation was obtained: O = 0.702 mm, which is in very good accordance with the actual opening recorded on-site.

Case B involved another investigated slab, where the tenant is CLB Brussels. The slab area was 26,000 m², with 40 m x 40 m jointless bay sizes and a thickness of 150 mm, using a C25-30 mix design with 35 kg/m³ of 60 mm length by 1.00 mm diameter hooked-end steel fibers. It was installed on a well-compacted sand base. The joint opening between the bays was 7 mm. The model presented here predicts a crack opening of 0.709 mm, which aligns well with the observed 0.65 mm crack opening recorded during the inspection.

Case C is the third investigated slab in Bornem, Belgium. The slab covers an area of 46,000 m², with 50 m x 50 m jointless bay sizes and a thickness of 180 mm, using a C30-37 mix design with 40 kg/m³ of 54 mm long by 1.00 mm diameter steel fibers with conical heads. The slab was installed over a grid of piles with 2.25 m x 2.25 m spacing. The joint opening between the jointless bays was 9 mm. The measured crack opening was up to 0.8 mm, which compares well with the model output of 0.723 mm. The calculated maximum crack opening is also in good accordance with the measured values.

Table 8.2 The minimum concrete coverage of the rebars according to ACI 357. The National Ready-Mixed Concrete Association Recommendations and the Euro Code

Exposition	Dry/general	De-icing salts	Marine
Maximum aggregate size	Minimum coverage	Minimum coverage	Minimum coverage
<20 mm	37 mm	50 mm	63 mm
>20 mm	57 mm	70 mm	102 mm
EuroCode: D, diameter of rebar	D + 10 mm	D +35 mm	D + 55 mm

Table 8.3 Field experience measurements of cracking and joint opening

	Case A- Poland (ground bearing slab)	Case B – Brussels (ground bearing slab)	Case C – Bornem (piled suspended slab)
L, mm (bay size)	36000 x 36000	40000 x 40000	50000 x 50000
L_f, mm (fiber length	35 mm	60 mm	54 mm
D, mm (fiber diameter)	0.75 mm	1.0 mm	1.0 mm
μ	0.06	0.06	0.12
K	0.05	0.05	0.1
V_f	0.32	0.45	0.51
β	2	3	2
ε_R	$\beta \times 10^{-4}$	$\beta \times 10^{-4}$	$\beta \times 10^{-4}$
λ	1.05	1.15	1.15
Computed Crack width, mm	0.702 mm	0.709 mm	0.723 mm
Measured crack width, mm	0.5-0.7 mm	0.65 mm	0.8 mm

8.13 CONCLUSIONS

The prediction of shrinkage crack opening in SFRC jointfree slabs on grade and in piled suspended slabs is now possible using the model proposed in this chapter.

The parameters used by the model include concrete shrinkage, fiber shape and geometry, the volume percentage of fibers, the k-value derived from the Westergaard plate bearing test, and the distance between construction joints. The model is applicable under the assumption that the slab has been laid according to the current state-of-the-art methods.

This model is based on practical measurements of cracking in completed projects over the last 25 years, incorporating the parameters as defined in the literature.

REFERENCES

1. B. Bissonnette, E. Attiogbe, M. Miltenberger, C. Fortin. "Drying shrinkage, curling, and joint opening of slabs on ground." *ACI Materials Journal*, Vol.104, 257–269, May–June 2007.
2. K. Bondy "Cracking in ground-supported post-tensioned slabs on expansive soils." *PTI iTechnical Notes*, Issue 6 August 1995.
3. K. Bondy. "Shrinkage-compensating concrete in post-tensioned buildings." *Structure Magazine*, p. 18 April 2010.
4. G. Garber. "Design and Construction of Concrete Floors", Dec.2019, CRC Press, Taylor and Francis Group.
5. A. Lambrechts, D. Nemegeer, I. Vanbrabant, H. Stang. *Durability of steel fiber reinforced concrete*, SP-212-42, p. 667.

6. M. Raupach, C. Dauberschmidt. "Investigations into the critical corrosion-including chloride content of steel fibres in artificial concrete pore solution." *Materials and Corrosion*, 53, 408–416, 2002.

7. E. Garcia-Tengua. "Using decades of data to rethink proportioning and optimization of FRC mixes: the optiFRC." *BEFIB*, 827–838, 2020. Rilem-FIB symposium on FRC.

8. X. Destrée and A. Lazzari. "Industrial Floor and construction Method." US Patent 4,640,648, 1983.

9. X. Destrée. "Structural application of steel fibre as principal reinforcing: Conditions -design-examples." *RILEM PRO*, 15, 291–298, Fibre Reinforced Concretes BEFIB 2000, Lyon/ France.

10. X. Destrée, "Twincone SFRC structural concrete, fiber-reinforced concrete modern developments." *Proceedings of the Second University-Industry Workshop, UBC*, pp. 77–86.

11. E. Alexandre, B. Bouhon. Jointless steel fiber-reinforced concrete slabs-on-grade and on piles, pp. 89–102, SP-268-8, 2010.

12. R. Ytterberg. "Shrinkage and curling of slabs on grade." *Concrete International*, 9,p 22–31, April 1987.

13. R. Meininger, "Drying Shrinkage of Concrete." Engineering Report RD3, National Ready Mixed Concrete Association", 1966.

14. X. Destree, Y. Yao, B. Mobasher. "Sequential cracking and their opening in steel-fiber- reinforced joint-free concrete slabs." *ASCE Journal of Material*.28, n°4, 04015158-4.p.1–11.

15. M. Bakhshi, B. Mobasher, C. Soronakom. "Moisture loss characteristics of cement-based materials under early-age drying and shrinkage conditions." *Construction and Building Materials*, 30, 413–425.

16. B. Bissonnette, E. Attiogbe, M. Miltenberger, C. Fortin. "Drying shrinkage, curling, and joint bissonnette: drying shrinkage, curling, and joint opening of slabs on ground." *ACI Materials Journal*, 30, 257–269, May–June 2007.

17. P.S. Mangat and M. Azari. "Shrinkage of steel fiber reinforced cement composites." *Materials and Structures*, N°21, 163–171, 1988.

18. T. Voigt, V. Bui, and S. Shah. "Drying shrinkage of concrete reinforced with fibers and welded-wire fabric." *ACI Materials Journal*, 101(3), May-June 2004.

19. B. Malmberg, A. Skarendahl, "Method of studying the cracking of fiber concrete under restrained Shrinkage." Swedish Cement and Concrete Institute, 173–179.

20. J. Silfwerbrand. "Design of steel fiber reinforced concrete slabs on grade for restrained shrinkage." *Rilem PRO39*, 975–984, 2004.

21. H. Westergaard. "Stresses in concrete pavements computed by theoretical analysis." *Public Roads*, 7, 25, 1926.

CONSTRUCTION AND BUILDING MATERIALS

K. Bondy, "Cracking in ground-supported post-tensioned slabs on expansive soils." *PTI iTechnical Notes*, Issue 6 August 1995.

H. Haynes, K. Bondy. "Contraction joints for residential post-tensioned slabs." *ACI, Concrete International*, p. 35, July 2020.

Chapter 9

SFRC Slabs vs. Self-Stressing S.F.R.C. jointless slabs (a Primekss invention)

9.1 A GAME CHANGER FOR SFRC SLABS

The slabs on ground were the first significant industrial application of SFRC as a complete replacement for wire meshes. This application began developing mainly in the early 1980s, starting in Belgium and Holland, and later spreading throughout Western Europe and beyond.

The game changer was the NATO full-scale test in Sanem (Luxembourg, 1983), as reported in the earlier chapter on full-scale tests of SFRC slabs. A 130 mm slab with 7 m x 7 m saw cuts and 20 kg/m^3 of Eurosteel fibers (1 mm diameter x 60 mm length, undulating) was selected over a 200 mm slab with a top mesh of 8 mm x 150 mm and 5 m x 5 m saw cuts.

We were awarded the 100,000 m^2 NATO project and completed it accordingly.

To summarize, before that NATO test, the SFRC slab market was almost nonexistent. All slabs were reinforced with wire meshes, except for a small number of SFRC slabs with a 30 kg/m^3 dosage rate of hooked-end fibers (0.6 to 0.8 mm x 60 mm length), applied at a dosage rate of 30 kg/m^3 in thicknesses almost identical to traditional slabs.

The NATO test slab design was much more economical and ultimately saved 7,000 m^3 of concrete. It was also significantly more user-friendly due to the complete omission of steel wire meshes, which otherwise had to be installed manually.

I am still proud today to have been the engineer who pushed for and demonstrated the feasibility of this new solution for the NATO slab.

At the time, during the acceleration of the Cold War, a new weapon storage warehouse site was planned in Sanem (Luxembourg), consisting of 20 warehouses, each measuring 5,000 m^2. The specified design was a 200 mm thick slab in B 25 concrete, reinforced with an 8 mm x 150 mm x 150 mm wire mesh placed at 1/3 of the slab depth. The loads were specified as 50 kN/m^2 UDL or a 100 kN axle load on two wheels, each with a 200 mm x 400 mm contact area, resulting in a 0.625 N/mm^2 contact pressure. The slab design included sawn cuts at 5 m intervals.

DOI: 10.1201/9781003188315-9

The base consisted of a 500 mm crushed slag layer at optimal compaction, placed above a lime-treated natural clay ground. The expected Westergaard k-value was 10 kg/cm³ (0.1 N/mm³ or 375 pci).

We proposed an alternative solution of 130 mm thickness in B 25 slab reinforced with 20 kg/m³ of steel fiber made from 1 mm diameter steel wire (1100 MPa tensile strength) and 60 mm length, featuring an undulated shape with six undulations of approx. 8 mm pitch and 1.0 mm wave depth at the fiber centerline, as shown in Figure 9.1. Sawn cuts were placed at 7 m intervals.

The offer was initially "freshly" received and regarded as a reckless offer by a young daring engineer. NATO called, however, for a technical meeting of almost 30 people where I was invited to present technically the solution before the Army, the Ministry of Defense, the Ministry of Public Works of Luxembourg, the Warehouses State Agency, the Road and Bridge Laboratory of Luxembourg and SECO, the technical controller, the OREX Lab, the University of Liège civil Engineering labs.

I remember of the most exciting moment was when the General in chief, at the end of the discussion, noticed that nobody around the table would agree with my SFRC slab offer. Then he said: *"This young man proposes here a solution to save 7000 m³ of concrete to fit in our tight budget. Who is against."*

Nobody dared to be!

The General in chief designated M. Pierre-Marie Dubois at Seco Controller office to be responsible to demonstrate if our solution was viable or reckless.

M.P-M. Dubois instantly decided on a full-scale slab loading test. The procedure involved testing a 7.7 m x 7.7 m slab with a thickness of 130 mm, reinforced with 20 kg/m³ of steel fibers. The test was organized on-site on the same 500 mm thick crushed slag base over lime-treated clay soil.

A positive result required the slab to sustain a 100 kN load applied to a 200 mm x 400 mm plate at any point without any detectable cracking, either on the top or bottom, as verified by ultrasonic electrodes.

The 100 kN load was gradually applied at the free corner, along the free edge, and at the center point of the slab. The objective was to ensure a safety factor of 2 on the elastic limit.

As detailed further in the earlier chapter "Full-Scale Test," the first-minute cracks were observed at loading intensities of 200 kN (free corner of the slab), 250 kN (free edge), and 320 kN (middle of the slab), which were 3 to 4 times higher than calculations derived from any Westergaard theory.

The SFRC Sanem NATO slab project was installed in 1983 at a thickness of 140 mm to account for tolerances in slab thickness.

The slab's concrete constituents included crushed granulated slags in fractions of 0–8 mm, 8–16 mm, and 16–25 mm, a by-product of local ARBED

Figure 9.1 Typical Eurosteel I mm x 60 mm 1200 MPa steel fiber.

steel production. The cement used was 320 kg/m³ of HF cement, equivalent to today's CEM III B 32.5. This mix design was clearly of a low carbon footprint, well ahead of its time.

The Eurosteel fiber production by ARBED, our licensed manufacturer and precursor to the current ArcelorMittal, was initiated to supply the approximately 300 tons of steel fibers needed to complete this NATO project.

The granulated slag was very porous and required a large quantity of water to achieve a water-saturated surface dry condition. Up to 200 kg/m³ of water was added to obtain suitable workability with a superplasticizer. It is likely that these aggregates provided internal curing, as the slabs remained almost "crack-free" and without curling throughout their lifespan. At that time, internal curing was completely unknown, even in the literature.

Today, the NATO slabs remain in good service condition and satisfactory after over 40 years of use.

Following the successful 1983 NATO reference project in Sanem, Luxembourg, several similar large-scale NATO projects were completed in SFRC during the following years in Belgium and Holland. This quickly initiated the day-to-day market for SFRC slabs in Western Europe.

The purpose of this chapter is not to repeat or rephrase the large amount of existing meaningful literature. Among the most reputable and widely used references worldwide are Technical Report No. 34 in the UK, the ACI 360 "Design of Slabs on Ground" by the American Concrete Institute, and more recently, the ACI SPEC-544.12-23 (Reference 1), where the fiber dosage rate is calculated as temperature reinforcement.

9.2 THE WESTERGAARD THEORY BACKGROUND

The entire theory behind traditional ground-bearing slab design is primarily derived from the 1925 Westergaard theory.

The Westergaard model was developed in the 1920s for the US Army to address how to quickly design the thickness of an aircraft pavement using a simple in-situ test value that could be applied anywhere the Army needed to operate.

It was a highly attractive method because a technician on-site, using the Westergaard test setup, could rapidly determine the required slab thickness. The 30 in. (762 mm) rigid steel disks were placed under a hydraulic jack to lift a 30 kN truck axle of that era. The ground beneath the 762 mm diameter round plate was subjected to a pressure of 0.07 N/mm² (70 kPa), which remains the standard today.

Professor Westergaard solved the Laplace plate equation using his hypothesis of a second member expressed as $-q/D$, where q is the contact pressure and D is the rigidity of the elastic slab.

His bearing coefficient K represents the ground-bearing body as an elastic liquid, assuming it lacks shear strength and does not provide friction.

According to Westergaard's equations, the slab's elastic flexion stress depends on $\ln(1/K)$, the point loading intensity, and the footprint size of the point loading.

When the K-value increases significantly, from 0.03 N/mm³ to 0.1 N/mm³, the maximum flexion stress of the slab under a central load decreases by only 22%. Consequently, the slab thickness can be reduced by just 11%, from 200 mm to 180 mm.

By Westergaard's method, the incentive to improve ground-bearing capacity is minimal, given that such improvement often costs much more than the equivalent savings of 20 mm of concrete, as in this example.

However, in reality, the ground is not an elastic liquid body. It provides significant shear strength through internal friction—unlike a liquid, which can be easily penetrated with a finger, a well-graded and compacted granular base resists penetration effectively.

As discussed in the earlier "Full-Scale Test" chapter, full-scale tests of slabs on grade, up to cracking and yielding, reveal a significant overdesign of slab thickness when the Westergaard method is applied. Westergaard's K-value hypothesis underestimates the first crack and collapse load of the slab on grade by a factor of 4 to 7.

This, however, is good news, as it means slabs are statically overdesigned, making them reliable and safe in service over time.

Yet, many slabs do not perform as expected. Numerous slabs show issues with serviceability, such as uncontrolled cracking, poor joints, and curling movements, leading to frequent disputes. In most cases, overall shrinkage is the primary culprit.

Figure 9.2 shows the Soum model of curling at the joint, as a function of D, the diameter of larger aggregates, and η, the relative humidity of the ambient air.

The slab lifts off up to 1 m away from the joint, leaving the slab edge cantilevered and unsupported, which eventually leads to severe cracking over time.

The Westergaard model doesn't deal with the slab shrinkage.

9.3 SLAB SHRINKAGE

The slab shrinkage is not dealt with by standard calculations but could be addressed on-site to ensure the optimal mix design is used and installed. The required slump increase should result only from the addition of superplasticizer admixtures.

The rule is to achieve the highest possible optimum unit weight of fresh concrete. Anything less than the optimum means more fine content, more water, more air, more filler, and thus more shrinkage.

The sawn cutting or shrinkage joints, placed at close distances such as 1 m, 4 m, 6 m, or 12 m and executed at an early age, are essentially man-made cracks that rarely remain controlled over time. These joints open and curl without preventing cracks in the long run.

JOINT CURLING

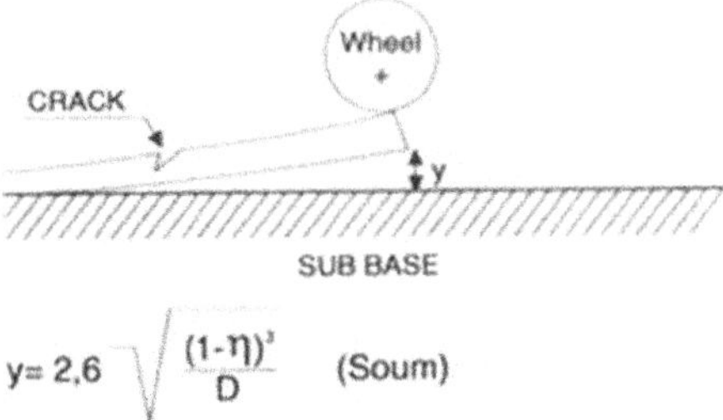

$$y = 2,6 \sqrt{\frac{(1-\eta)^3}{D}} \quad \text{(Soum)}$$

η = percentage water content in air with reference to saturation
D = size of aggregate

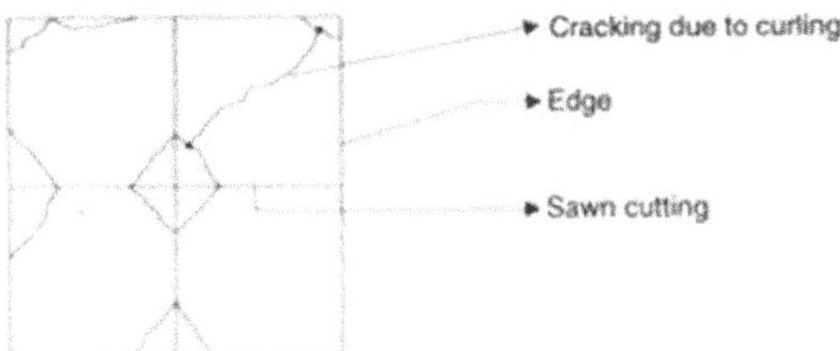

Figure 9.2 Typical cracking induced by the curling of joints

Condition	Potential Increase in Shrinkage	Cumulative Effect
Temperature of concrete	8%	1.00 x 1.08 = 1.08
6"-7" vs. 3"-4" Slump	10%	1.08 x 1.10 = 1.19
Excessive haul in transit mixer	10%	1.19 x 1.10 = 1.31
Use of smaller aggregates	25%	1.31 x 1.25 = 1.64
High shrinkage cement	25%	1.64 x 1.25 = 2.05
Excessive "dirt" in aggregate	25%	2.05 x 1.25 = 2.56
High shrinkage aggregates	50%	2.56 x 1.50 = 3.84
Admixtures	30%	3.84 x 1.30 = 5.00
Total increase	**Summation = 183%**	**Cumulative = 400%**

Figure 9.3 The cumulative effect of parameters on shrinkage according to ACI knowledge.

In reality, cracks and failures are mostly caused by the detrimental shrinkage generated by poor-quality materials, poor mix designs, or poor slab installation, even though all of these may comply with standard codes of practice.

On-site, rely only on what you measure to take corrective actions. Monitor the "shrinkage content" of the concrete slab by checking every parameter from the concrete plant to on-site installation.

In Figure 9.3, taken from an ACI presentation, the effect of various parameters related to concrete constituent materials and supply on total shrinkage is shown.

The cumulative effect is equal to four times the reference concrete shrinkage! Adding to this, factors like poor curing conditions with high winds or air currents inside a building, high ambient temperatures, excessive sun exposure, salts and shells in concrete, flaked aggregates, and undocumented set accelerators can increase the shrinkage by 400%, doubling it to 800% or more.

No techniques or specialty products can turn bad concrete with poor installation into good concrete!

By 1985, the use of SFRC in slabs on grade had become almost routine. However, the standard wire mesh slab is still widely used today, mainly because it is perceived as a cheaper solution. What could be cheaper than a light slab with chicken mesh in the mud?

Still, some SFRC slabs have not performed as expected, primarily due to excessive shrinkage that the fibers cannot control, even though these slabs comply with design standards or guidelines.

Today, cracks and bad joints still plague the slab industry, as concrete shrinkage exceeds the level observed 40–80 years ago, as outlined in Reference 2, by Burrows.

Shown in Figures 9.4 and 9.5 is a typical German automotive factory slab from 2023, designed strictly according to German design guidelines for SFRC slabs, where the reinforcement consists of wire meshes combined with 35 kg/m³ of steel fibers. Figure 9.6 shows the installed wire meshes and Figure 9.5 shows the finished slab while the job site is still under

Figure 9.4 Typical German slab with wire mesh plus steel fiber.

Figure 9.5 Bad joints following the German standard slab design.

construction, well before the slab has reached its industrial service condition. You can already observe the poor behavior of the construction joint.

Construction joints, regardless of their load transfer capacity, open to relieve stresses and curl, leading to cracking and chipping of concrete at the arris.

Frequently, at the joints, slabs are no longer in good service condition due to vertical movements and no longer maintain satisfactory flatness or levelness. The dynamic actions of moving trucks contribute to continuous degradation.

The first and most important factor to address in slab design is shrinkage throughout the entire lifespan of the slab.

9.4 SFRC JOINT FREE FLOORS

In 1984, I, together with M. A. Lazzari, my boss at the time, was the inventor of the SFRC jointfree floor patent, granted in Europe (Eurosteel Patent Number 0137 024), the USA (Patent Number 4,640,648), and many other countries.

This was indeed a true invention, as steel fibers were not given any recognition for reinforcing concrete slabs at the time, and the joint-free floor concept, regardless of the type of reinforcement, was unknown in standards and guidelines. Many slabs were still cast in long strips or even in a checkerboard pattern, depending on the location. Clearly, the man of art had no idea how to construct a jointfree floor.

By the Eurosteel S.A. patent, the "man of art" could construct it, as outlined in the claims shown in Figure 9.6.

We claim:

1. A process for making a totally independent and
continuous slab of fiber-reinforced concrete devoid of
seams, which comprises the steps of:
   providing a foundation course;
   compacting said foundation course to a k value of
      Westergaard of at least 5 kg/cm³;
   leveling said foundation course with a tolerance of at
      most ±1 cm with respect to a reference level;
   placing a plastic sheet on said foundation course;
   placing a compressible mattress adjacent fixed bound-
      aries which are to limit or traverse said slab so as to
      separate said slab from said fixed boundaries;
   placing complementary reinforcements adjacent said
      fixed boundaries;
   casting on said foundation course, adjacent said fixed
      boundaries, a limited shrinkage concrete reinforced
      with fibers;
   allowing said concrete to set.

2. A process as claimed in claim 1, in which said
concrete contains reinforcing fibers in a proportion
comprised between 25 and 30 kg/m³.

3. A process as claimed in claim 1, in which said
concrete contains reinforcing steel fibers in a proportion
comprised between 25 and 30 kg/m³, said fibers having
a diameter between 0.5 and 1 mm, a length between 40
and 60 mm and anchoring means selected from the
group consisting of ondulations, corrugations, hooks
and bulges.

4. A process as claimed in claim 1, wherein the
shrinking of said concrete is limited by reducing its
content of cement and water.

5. A process as claimed in claim 4 wherein a Portland
type of cement is used in said concrete.

6. A process as claimed in claim 1, wherein the slab is
not to be subjected to charges exceeding 3000 kg/m².

7. A process as claimed in claim 1 wherein said setting
of said concrete is allowed for at least six weeks.

* * * * *

Figure 9.6 Claims of the year 1984 patent: EUR 0 137 024 and USA 4,640,648.

Note that the provisions of this patent became public in 2001, and in France, the official DTU standard for the design and installation of logistic slabs adopted a very similar approach in 2004. Today, most large-scale logistic slabs are constructed accordingly.

For slabs designed to carry significant loads along with frequent vehicular traffic, the best option is to go "joint free" with the number of construction joints between adjacent bays minimized. Steel fiber reinforcing, combined with a limited matrix "shrinkage content" and reduced friction from the grade, is a proven way to achieve "jointfree" floors.

In 1983, I authored the patent for such a technique, enabling the "Man of Art" at the time to construct, for example, a 150 mm thick slab with 30 kg/m³ of 1 mm x 60 mm steel fibers in a B 25 mix. This mix had a W/C ratio smaller than 0.55, included a superplasticizer admixture, and was laid over a k=0.03 N/mm³ base. The bay size was approximately 3000 m² between construction joints, designed to support a 30 kN/m² UDL intensity or 50 kN point loadings.

In the early 1980s, a 30 kg/m³ dosage rate of 1 mm x 60 mm steel fibers was sufficient reinforcement for a 2000 m² bay size, as outlined in my patent and common practice at the time.

Figure 9.7 Chipping-out of regular SFRC at a sophisticated sinusoidal-shaped joint.

However, by 2024, cements are no longer what they were in 1982. Modern cements often generate much more strength and shrinkage due to changes such as a higher Blaine number (around 5000 cm²/g).

Table 9.1, from the Iowa State University Institute of Transportation (2014), summarizes parameters influencing drying shrinkage, including cement composition, W/C ratio, high-range water-reducing admixtures, aggregate grading, curing practices, and more.

Table 9.1: Iowa State University Institute of Transportation, on factors affecting shrinkage (2014)

Today, a suitable mix to meet joint-free floor requirements should have a water content limited to a maximum of 0.55, with 0.50 being more recommended. The cement content should be around 330 kg/m³ (avoiding more than 30% SCM), with a 22 mm maximum aggregate size, and 30 to 50 kg/m³ of steel fibers with diameters ranging from 0.75 mm to 1 mm and lengths between 50 to 60 mm. Steel fiber aspect ratios should ideally range from 50 (minimum) to 60 (maximum).

Supplemental cementitious materials (SCM) should not be counted as Portland Cement (Type I) in W/C calculations but instead weighted at 0.3 to 0.6, depending on their "pouzzolanicity" or hydraulicity.

High percentages of SCM can promote cracking, especially under winter conditions, as strength development slows while shrinkage due to water volume loss continues.

Table 9.1 Summary of influence factors, on concrete shrinkage

Categories	Factors	Specific factors	General findings and conclusions on shrinkage
Paste quality	Cement characteristics	Chemistry	Higher sulfate and gypsum content may reduce shrinkage
		Fineness	Through changes in the rate of hydration and water demand, finer cement leads to higher shrinkage
	Supplementary cementitious material	Fly ash	Class F fly ash potentially decreases shrinkage; Class C fly ash increases shrinkage with increasing dosage
		Slag cement	Conflicting results, depending on paste content; overall, comparable to ordinary PCC mixtures
		Silica fume	Conflicting results; depending on w/cm, curing period, and paste content; limited replacement may reduce long-term shrinkage
		Ternary mixtures	Normally, adverse effect can be diminished (i.e., PC+slag cement+silica fume) or even reduced (i.e., PC + slag cement + F fly ash)
	Chemical admixtures	Superplasticizers	May increase concrete shrinkage with a given w/cm ratio and cement content; dependent on admixture chemistry
	Air content		No significant effect on the magnitude of drying shrinkage if it is less than 8%
Paste quantity	w/cm		Influence is relatively small with a constant paste content for w/cm > 0.40
	Paste content		For a given w/cm ratio, shrinkage linearly increases with increasing paste content/volume
Other factors	Aggregate	Type	Influenced by mixing water demand and the stiffness of the aggregate

(Continued)

Table 9.1 (Continued)

Categories	Factors	Specific factors	General findings and conclusions on shrinkage
		Gradation	Indirect influence through the change of water demand and paste content; increased size results in decreased paste content, so decreased shrinkage
		Fines	Most fines increase shrinkage
	Curing	Duration	Increasing curing period reduces the amount of unhydrated cement, resulting in reduced overall shrinkage
	Environment	Relative humidity	Higher relative humidity leads lo lower shrinkage
		Temperature	High temperatures accelerate moisture loss leading to higher shrinkage
		Wind	Similar to temperature effect
	Construction		Improper addition of water during finishing, and poor curing will increase shrinkage
	Geometry		Increasing the volume-to-surface area ratio decreases shrinkage

In case of a ternary cement with a content of 350 kg/m³, where 175 kg half is a Type I CEM and the other half consists of SCMs, together with 180 kg water, the W/C ratio can be calculated as follows:

$$180 / (175 + 0.5 \times 175) = 180 / 262 = 0.68 \text{ instead of } 180 / 350 = 0.51$$

It is always very informative to measure and record the volume weight (or density) of the concrete mix upon arrival at the jobsite. Dense concrete typically ranges from 23.8 kN/m³ to 24 kN/m³ or even higher with heavier aggregates. Lower figures often indicate excessive water content or too high a mortar content, leading to increased shrinkage, even if the compressive strength meets the 30–37 MPa level.

A jointfree slab bay size of 1000 m² is a minimalist example but can extend up to 3000 m², with approximately 50 m between day joints.

The day joints are armored in steel to function theoretically as zero-moment joints with full shear transfer. However, most of these joints perform poorly, with openings exceeding 10 mm. This leads to concrete chipping along the joint and spalling of the concrete arris against the steel joint surface, as shown in Figure 9.7

A study we conducted at the University of Pas de Calais - IUT Bethunes (France) in 1992–1994 showed that 40 kg/m³ of 1 mm x 60 mm undulating steel fibers (Eurosteel-Silidur type) reduced the curling of a free corner by 40% for a 150 mm slab in C 30-37.

The study also revealed that a superplasticizer (high range water reducing admixture) used to increase workability without reducing the water content increases the shrinkage and curling of a slab.

For all types of concrete slabs, carefully monitor the actual water content to ensure it remains within the specified limit, allowing the superplasticizer to improve and speed up installation without adverse shrinkage effects.

9.4.1 Shrinkage in thick slabs

The amount of drying shrinkage decreases with the depth of the slab. Shrinkage is maximum at the surface, where vapor water loss occurs most rapidly, denoted here as s_{max}.

At depths of 300 mm and beyond, shrinkage is almost negligible, with any increase over time being extremely slow.

The shrinkage profile from the surface to a depth of 300 mm is nearly parabolic. For example, if s_{max} represents the surface shrinkage and shrinkage is zero at a depth of 300 mm, then we have:

A = area above the parabola from surface to H depth and B = the area under the parabola so that A + B area is a rectangle.

$$\text{Area A} + \text{B} = s_{max} \times H; \text{area B} = (2/3)s_{max} \times H \text{ and area A} = s_{max} \times H - (2/3)s_{max} \times H$$

So the area A = $(1/3)\, s_{max} \times H$

For example: what is the total possible shrinkage S_T of 1 m deep slab when the surface shrinkage is of 4 x 10⁻⁴? $S_T = (1/3)\, 4.10^{-4} = 1.33 \times 10^{-4}$ in μstrain

If such a shrinkage is fully restrained, it should result in a tensile stress:

$f_s = 1.33 \times 300 \times 1000 = 399,000$ N, so that a D section top mesh is needed; $D = 399000 N / 500 (N/mm^2) = 798$ mm² or 12 mm rebarswith 150 mm spacing in both directions of a weight of 12 kg/m².

$f_s = 1.33 \times 300 \times 1000 = 399,000\, N$, so that a D section top

mesh is needed; $D = 399000 N / 500 (N/mm^2) = 798\, mm^2$ or

12 mm rebars with 150 mm spacing in both directions of

a weight of 12 kg/m².

Instead of a top mesh, steel fibers should be f_{r1} = 1.33/0.45 = 2.96 N/mm² (EN 14651), typically a minimum of 25 kg/m³ of HE 90/60 steel fibers per cubic meter.

In case of a 1 m depth raft, M_{Rd} = 800 kNm/m, which should be needed in traditional reinforced concrete, a D section of rebars:

D = (800,000 Nm/m) / 0.9 x 850 mm x 435 N/mm² = 2400 mm²/m or 25 mm rebars at 200 mm distance apart in top and bottom most likely thus 77 kg/m³ plus the overlapping and the supporting chairs, hence ca. 90 kg/m³ reinforcing bars.

With steel fibers, it becomes:

$$F_f = 6 \times 800,000 / 1000^2 = 4.8\, N/mm^2.$$

$f_{.R3}$ > η_{det} x f_{rd} / 1.5 = 2 x 4.8 / 1.5 = 6.3 N/mm² where η_{det}=2 (EN-SS 812310) is the statically indeterminate factor in the EN-SS 812310 standard or the design in fiber-reinforced structures in case of two-way slabs.

To allow for easy placement, it is recommended to use a walking mesh at 150 mm depth so that the workers will not sink in the 1 m depth. The walking mesh is in general a 8 mm diameter x 150 mm spacing of 335 mm²/m section (335/2400 = 0.14) that offers however a M_{Rd} = 112 kNm/m.

The f_{rd} of the fibers can be reduced by 112/800 = 14% as well so that $f_{rd} = $ 0.86 x 6.3 = 5.42 N/mm² average value of f_{r3} or f_{r3m} (EN 14651) in a C30-37 mix. The dosage rate of steel fibers can be reduced to 40 kg/m³ thus a steel reduction factor 90/40 = 2.25!

9.4.2 SFRC equivalent design to the traditional with wire meshes

As many engineers continue to design traditional reinforced concrete slabs on grade, a good first step is to make an equivalent design including SFRC. It is indeed quite simple to do:

H_T := 200 mm   Traditional thickness of slab

$u := 1\, \dfrac{N}{mm^2}$ Unit number

c := 30 mm   Concrete coverage of wire mesh

$H_{eff} := H_T - c = 170\, mm$ Effective depth

$f_{rl} := 3.5\, \dfrac{N}{mm^2}$ EN14651 post cracking flexion
                                               strength

$f_{r3} := 4\, \dfrac{N}{mm^2}$

$$f_L := 4.5 \, \frac{N}{mm^2}$$

EN14651 Limit of Proportionality in flexion

$$H_{SFRC} := H_{eff} \cdot \left(\frac{2 \cdot f_L}{2 \cdot u + f_{r1} + f_{r3}} \right)^{\frac{1}{2}} = 165.466 \, mm$$

Is the SFRC Slab equivalent thickness

$$\eta_{det} := 2$$

(from EN-SS812310) is the Structural Indeterminacy factor

$$\gamma_c := 1.5$$

(from SS812310), SFRC is the material factor

$$M_{Rd} := \frac{\eta_{det} \cdot f_{r3} \cdot H_{SFRC}^2}{6 \cdot \gamma_c} = 24.337 \, kN \cdot \frac{m}{m}$$

Is the SFRC Resisting moment expression and intensity

$$f_y := 435 \, \frac{N}{mm^2}$$

Steel rebars tensile strength

$$v := 1 \, m$$

$$S_y := \frac{M_{Rd} \cdot v}{0.9 \cdot f_y \cdot H_{eff}} = 365.665 \, mm^2$$

Traditional steel section to provide the same resisting moment

Conclusions: a 165 mm slab in C30-37 with 35 kg/m³ of HE1/50 steel fibers is

equivalent to a 200 mm slab with top and bottom meshes of 9 mm x 150 mm (424 mm² > 365 mm²)

thus 424 mm²/m in top and bottom of a unit weight of 6.3 kg/m² each.

In the traditional solution, 2 x 6.5 kg/m² steel rebars are needed; thus, 15 kg/m² overlapping is included or 15 kg/m² /0.2 m = 75 kg/m³ instead of 35 kg/m³ steel fibers (a reduction factor of 2.14). Moreover at 35 kg/m³ of HE1/50 steel fibers, with maximum W/C< 0.55, a jointfree slab up to 1500 m² area is possible. The application of steel fibers instead of wire meshes is significantly more beneficial, economical, and sustainable than traditional methods. I used these calculations in the early 1980s to initiate and accelerate the SFRC flooring market in Belgium and Holland. At that time, the structural indeterminacy factor was unknown, but I relied on the post-cracking flexion strength of SFRC observed during tests of square plates with 500 mm x 500 mm clear spans and 100 mm thickness in flexion. In these cases, the flexion strengths were nearly double those observed in small prismatic specimens, with a scatter of only 12% at most, compared to up to 50% in smaller specimens. As shown in Figure 9.8, a 20 kg/m³ steel fiber dosage of a 1 mm x 60 mm Eurosteel undulating type provides a full plastic plateau, similar to the NATO full-scale test of 1983.

The reference is taken out of BFT magazine n°11/1985 in p.747 under Reference 3.

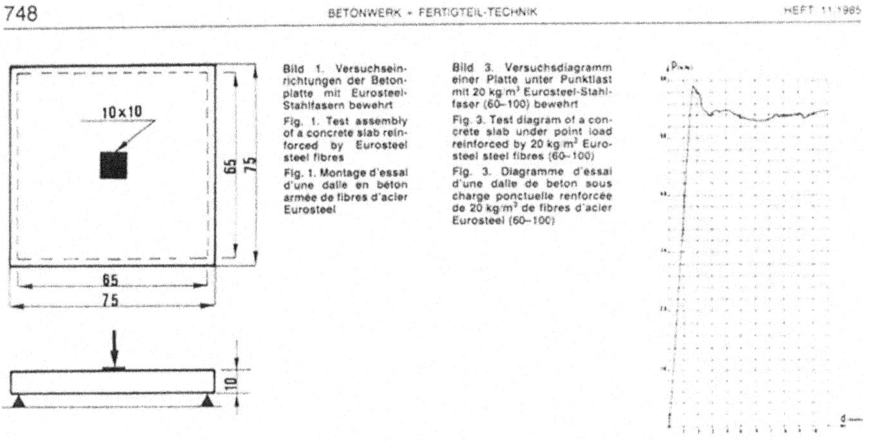

Figure 9.8 Small square slab flexion test in 1984.

Statically indeterminate square and round slab flexion testing later became part of EN 14488-5 for shotcrete, as small plates could be "shot" more easily than small beams. The precursor of this method was the SIA 162 and EFNARC standards. This testing originated in Western Canada for shotcrete applications, particularly for strengthening old tunnels to prevent falling blocks.

Since 1984, I have favored this method using small square plates, as they are much easier to produce on-site and far more stable during testing than small beams in flexion.

In reality, ground-bearing slabs benefit from the plasticity and friction of a quasi-elastic or elasto-plastic grade and the ground beneath them.

Referring to EN 14651 strength numbers to design slabs on grade, which are defined solely by a K-value typical of a liquid material (as seen in TR 34, 4th Ed.), remains difficult to justify from both economic and sustainable perspectives.

9.5 PRIMEKSS, THE SELF STRESSING SFRC SOLUTION AGAINST SHRINKAGE

The observation of full-scale tests of slabs on grade, which are vastly overdesigned by the Westergaard method, shows that eliminating shrinkage cracking and curling movements opens the door to halving the typical TR34 or ACI 360 SFRC slab thickness.

Without the effects of shrinkage, the slab becomes highly stable and less dependent on the time factor. It maintains full base support throughout, ensuring that no cantilever conditions develop near traditional joints, where conventional SFRC slabs tend to curl and lose support regardless of the type of construction joints used.

At the 2007 ACI Puerto Rico Convention, while enjoying some spare time on a small Caribbean Island off the convention coastline, I advised Janis Oslejs, CEO and founder of Primekss, to combat shrinkage by strongly incorporating Type K Cement slurry on-site. This approach allows for significant thickness reductions in SFRC slabs to be safely achieved.

Type K cement generates ettringite volume expansion in water-cured concrete. This physical expansion is restrained by the steel fiber reinforcement and the friction against the grade.

Starting with the knowledge of the ACI 223 Committee, as referenced in 4, we introduced calcium sulfo-aluminate clinker into the mix. This reacted with the C3A in the cement to form ettringite at an early age, typically within the first seven days after mixing.

Ettringite formation in concrete is known for its substantial volume expansion before complete hardening, effectively compensating for shrinkage contraction. At this early stage, the concrete slab, still relatively "soft," can accommodate the expansion without bursting damage, provided that the expansion is restrained by internal reinforcement and friction on the grade.

I had visited Type K slabs in the USA as early as the mid-eighties, during the time I was developing the joint-free flooring technique for SFRC slabs. At that time, I was not convinced, as I also observed significant cracking and joint opening. I recall visiting the Dayton (OH) Air Base Museum to inspect a Type K slab, which was far from convincing when compared to our patented SFRC joint-free slabs of that period.

Around 2005, I revisited more Type K slabs, and the results were better than what I had seen 20 years earlier. Between Chicago and Saint Louis, I visited several Albertson's "Do It Yourself" superstores, where 150 mm-thick slabs with wire mesh and Type K shrinkage-compensating cement were installed. These were convincing, showing minimal cracking, joint openings up to 20 mm, and no curling along the joints.

In Europe during the mid-1980s, cements were increasingly binary or ternary, with high C2S content and high Blaine numbers to achieve early strength despite reduced Portland clinker content. However, these newer formulations contributed to more cracking and curling compared to the lower Blaine and lower-grade cements of 1985.

The Albertson's slabs were wire-mesh reinforced primarily to restrain the chemical expansion. We soon discovered that steel fibers at 35–40 kg/m^3 dosage rates could restrain the expansion even more effectively. This is attributable to the high specific surface area (m^2/kg) of steel fibers, approximately 10 times greater than that of wire mesh.

As an R&D consultant to Primekss, I introduced and continued developing the patented PrimeXcomp slab concept over the years. It became clear that simply adding Type K cement to the mix was insufficient. We needed to make the expansive reaction more adaptable to local ready-mixed concrete compositions, cement properties, temperature variations, and climate conditions.

At Primekss, we developed a more robust process tailored to the diverse needs of regions such as Scandinavia, Texas, Alberta, Minnesota, the Middle East, South Africa, and beyond.

9.5.1 How to halve the thickness as per TR 34 or ACI 360 of a slab on grade

- Halving the thickness down to 100 mm or thinner will require a base that does not deform during jobsite activities and avoids rutting by wheels. The grade must remain undeformable under jobsite traffic to ensure that the tolerance on thickness, including base and slab surface finishing tolerances, does not exceed 15 mm, resulting in a slab thickness of at least 75–85 mm. At a K-value of 300 pci (0.08 N/mm^3), this condition is met.
- Omit all types of joints, as each joint becomes a source of trouble and is difficult to repair.
- Target minimal concrete shrinkage for the plain concrete supplied. Specify and ensure that the concrete, as delivered, is as dense as possible with the available constituent materials: a target minimum of 24 kN/m^3 (150 lbs/ft^3) with normal weight aggregates of approximately 2600 kg/m^3 (163 lbs/ft^3) density. Any fresh concrete density below 23.80 kN/m^3 indicates higher water or air content, leading to increased shrinkage.
- Implement the PRIMEKSS patented process by adding and mixing, on-site, the expansion-generating binders in slurry form along with liquid chemical admixtures to enhance mix flow, control setting time, and ensure robust expansion performance under summer or winter conditions.
- Use a fiber blast machine to blow and mix steel fiber reinforcing (ASTM Type 1) on-site at a 0.45% volume fraction dosage rate (minimum of 35 kg/m^3) with an aspect ratio of no less than 50 and no greater than 60.
- Cure the slab under specific water-saturated blankets for at least 7 days.

9.5.2 Developing further the restrained expansion

The PRIMEKSS SIA and Primekss North America companies further developed the Type K technique by incorporating a novel and proprietary formula of calcium sulfate hydrate and semi-hydrate subjected to precise heating treatment and grinding. Together with yé'elemite, adequate dolomite, and the PRIMEKSS CPEA proprietary liquid admixtures, this formula propels the expansion further, creating a net compression stress that becomes permanent in the section, even after shrinkage completion, resulting in almost zero shrinkage.

The Primekss expansion binder formula significantly increases the robustness of the expansion reaction, making it less dependent on external temperature, time, and water content prevailing on-site.

Yé'elemite can occur naturally but is also manufactured in kilns by burning limestone, bauxite, and anhydrite together at a temperature of 1250°C, then grinding it to a suitable particle size. During hydration, in the presence of calcium and sulfate ions and requiring a high quantity of reaction water, it forms the swelling crystal ettringite.

Ettringite is a fibrous mineral that is also responsible for a fraction of the total strength in the cementitious materials. The stoichiometric equation of the Yelemite hydraulic reaction is shown in References 5 (Lea's Chemistery-1988) and 6(A Neville)

$$3 \cdot CaO \cdot 3Al_2O_3 \cdot CaSO_4 + 8 \cdot CaO_4 + 6 \cdot CaO + 96 \cdot H_2O \qquad (9.1)$$
$$\rightarrow 3 \cdot \left\{ Ca_6 \left[Al(OH)_6 \right]_2 (SO_4) 3 \cdot 26 \cdot H_2O \right\}$$

The expansion is caused by quite a high number of water molecules as shown in (1) (indeed, 96) and was bound to form ettringite, the hydrated product of the reaction.

As recommended in the ACI 223 (Reference 4), a water curing process under burlaps is used for 14 days.

The yé'elemite-generated expansion develops during the first 7 days of hardening, while the E-modulus is still small, ensuring that the concrete does not burst. This contrasts with the long-term formation of ettringite, which occurs when the tri-calcium aluminate (C_3A) from Ordinary Portland Cement (OPC) reacts with external sulfate ions in hardened Portland cement concrete.

The expansion can be measured on laboratory prismatic specimen according to ASTM C 157, and typical values are shown as function of the age of

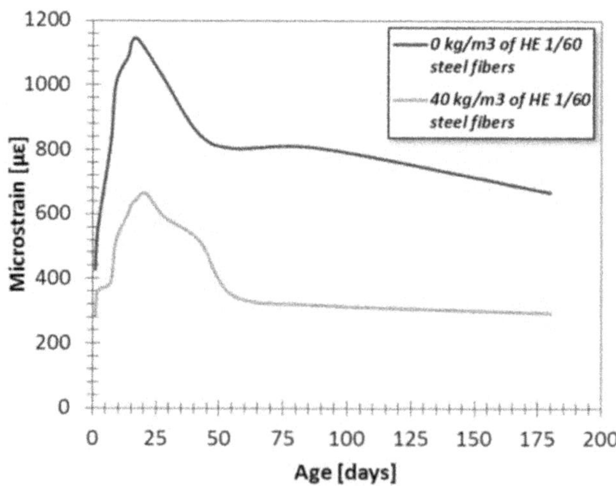

Figure 9.9 Expansion followed by shrinkage of a plain mix and of a SFRC mix in function of the age of the concrete samples.

Figure 9.10 Expansion and restraint of SFRC measured on small slab specimen.

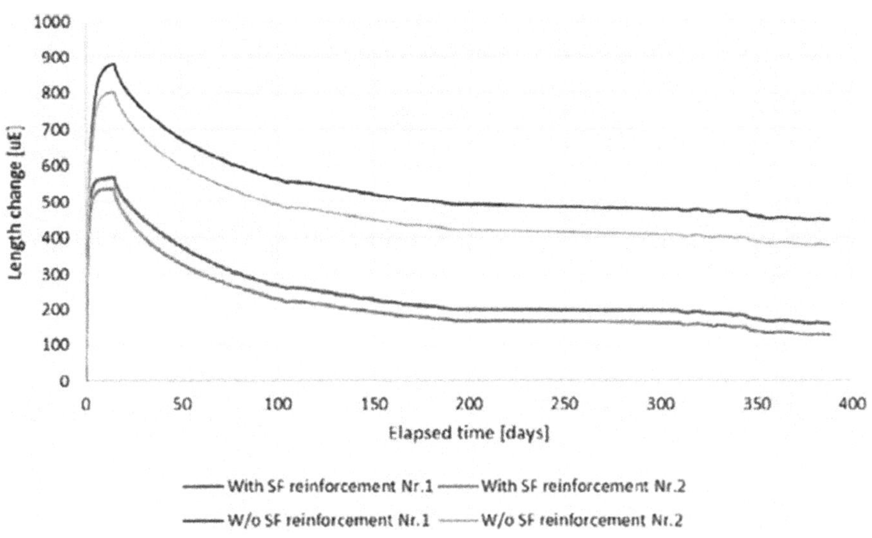

Figure 9.11 Expansion and restraint with and without steel fiber reinforcing of small slab specimens.

concrete, starting from day 1, in the figure 9.9 and 9.10 for different steel fiber reinforced concrete mixtures A to D with an increasing early expansion A to D.

The abscissa in Figures 9.9a is given as the inverse of the square root number days of age in order to be able to extrapolate to an infinite number of days at zero abscissa and also to obtain the permanent expansion.

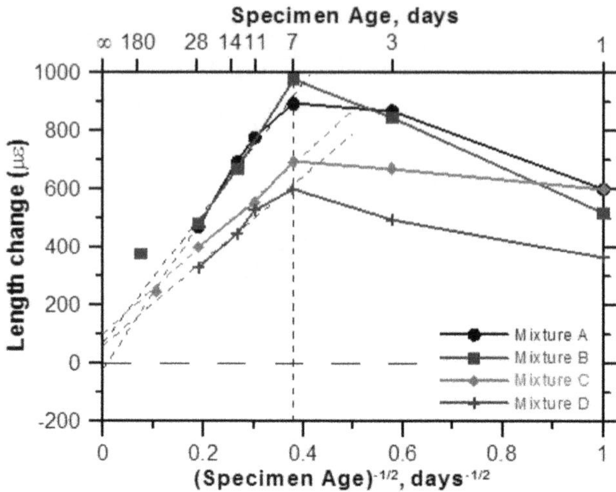

Figure 9.12 Zero shrinkage concrete deformations in function of age in inverse square root of number of days up to infinite time at zero abscissa.

The internal restraint provided by the 40 kg/m³ dosage rate of steel fibers and the friction on the grade is able to contain the initial expansion. Without steel fiber, it is from an expasion of 1000–1200 µstrain down to 75–125 µstrain after completion of the shrinkage, at infinite time.

The expansion and internal restraint by the steel fibers at 40 kg/m³ dosage rate is also measured using two vibrating wire strain gauges installed in X and Y directions on a small slab sample of 600 mm x 600 mm x 100 mm thickness, as shown in Figures 9.10 and 9.11. The scatter of the results with small slabs is significantly reduced. The whole test is reported by R. Cepuritis Dr. Ph.D. CTO. of Primekss, in Reference 7.

The slab tests show that the residual expansion at 400 days is of 1/3 of the 7 days expansion so that the permanent expansion will end up at approx. 100 µstrain, somewhat more than measured on the ASTM C 157 prismatic specimens as shown in Figure 9.12.

The resulting permanent compression stress over the section is of σ_{cp}/γ_c = 1.33 to 2.33 N/mm² where γ_c = 1.5 as the material coefficient. It is indeed a chemical post-tensioning that develops into the slab.

As a result, the slab is much stiffer, as if it was much thicker in traditional. This is combined with the steel fiber reinforcement, totally protected against concentrated point loading punch-out.

In 2023, the Primekss technique, with 15 years of site experiences and further refinements, has resulted in over 15 million m² of completed PrimXcomp slabs.

The PrimXcomp has also been used very successfully in ground-suspended slabs on piles, with about 6 million m² completed since 2011, continuing through to 2024. PrimXcomp concrete is often described as zero-shrinkage concrete as the slab shrinkage becomes no longer detrimental.

In Reference 7, in the case of a piled slab supported on a 3 m x 3 m and 4 m x 6 m grid of piles, it is reported that

> Primekss was able to demonstrate through the extensive testing of local materials (part of the PrimeQuality QA Process) to ACREO and Fairhurst (engineers) that the adoption of an efficient PrimX self-stressing steel -fibre-reinforced-concrete (PrimX SS-SFRC in short) solution would reduce the consumption of both concrete and reinforcement to achieve a net embodied carbon saving of 55 %. It has been demonstrated by comparing the traditional plain reinforced proposal of 260 mm concrete thickness with 180 kg/m^3 reinforcing bar to the 190mm chemically post-tensioned solution with 50 kg/m^3 steel fiber.

The addition of Primekss proprietary binders, cementitious additives, and CPEA liquid admixtures to SFRC controls the concrete shrinkage at lifetime, as shown in the previous figures.

In Figure 9.12, the abscissa is presented such that 1-day is represented as 1, further at 7 days as 0.38, at 14 days as 0.27, and an infinite number of days as 0. At infinite time, the resulting expansion is approximately 75 µstrain, which is about 25% less than that observed with the slab specimen type.

The main advantage of the zero-shrinkage concept is that jointless, crack-free slabs become feasible and fully accessible worldwide, thanks to the robust Primekss concrete process, regardless of weather conditions or the nature of local cement types. With steel fiber reinforcing, the tensile strength of the slab concrete becomes a reliable 3D property that designers can depend on for flexion, shear, punching-out, fatigue, and dynamic actions.

The elimination of curling along the edges ensures that the slab remains in full and permanent contact with the grade, removing negative moment cracking along joints and edges as a loading case. This allows forklift trucks to operate smoothly without encountering bumps at each joint.

A typical permanent compression stress of up to 3.0 MPa develops in the slab section when movement is restrained by friction from the granular base.

In Figure 9.12, at zero abscissa, the change in length is of around 75 µstrain so that µ x E = 1.5 MPa.

Any polythene sheeting under the slab is no longer necessary, though I must admit that some owners still prefer it. In concrete slabs, some form of surface cracking is inevitable, even in post-tensioned slabs and does not affect their performance.

Without the plastic sheet underneath, the construction joint opening is almost zero after several months or more, but some micro-cracking (less than hairline) may appear. With the plastic sheet underneath, there may be limited joint opening, but the surface is almost crack-free.

I repeat here: any customer who wants to exclude all possible cracking should not use concrete and should instead opt for a different material—though I still do not know what that material could be.

The PrimX-Self Stressing SFRC application, for both slabs on grade and suspended slabs on piles, eliminates detrimental cracking—a rare advantage, regardless of the technique used.

I must say I'm quite proud of such an achievement as it results from 45 years of R&D activity against slab shrinkage and for steel fiber reinforcing of concrete.

More than with any other organization I have been involved with, I could focus during the last 20 years with a brilliant team we hired and trained to become the hart and the bones, and the R&D team of the Primekss company.

9.6 DESIGN OF PRIMXCOMP JOINT FREE SLABS ON GRADE

When the Kw of Westergaard is increased from 0.03 N/mm³ to 0.08 N/m³, keeping all other parameters constant, the moment and the maximum permissible point loading intensity are increased by almost 50%.

Hence, a zero-shrinkage fiber-reinforced concrete slab of 100 mm thickness on top of a base showing K = 0.08 N/mm³ becomes suitable in case of point loading intensities of up to 120 kN.

The SFRC needed for the application shows a f_{r3} (EN 14651 average post-cracking flexion strength of concrete) ranging from 4 to 5 N/mm².

When compared to traditional slabs, 80–100 mm of concrete thickness is saved. As shown in Figure 9.13, a design diagram summarizes the thickness required as a function of the statistical point loading intensity for a steel fiber-reinforced zero-shrinkage concrete slab by Primekss.

The diagram has been calculated for Kw = 0.08 N/mm³ (equivalent to 300 pci) and indicates the load intensity of a single leg in a back-to-back configuration, as shown in Reference 8.

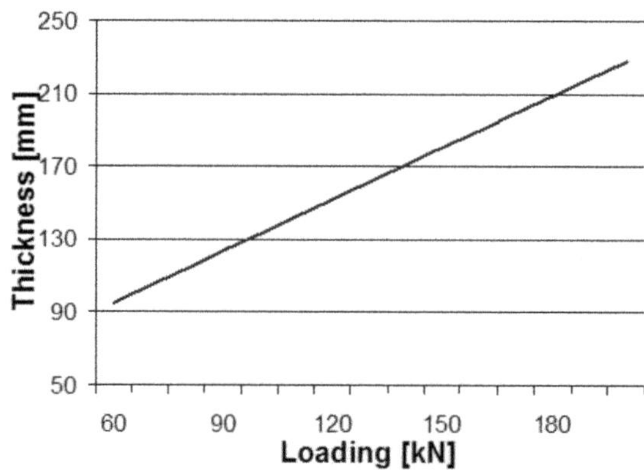

Figure 9.13 Design diagram for Primekss Zero-Shrinkage floors.

The thickness as a function of the. single Point Loading Intensity of a back-to-back leg case. When K_w = 80 MPa/m

9.6.1 Practical design

Figure 9.13 is the resultant from the following equation to obtain the PrimeXcomp slab thickness H (mm) as a linear function of Q (kN) of the single leg intensity of a back-to-back case:

$$H\left(80\,MPa/m\right) = 0,45 \times P + 48, \text{where } P = 2Q(kN) \tag{9.2}$$

We will see further that this formula leads to traditional SLS loads at maximum intensity.

When the value of Kw is no longer 80 MPa/m, instead a K2 value, the thickness is given by the following expression:

$$H\left(k2\right) = H\left(80\,MPa/m\right) \times (80/k2^{1/4} \tag{9.3}$$

For example, a total back-to-back loading intensity of P=120 kN on a Kw value of 80 MPa/m requires H (80 MPa/m) = 0,45 x 120 + 46 = 100 mm PrimXcomp thickness.

When the K-value drops to 30 MPa/m (100 pci), the PrimXcomp slab needs to be thicker.

In my opinion, the best EN 14651 performing hooked-end fibers in regular concrete are with length 0.8 mm x 60 mm and with a wire tensile strength of 1200 MPa. However, the aspect ratio of 60/0.8 = 75 is more prone to fiber-like appearance in the finished surface.

To get the best EN14651 performance, 60 mm length is preferred.

HE 0.9/60 1200 MPa steel fiber is generally close in EN 14651 performance to the HE+1/60.

Higher EN 14651 pure performance could be obtained by more slender fibers with higher aspect ratios but we could however observe more fiber showing in the surface of slab.

More slender fibers are also more costly in kg/m³ so that some manufacturers would propose more slender fibers at a much smaller dosage rate, sometimes down to 15 kg/m³, for jointfree slabs or proposed to combine them with a light top mesh. Sometimes the slender fiber unit price becomes so high that a mesh becomes cheaper.

HE 0.9/60 1200 MPa is generally nearly similar in EN 14651 performances to HE+1/60 steel fibers.

To get the best EN14651 performance, the 60 mm length of fiber is preferable.

9.6.2 Effect of better k-value

$$H(30\,MPa/m) = 102\,mm + \left(80/30\right)^{1/4} = 100 \times 1,28 = 128\,mm$$

When using a PrimXcomp slab, a full continuity is obtained across the construction joints, and the only loading case to verify is that of the center point loading. The edge and corner cases of point loading are not considered anymore.

The design is then very simple, limited to one case and following the law here in the design diagram shown in Figure 9.13.

Consequently, there is no need of running the overly complicated edge thickening or reinforcing like in traditional methods.

9.6.3 A full-scale loading test of a Primekss slab close to a construction joint

The purpose of the test was to demonstrate that no reinforcing or thickening of the Primekss slab is needed alongside the construction joint.

The slab in Sweden at Rosenbergs was 130 mm thick, which was the minimum thickness accepted by the owner's consultants, with a dosage rate of 35 kg/m³ of HE 90/60 steel fibers made from the typical compression self-stressing Primekss concrete. The grade was k = 0.08 N/mm³, determined by the Westergaard plate test.

As usual, the mix design included 300 kg/m³ CEM II L 42.5 cement, along with the proprietary Primekss binders and liquid CPEA admixtures, to obtain the standard laboratory ASTM expansion, which stabilizes at an infinite time within 80–100 µstrain restrained expansion.

The slab of 130 mm was accepted by the owner and consultants to carry a leg load of 80 kN in back-to-back case at 150 mm distance to the joint, together with a 60 kN/m² UDL.

The first striking observation by the owner engineers was a "completely closed construction joint of the Alpha type. This is a benefit resulting from the chemical self-stressing effect. Indeed, the more the construction joint opens, the weaker it is and the more reparations will be needed during its life span.

The loading was organized by stacking, as shown in Figure 9.14, with up to 32 thick steel plates weighing 1.92 tons each, resting on four legs, two of which were positioned 150 mm from the joint.

The maximum possible experimental leg load attained was 154 kN per 100 mm x 100 mm base plate in the back-to-back case, as shown in Figure 9.15.

At 2 x 80 kN in back-to-back loading, the slab showed no cracks or visible deformations.

The loading process continued up to a maximum of 616 kN (4 x154 kN). At (2 x 154 kN) in back to-back loading at 150 mm from the joint, initially, no cracks were observed. After 15 minutes, a hairline crack of 0.2 mm opening appeared but closed back to zero after unloading. At this stage, the crack opening didn't even attain the 0.5 mm of the f_{r1} (EN 14651) limit for the service condition (SLS).

The maximum deflection at the construction joint was 2 mm.

The test team of Invator laboratory didn't expect such a high resistance, and no additional steel plates were available on-site to increase the loading intensity beyond 616 kN. This prevented reaching the ULS limit, defined by

Figure 9.14 Stacking up of the 32 steel plates of 19.2 kN own weight each.

Figure 9.15 The back-to-back loading at 150 mm distance to the construction joint and seen just behind the legs.

1.5 x SLS load limit. The 3 mm crack opening at the f_{r3} flexion stress level at the ultimate state(ULS) was thus never attained.

The SLS condition of the SFRC is generally by standards defined by a 0,5 mm crack opening at f_{r1} flexion stress level, which was indeed by far, never attained here.

Deriving it from the test, one can estimate prudently the maximum service point loading intensity, at 125 kN instead of 80 kN by contract.

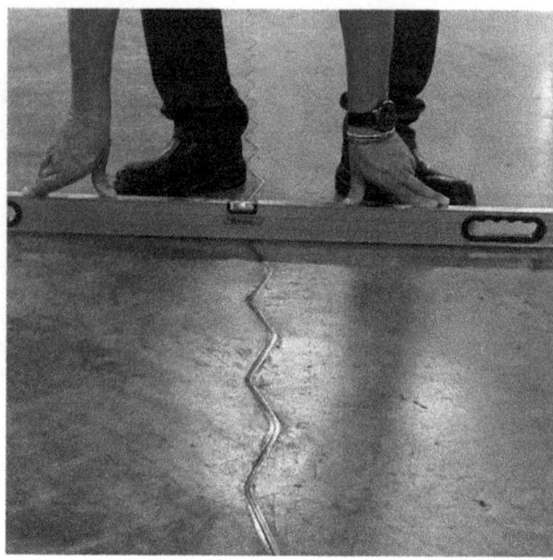

Figure 9.16 No curling of a Primekss slab after 4 years of full service (South Africa).

The slab could be however thinner to resist the 2 x 80 kN point loading in back-to-back case.

The reduction of thickness can be of $(125/80)^{1/2} = 25\%$, thus 100 mm nominal thickness.

Note that the Primekss design diagram derived from the Vaasa Full-scale test of 2009 in Reference 8 shows 115 mm thickness for 80 kN leg loading in back-to-back case, and thus 110 mm indeed under 2 x 125 kN. The full-scale test procedure at Roserbergs confirmed that the Primekss design diagram in Figure 9.12 and the formula are quite conservative.

Indeed, H = 0.45 x (2x 80) + 48 = 120 mm

$$H = 0.45 \times (2 \times 135) + 48 = 170\,\text{mm}$$

The end user and owner could have opted for a thinner Primekss slab, reducing the thickness to 110 mm instead of 130 mm.

The thickness tolerance at Primekss is maintained within a 15 mm variation, ensured by systematic quality control during slab installation.

The Primekss self-stressing SFRC jointfree slab system offers the following-decisive advantages:

– Elimination of any detrimental shrinkage, thus total crack control at life and no curling of joints at all. -Full-service ability at whole slab life with invariable surface tolerances as shown over a construction joint is shown in Figure 9.16.
– No surface grinding is needed. Unlimited joint distance nor shape and size limitations.
– Up to 50 % slab thickness reduction without traditional reinforcing; Very significant savings as well as significant time savings on the job site.

Figure 9.17 A Primekss slab at a construction joint.

- Primekss supplies a 10 or 15 year full usage guarantee because the Primekss Self-Stressing SFRC floors are quasi-fully continuous, shrinkage free, smooth, and thus free of maintenance at life like these shown in Figures 9.17 and 9.18 (such as a 90 mm thickness PrimX slab in South Africa after 4 years of usage). Observe that in both figures, there is no joint opening after the operations on it during 4 years.

Once, many years ago, I heard at an ACI session that all slabs are good between cracks and joints.

Primekss self-stressing SFRC are even better as the construction joints are movement-free, and cracks are minimal, limited to the very surface of the slab, as demonstrated by the 90 mm slab shown in Figure 9.16.

9.7 PRIMXCOMP PILED SLABS

Two full-scale tests have been conducted on PrimeComposite slab-on-piles concrete floors: one in Klaipeda (Lithuania, 2011) and another in Gothenburg (Sweden, 2014) for the Tingstad project.

The Klaipeda slab was of a 210 mm thickness with a dosage of 50 kg/m³ of Twincone steel fibers, placed on a pile grid of 4 m x 4 m with 1 m x 1 m pile heads. It was designed to withstand a loading of 30 kN/m² loading. The full-scale load test, shown in Figure 9.19, involved imposing a 30 kN/m² loading intensity on a 100 m² area for a duration of 3 months. After 3 months, the maximum recorded deflection was 1.5 mm, even though the ground underneath had settled, losing contact with the PrimeXComp suspended slab.

The slab at the Tingstad project was designed with a thickness of 220 mm, deepening to 250 mm over the pile heads. The 300 mm diameter piles

Figure 9.18 Primekss slab of 90 mm thickness.

Figure 9.19 Full-scale loading test of the Klaipeda (Lithuania, 2011) PrimXComp slab-on-piles.

featured 1 m diameter heads and were spaced on a 4.0 m x 4.7 m grid. The slab was reinforced solely with 55 kg/m^3 of HE+ 1/60 steel fibers, serving as the structural reinforcement to meet the 40 kN/m^2 service load requirement.

The full-scale testing at Tingstad (Sweden) project, shown in Figure 9.20 was carried out by the Swedish Cement and Concrete Research Institute

Figure 9.20 Full-scale loading test of the Tingstad project (Gothenburg, Sweden, 2014) PrimXComp slab-on-piles.

(CBI). The full-scale test procedure involved a distributed load of 44.8 kN/m² over the loaded area of the slab. Load was applied and held constant for 8 days. There was a 21 mm gap underneath the slab such that it was under fully suspended elevated conditions.

During loading application, deflections of the slab were very limited and there were no signs of distress or structural failure (e.g., no excessive and permanent deflections, significant cracking, development of yield lines, etc.). The average pile settlement was 0.95 mm after 8 days of loading, and the maximum differential mid-span deflection of the slab, calculated as the mid-span settlement minus the average pile settlement, was 2.3 mm.

These results proved that the combination of Primekss' concrete technology and HE+1/60 steel fibers at a high dosage rate have created a PrimeComposite slab (PrimXcomp) that provides a very high stiffness. The slab easily supported full-scale load testing, proving that the slab has the required load-bearing capacity and that the design assumptions for the slab are correct.

Primekss self-stressing SFRC slabs from 90 mm thickness and up to 4000m²of bay size between construction have been built and used for years in Scandinavia, Baltics, South Africa, and the USA. Note that the construction joints remain closed and don't show any curling after these years.

9.7.1 A Simplified rapid method of design

The PrimeXComp structural slab thickness can then be given by the following experimental formula found in Reference 9 by Destrée and Cepuritis:

$$H = 0.69 \cdot \left(R_p / \left(f_{R,3} + \sigma_{cp} \right) \right)^{1/2}, \tag{9.4}$$

where R_p is the total unfactored pile reaction, $f_{R,3}$ is the flexural strength of the SFRC according to EN 14651, and σ_{cp} is the permanent PrimeXComp post-tensioning compression stress. When pile heads are used, as calculated above, H has to be decreased by 20 mm thickness.

Then for the above-described Tingstad project full-scale testing case: $f_{R,3}$ = 5 N/mm², $\sigma_{cp}/1.5$ = 2.0 N/mm², R_p = 4.0 m · 4.70 m · (40 + 5.28) kN/m2 = 851254 N = 851.54 kN.

$$H = 0.69 \times \left(851254 / 5 + 2\right)^{0.5} = 240\,mm.$$

Construction joints are installed at ¼ of the span between the piles. Example of a case: Project NCC/Göteborg:

3.80 m x 3.80 m; 40 kN/m² UDL ; 250 mm; f_{r3} = 4.5 N/mm², steel fibers at 45 kg/m³ dosage rate, no chemical post-tensioning here, thus a regular SFRC solution without the chemical post-tensioning.
H = 0.68 x (46 000 x 3.80 x 3.80 / 4.8) $^{0.5}$ = 253 mm rounded at 250 mm
Idem with p-t at σ_{cp} = 2.50/1.5 = 2 N/m²
H = 0.68 x (46 000 x 3.80 x 3.80 / 6.47) 0.5 = 218 mm rounded at 220 mm.

The chemical post-tensioning can save here, 30 mm of slab thickness and ensures the optimal shrinkage cracking control.

9.7.2 Comparative Full-Scale Tests (Fall 2023 at Jelgava University, LV.)

The comparison was done between regular SFRC and PRIMEKSS Self-Stressing SFRC, both at 50 kg/m³ of HE 90/60 steel fibers.

Two adjacent slabs of 150 mm thickness and of ca. 300 m² area each thickness have been built as suspended 2 m above the ground level and supported by a grid of columns of 300 mm diameter, 3 m spans center to center, in both ways as shown in Figure 9.21.

The two slabs have been loaded by UDL to full SLS service conditions, to ULS and then further to collapse. The UDL footprint of the big bags stacked-up was an L-shaped distribution of loads as shown in Figures 9.22 and 9.23, in order to activate the highest possible two-way moment and shearing over the piles.

The L-shaped loading pattern is shown in Figure 9.23
The final collapse patterns are shown in Figure 9.24:

The self-stressing Primekss SFRC is shown like a fan pattern from the top while the regular SFRc pattern is shown from the bottom and denotes a more classical yield line pattern from edge to edge.

The quasi-fan pattern observed in the top of Figure 9.22 with SS-SFRC results from a stiffer slab compared to the regular SFRC slab shown at the bottom of Figure 9.22.

Figure 9.21 Aerial view of the two slabs tested.

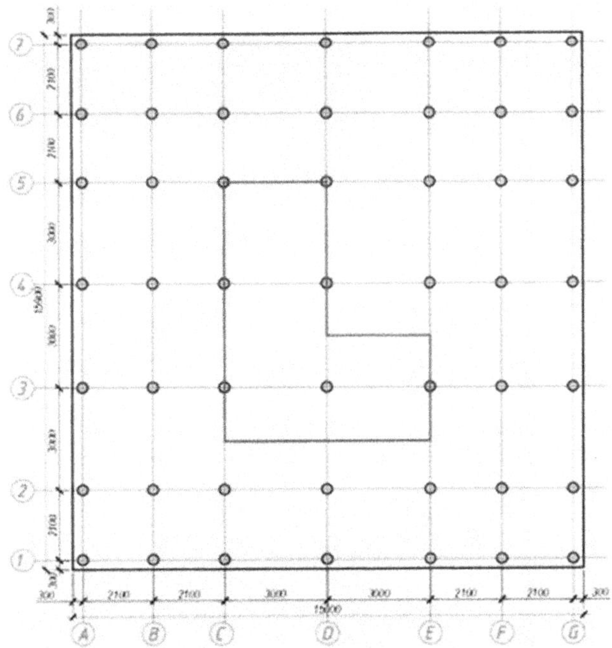

Figure 9.22 L-shaped UDL footprint.

Figure 9.23 Aerial view of the UDL.

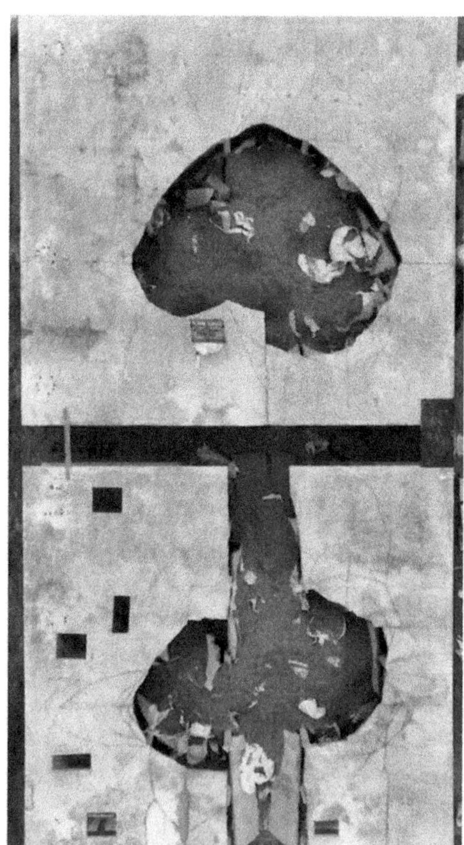

Figure 9.24 Aerial view of the collapse pattern of both slabs (SS-SFRC at top/ SFRC and at bottom).

Table 9.2 Summary of observations and conclusions about the structural full-scale tests

		SFRC	*Self-stressing SFRC*
1	Cracking in number	N	N/3
2	Cracking in opening	O	O/2
3	Deflection at 17 kN/m²	3 mm (S/1000)	2.5 mm (S/1200)
4	Deflection at 25 kN/m² after 1 week loading	6,5 mm (= L/462) (> L/500)	4.5 mm (= L/667) (< L/500)
5	Deflection at 17 kN/m²	1.8 mm	1.1 mm
6	Permanent deflection after unloading of 25 kN/m²-1 week	3.5 mm	2 mm
7	**Collapse Loading intensity**	44 kN/m²	61 kN/m²
8	**SLS allowable loading**	21 kN/m²	27 kN/m²

Table 9.2 is an overview of the observations and the conclusions for designing.

Line 5 shows the deflection after 1 day of loading, while Line 3 represents the deflection after 1 week. The number of cracks in the Self-Stressing SFRC (SS-SFRC) is one-third that of regular SFRC, and the crack openings are half as wide. The SS-SFRC is significantly stiffer than the regular SFRC.

The number of cracks, their locations, and their openings were recorded using 800 linear meters of internal optical fibers installed at the top, half-depth, and bottom of each slab. Prof. Dr. G. Fischer of TU Denmark developed this system for Primekss during the two full-scale suspended slab tests.

No shear collapse has been observed but well some micro cracks around the piles at ULS on the surface of SFRC slab only.

We have compared SFRC vs. SS-SFRC in observed facts (at 50 kg/m³ HE90/60 steel fiber dosage rate):

- L2 and L3 are precisely re-calculated by dividing the number of big bags by the loaded area so that Line 2 = 17 kN/m² and Line 3 = 25 kN/m².
  Line 2 and Line 3 intensities don't include the own weight of the slab.
- Conclusions for designing: Line 3 shows 20% less "δ" deflection for SS-SFRC. SS-SFRC can be thinner than SFRC by 8%
- Line 4 shows that the SS-SFRC can carry more loading to obtain 6 mm deflection (L/ 500) thus more loading by 8% thus 27 kN/m² instead of 25 kN/m³ (4.5 mm x 1.08 +1 mm extra creep = ca. 6 mm = L/500).
- The line 5 shows that SFRC can carry more loading than 17 kN/m². ArcelorMittal TAB-Structural calculation sheet allows for such an SFRC slab, ca. maximum 21 kN/m². Under the same loading intensity, based upon piled slab loading, the SFRC could be 13% thinner.
- Practical conclusions for design are: SS-SFRC is thinner than SFRC by 8% as a minimum.

- The maximum SLS loading (line 8): The maximum allowable SLS loading intensity for the SFRC slab is of 21 kN/m² and for Primeks self-stressing SFRC of 27 kN/m².

A complete report of this full-scale test will be prepared by the Technical University of Riga (Prof. Popescu), M. Suta, an engineer at Primekss, and Ph.D. student at Riga TU. The investigation by laser optical fibers has been carried out by Prof. Dr. G. Fischer of TU. Denmark, and Prof. Dr. Tjalsten of the University of Lulea was in charge of providing technical advice for the full-scale tests.

Note that a TR34 4th Ed. compliant design would have required a 200 mm slab thickness for the same SLS loading compared to the 150 mm used here.

9.7.3 Jelgava FST back calculation following the ACI544-6R15

The calculation verification is done according to ACI 544 6R15 of the American Concrete Institute.

$$t := 150 \cdot mm$$

Slab thickness

$$f_{ck} := 35$$

Cylinder strength (characteristic)

$$f_{ckc} := 45$$

cube strength

$$f_{ctm} := 0.3 \cdot \left(f_{ck} \right)^{\frac{2}{3}} = 3.21$$

Tensile strength of plain concrete

$$\omega := \frac{f_{ck}}{f_{ctm}} = 10.904$$

$$f_{r3} := 5.000$$

$$\sigma_{cr} := f_{ctm}$$

50 kg/m³ HE+-90/60 EN 14651 - characteristic typical value of EN 14651 flexion strength at ULS

Ratio of strengths

$$\mu := \frac{f_{r3}}{3.104 \cdot f_{ctm}} = 0.502$$

ACI 544 8R16

$$u := 1 \cdot \frac{N}{mm^2}$$

$$m_{Rdaci} := \frac{3 \cdot \omega \cdot \mu}{\omega + \mu} \cdot f_{ctm} \cdot \frac{t^2 \cdot u}{6}$$
$$= 17.324 \, kN \cdot \frac{m}{m}$$

(Eq 6 4 1b and H 8)

The calculation verification is done according to ACI 544 6R15 of the American Concrete Institute.

$$\gamma_f := 1.5$$

$$\sigma_{cp} := 1.3 \cdot \frac{N}{mm^2}$$

Material factor
PrimeComposite
permanent compressive
stress due to the chemical
post-tensioning

$$m_{RPC} := \frac{3 \cdot \omega \cdot \mu}{\omega + \mu} \cdot f_{ctm} \cdot \frac{t^2 \cdot u}{6} + \frac{\sigma_{cp} \cdot t^2}{6 \cdot \gamma_f}$$

$$= 20.574 \, kN \cdot \frac{m}{m}$$

SFRC resisting moment

$$\gamma_c := 24 \cdot \frac{kN}{m^3}$$

Concrete volume mass

$$q := 27.00 \cdot \frac{kN}{m^2}$$

Unfactored UDL at SLS

$$W_G := \gamma_c \cdot t = 3.6 \, \frac{kN}{m^2}$$

Slab unit weight

$$D := 300 \cdot mm$$

Pile diameter

$$L_x := 3.00 \cdot m$$

Spans

$$L_y := 3.00 \cdot m$$

$$\lambda_{DL} := 1.35$$
$$\lambda_{LL} := 1.5$$

Load factors on the own weight
variable load factor

$$\lambda_Q := \frac{(\lambda_{DL} \cdot W_G + \lambda_{LL} \cdot q)}{W_G + q} = 1.482$$

Average load factor

$$L_{rx} := L_x - D - t = 2.55 \, m$$
$$L_{ry} := L_y - D - t = 2.55 \, m$$

Net spans (ACI 544 6R15)

$$\Phi_P := 0.90$$

Reduction factor of resisting
moment

$$M_{Px} := \left(\frac{(\lambda_{DL} \cdot W_G + \lambda_{LL} \cdot q) \cdot (L_{rx})^2}{16} \right)$$

$$= 18.435 \, \frac{1}{m} \cdot kN \cdot m$$

Design moment of a span in the
middle of the slab (eq.7.5.3c
and 6.3 11.1m)

$$\frac{M_{Px}}{\Phi_P \cdot m_{RPC}} = 0.996$$

< 1, OK

punching shear verification:

$$V_{pc} := \left(2 \cdot \pi \cdot \left(\frac{3 \cdot t + D}{2}\right) \cdot t \cdot 0.66\right) \cdot \left(\mu \cdot \sigma_{cr} + 0.175 \cdot \frac{\sigma_{cp}}{u}\right) \cdot u = 428.814 \, kN$$

Shear resistance at critical perimeter (eq.7 6a)

$$R := \left(\lambda_{DL} \cdot W_G + \lambda_{LL} \cdot q\right) \cdot L_x \cdot L_y = 408.24 \, kN$$

Total factored reaction

$$R_{SLS} := \frac{R}{\lambda_Q} = 275.4 \, kN$$

Total unfactored reaction

$$\frac{R}{V_{pc}} = 0.952$$

<1, OK

Deflection

Long-term modulus of elasticity PC DC (after 1-week L3 loading)

$$E := 13500 \cdot \frac{N}{mm^2}$$

$$\delta := \frac{0.185 \cdot \left(\left(q - \frac{0.77 \cdot \sigma_{cp}}{f_{r3} \cdot u} \cdot q\right) \cdot L_x \cdot \left(\frac{L_{rx}}{t}\right)^3\right)}{E}$$
$$= 4.362 \, mm$$

Eq J.2 (ACI 544-6R15) deflection: A new deflection expression to take into account of the self-stressing stress Span to deflection ratio > 500. OK! (EuroCode 2 limit) The slab is a lot stiffer! average loading factor

$$\frac{L_x}{\delta} = 687.813$$

$$\lambda_M := \frac{\lambda_{DL} \cdot W_G + \lambda_{LL} \cdot q}{W_G + q} = 1.482$$

Ultimate loading intensity

$$P_{ULS} := \lambda_M \cdot \frac{(W_G + q)}{M_{Px}} \cdot (m_{RPC}) = 50.625 \, \frac{kN}{m^2}$$

Experimental value

$$P_{Collapse} := 61 \cdot \frac{kN}{m^2}$$

$$\frac{P_{Collapse}}{P_{ULS}} = 1.205$$

So, we obtain here:
  UDL under-SLS= 27 kN/m²
  ULS = 50.6 kN/m² -1.48 x 3.6 kN/m² = 45 kN/m²
  UDL at collapse = 61 kN/m²

9.7.4 A detailed method of design of the Primekss Self-Stressing SFRC

As the section undergoes a permanent compression stress due to the restrained expansion caused by the friction on the grade, pile heads, and the internal friction from the steel fibers, the resulting gain in the resisting moment is significant, being increased by a factor α:

$$\alpha = \left(f_L + 0.8 \times \sigma_{cp/\gamma c} \right) / f_L, \tag{9.5}$$

as defined in Reference 9

Where f_L is the first crack strength in bending as shown in the reference.

Using a C 30/37 mix of f_L = 4.5 N/mm², we calculate α = (4.5 + 0.8 · 2.5/1.5)/4.5 = 1.30. The cracking moment is increased by 30% under the influence only of the restrained expansion.

Thus, a traditional steel fiber reinforced concrete piled slab of 200 mm thickness shall be replaced by 200mm/$\sqrt{1.30}$ = 175 mm thickness PrimXcomp slab thanks to the restrained expansion and this is for the same loading capacity, resulting in a stiffer slab, attributable to the absence of cracking thanks to the chemical post-tensioning.

We can now evaluate the contribution of the steel fiber reinforcing and therefore the ULS post-cracking flexion behavior to be taken into account.

The resisting moment of the section at ULS is derived here in (9.6), from the expression found in Chapter 5.6.4. "Constitutive laws" of the FIB Model Code 2010.

$$M_u = \left(0.5 \cdot f_{R3} - 0.2 \cdot f_{R1} \right) \cdot h^2 / 2 + \left(f_{FTs} - 0.5 \cdot f_{R3} + 0.2 \cdot f_{R1} \right) \cdot h^2 / 6, \tag{9.6}$$

where f_{FTs} = 0.45 f_{R1} and all f_{Ri} are average values of the EN 14651.

Under a worked-out and a simpler expression, it then becomes M_u = $(f_{r3} + 0.05 \cdot f_{r1}) \cdot h^2/6$, where f_{R1} and f_{R3} result from EN 14651 testing.

Taking into account the structural indeterminacy or the structural ductility factor η_{det} when top and bottom yield lines, in both OX and OY directions, develop:

$$\eta_{det} = 2, \text{for } 2-\text{ways slabs} \left(SS\,812310 \right) \tag{9.7}$$

and derived from the EN-SS 812310, a design standard for steel fiber reinforced concrete design of structures in Reference 10.

Hence, the expression of M_{rd}, the resisting moment becomes:

$$M_{rd} = \eta_{det}\cdot\alpha\cdot(\alpha f_{r3} + 0.05\cdot f_{r1})\cdot h^2 / (6\cdot\gamma_f),\tag{9.8}$$

where $\gamma_f = 1.5$ is the material factor of the fiber-reinforced concrete.

When the statical indeterminacy is taken into account in the yield line pattern, like in Fig. 9.24, η_{det} shall be ignored; otherwise, it is taken into account twice.

The external design moment expression in case of a q uniformly distributed loading, w being the self-weight of the slab and L_{rx}, the net span or the distance between the negative moment yield lines:

$$M_d = (1.35\cdot w + 1.5\cdot q)\cdot L_{rx}^2 / 16,\tag{9.9}$$

so that the moment equilibrium condition is:

$$M_d = (1.35\cdot w + 1.5\cdot q)\cdot L_{rx}^2 / 16 < \alpha\cdot(f_{r3} + 0.05\cdot f_{r1})\cdot h^2 / (6\cdot\gamma_f),\tag{9.10}$$

The net span as shown in Figure 9.25 is used for the yield moment calculation.

Let's check that, under its most onerous service loading intensity, the slab will not crack under flexion:

$$Max(M_{SLS}) = (w + q)\cdot L_{rx}^2 / 12,\tag{9.11}$$

Is the expression of the peak moment above a pile in an edge span, so that the maximum flexion stress is calculated here:

$$f_{max} = 6\cdot max(M_{SLS}) / h^2 \text{ shall be smaller than } f_L + 0.8\cdot\sigma_{cp}.\tag{9.12}$$

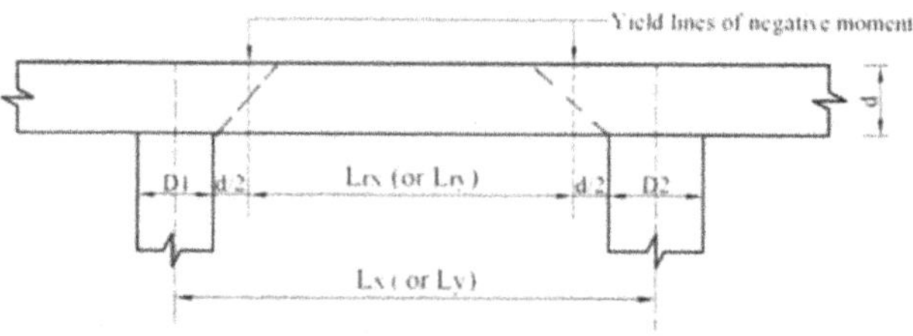

Figure 9.25 Determination of the location of the negative moments yield lines.

A numerical example is needed to understand and compare.

Consider a slab of 210 mm thickness, above a pile grid of 4 m x 4 m, where each pile is provided with a head of 1 m diameter. The slab is constructed using C30/37 concrete reinforced with 45 kg/m³ of HE +1/60 steel fibers (1 mm diameter, 60 mm length, with hooked ends, made from a constituent wire of 1500 N/mm² tensile strength).

Let's try a design for q = 50 kN/m² uniformly distributed loading:

$$L_{rx} = L_{ry} = 4m - 1m - 0.21m = 2.79m, \tag{9.13}$$

as defined in ACI 544 6R15 with the application of the EN 14651 flexion strength of a class d (5.6-1, FIB MC 2010), f_{r1} = 5.0 N/mm² and f_{r3} = 6.0 N/mm². Typically with 50 kg/m³ of HE+1/60 steel fibers, the design moment is:

$$M_d = 39.90 \, kNm/m,$$

and the resisting moment is:

$$M_{rd} = 41.58 \, kNm/m,$$

and thus:

$$M_d / M_{rd} = 0.96 < 1.$$

It provides an acceptable solution when using the yield line method with the shortest possible pattern of yield lines.

On the other hand, the SLS verification:

$$Max(M_{SLS}) = (w + q) \cdot L^2_{rx} / 12 = 35.70 \, kNm/m,$$

and hence we derive the flexion stress,

$$f_{SLS} = 6 \cdot max(M_{SLS}) / h^2 = 4.86 \, N/mm^2,$$

and

$$4.86 \, N/mm^2 < f_L + 0.8 \cdot 2 \, N/mm^2 = 4.5 + 1.6 = 6.1 \, N/mm^2,$$

so that the slab is not cracked in flexion under the most onerous service loading intensity.

The shear resistance is now checked around a pile head where the critical shear section Σ_c:

$$\Sigma_c = \Pi \cdot (d_p + 1.5 \cdot h) \cdot h = 0.86\,\text{m}^2. \tag{9.14}$$

The pile reaction being R_{SLS},

$$
\begin{aligned}
R_{-SLS} &= \left(4\,\text{m} \cdot 4\,\text{m} - (1.31 \cdot 1.31)\,\text{m}^2\right) \cdot \\
&\quad \left(1.35 \cdot 5.04\,\text{kN}/\text{m}^2 + 1.5 \cdot 50\,\text{kN}/\text{m}^2\right) \\
&= 1168\,\text{kN}. \quad (9.15)
\end{aligned}
\tag{9.15}
$$

The critical shear stress is calculated:

$$\tau_d = 1168\,\text{kN}/0.86\,\text{m}^2 = 1.36\,\text{N}/\text{mm}^2,$$

while the shear strength being τR:

$$\tau_R = \beta \cdot \left(f_L + 0.8 \cdot \sigma_{cp}\right) + 0.12 \cdot f_{r3}/\gamma_f, = 1.50\,\text{N}/\text{mm}^2, \tag{9.16}$$

where

$$\beta = \left(0.18\,f_{ck}^{1/3}\right)/\left(0.30\,f_{ck}^{2/3}\right) = 0.193. \tag{9.17}$$

Thus, $\tau_R > \tau_d$, verified.

The governing issues in these expanding yé'elemite mixture-modified steel fiber-reinforced concrete slabs are the ultimate verification and the maximum flexion stress under the service conditions, rather than the traditional shrinkage induced crack opening. The stiffness of the section is significantly enhanced, resulting in a very small deflections, typically in the range of the span/1000 to span/3000.

The expression (J.2) of the ACI 544 6R15 is suitable to calculate the deflection:

$$\delta/L = 0.185 (q + w))/\left(E\,(L/h)^{1/2}\right) \tag{9.18}$$

so that here in the example: $\delta/L = 1200$, where $E = 20.000\,\text{N}/\text{mm}^2$, a long-term value.

As we have observed seen in Figure 9.10, when the peak expansion is reached and that the watered burlaps have been removed, the slab will start to lose vapor water by evaporation of which shrinkage will result.

Although the slab will enjoy a net compression stress in the slab section, after complete drying, some minute cracking is inevitable but is not detrimental since these cracks don't open over the point that chipping-out of the crack edge will start. These minute cracks are limited in length and localized.

The construction joints can still open slightly, from as little as1 or 2 mm to more up to 5 mm, ensuring that no stepping or movement develops. It ensures the smoothest ride of any vehicular traffic throughout the slab's lifespan.

9.8 INSTALLATION OF PRIMXCOMP SLABS

We will only list here the most important tasks:

- To report every observation, recording, measurement, difference, and corrective action.
- To define the final mix design to receive the low shrinkage content plain concrete supplied by the batching plant.
- To check the reactivity of the local cement with the PrimXcomp proprietary constituents, mineral, hydraulic, and organic.
- Checking the defined concrete expansion and shrinkage under the ASTM C 853 standard procedure.
- Checking all preparatory works needed before installing, including the base bearing capacity, and its stable levelness that doesn't degrade under the jobsite traffic.
- At the arrival of concrete trucks, check all physical parameters including its fresh density, which needs to be more than 23.80 kN/m^3, and water content. The slump test is within 50 mm to 75 mm
- The whole Primekss system is installed on-site where all additions are included in a slurry that is pumped into the truck mixer.
- The steel fibers are blown separately at high velocity (25–30 m/s) over the whole surface of fresh concrete in the truck mixer at full speed of rotation.
- Introduce the binder slurry then include all proprietary additions of PrimXcomp materials by means of a pressure air pump into the truck mixer at full speed of rotation.
- Introduce the steel fibers separately at high velocity over the whole surface of mixed concrete in the drum of the truck mixer, by means of a steel fiber blast machine.

The placing of the PrimXcomp concrete is achieved by concrete dumpers with low pressure tires, or with a concrete pump of 125 mm hose diameter and in some case when the base is indeformable, by direct delivery out of the truck mixer.

Table 9.3 Equivalent design of SFRC of HE1/50 fibers vs. traditional steel wire meshes

Thickness (mm)	Wiremesh A193-A2S2-A393	Thickness Dosage (kg/m³)	Thickness Dosage (kg/m³)	UDL intensity (kN/m·j)	Rack leg (kN)	Forklift total weight.(kN)	Truck axle (kN)
120	1A193	120 - 20		30			
120	1A252	120 - 25		30			
140	1A193	125 - 20		35		45	50
140	1A252	125 - 25		35		50	60
150	1A193	130 - 20		40	20	60	60
150	1A252	130 - 25		40	20	60	65
150	2A193	130 - 35	140 - 25	45	25	65	65
150	2A252	130 - 35	150 - 25	50	30	70	75
160	1A193	150 - 20	130 - 25	45	30	75	75
160	1A252	150 - 25	140 - 30	45	30	75	75
160	2A252	140 - 35	150 - 25	50	35	80	85
160	2A252	140 - 30	160 - 20	55	35	85	90
180	1A393	150 - 25	140 - 35	50	35	80	85
180	1A252	170 - 20	160 - 25	55	40	185	85
180	2A193	160 - 30	140 - 40	50	35	85	100
180	2A252	160 - 35	150 - 40	60	45	90	120
200	1A252	180 - 20	160 - 30	60	50	90	95
200	1A393	180 - 25	170 - 25	60	50	90	100
200	2A252	160 - 40	180 - 30	70	55	95	130
200	2A393	190 - 40	190 - 30	70	60	100	130
220	1A252	180 - 20		70	60	100	130
220	1A393	200 - 25		70	65	100	130

220	2A252	190 - 35		70	70	110	130
220	2A393	210 - 45	200 - 35	70	75	110	130
240	1A252	200 - 20					
240	1A393	220 - 25					
240	2A252	210 - 30	180 - 45				
240	2A393	220 - 45	200 - 45				
260	1A252	220 - 20					
260	1A393	220 - 25					
260	2A252	220 - 35					
260	2A393	230 - 35	220 - 40				
300	1A252	250 - 20					
300	1A393	260 - 25					
300	2A252	250 - 30					
300	2A393	270 - 40	260 - 45				

To be watertight the SFRC rafts, water tightness specialty techniques application should be specified. SFRC ground bearing rafts are installed on a ground global reaction coefficient of MIN k = 30 MPa/m

9.9 BASIC EQUIVALENCE OF A WIRE MESH SLAB TO A BASIC STEEL FIBER REINFORCING

In the following traditional Table 9.3, the reader can find the equivalent steel fiber reinforcing to replace one or two layers of steel wire meshes (at the top and bottom of the slab) using a basic steel fiber product that is widely available, user-friendly, and competitively priced: the type HE1/50 -1100 MPa, hooked ends of 1 mm diameter x 500 mm length. of 1100 MPa tensile strength.

The replaced steel wire mesh is of types are A193, A252, or 3A93 mm² sections per meter run. The table is meant to be used rapidly by concrete plants operators or concrete contractors. Typical possible loading conditions for UDL, racking legs, and forklift trucks are mentioned. A Note that thickness of more than 240 mm is typically of ground-bearing rafts.

REFERENCES

1. "Performance-Based Fiber-Reinforced Concrete for slabs-on-Ground and Overlays- Specification (ACI SPEC-544.12-23)"
2. R. Burrows. "The Visible and invisible Cracking of Concrete." December 01-1998, ACI.
3. M. Sahloul, X. Destrée. Practical Investigations into steel fiber reinforced industrial floors. Beton+Fertigteil Teknik, Heft11/1985, 747–750.
4. ACI 223R10: "Guide for the use of shrinkage compensating concrete."
5. *Lea's, Chemistery of Cement and Concrete*, 4th edition, Arnold and John Wiley, 1988, p. 826–827.
6. A.M. Neville. *Properties of Concrete*, 4th Edition, 1996, pp. 446–449, John Willey and sons.
7. R. Cepuritis, K. Grinspons, D.A. Martin. "Chemically post-tensioned steel fiber reinforced industrial flooring slabs." *Floors and Screeds*, 8–10, Concrete (UK) September 2023.
8. Reducing CO2 Emissions of Concrete Slab Constructions with the Prime Composite slab system 9Xavier Destrée, Brad J. Pease. ICCS 2013, ACI-FIB, Tokyo.
9. Chemically Post-tensioned Ultrathin Joint Free Fibre Reinforced Concrete Slabs with Zero Shrinkage on Grade and on Piles, X. Destrée, R. Cepuritis, PrimekssLabs, FIB 2018, Melbourne.
10. EN-SS812310 is a Swedish Standard: Fibre Concrete - Design of Fibre Concrete Structures.

Chapter 10

SFRC design standard documents and slabs

10.1 RELEVANT DOCUMENTS

In general, , we recommend our reader to read in detail the ACI 544-4R18, "Guide to Design with Fiber-Reinforced Concrete" where testings, designs, applications construction, and solved problems are explained.

There are a number of documents we should know as far as SFRC structural slabs are concerned. These documents are listed below.

(0) The EN 14651 flexion test on small beam specimen.
(1) The EC2 Annex L for SFRC is still a draft (2023) but could be applicable, in a likely further modified version, from 2025
(2) The EN-SS812310 "design standard for SFRC structures," a Swedish standard.
(3) The DIN 1045 DAFSTB, a German standard document for the design of the SFRC
(4) The Technical Report no.34, fourth edition (The Concrete Society, UK)
(5) The ACI 544-6R15, regarding the design and construction of SFRC slab structures design
(6) The FIB-Modelcode 2010 and the FIB Bulletin no. 105 are devoted to applications and examples of calculations.

At the end of this chapter, slab-related calculation examples of the seven documents mentioned above are presented.

10.1.1 The EN14651

We have discussed this standard extensively in earlier chapters, so we will not repeat those comments here.

To summarize, it should be noted that the biases of the EN 14651 standard method are numerous.

Its results are affected by considerable and unpredictable scatter. This scatter arises not only because many parameters related to sampling, manufacturing, compacting, placing, and vibrating are imprecise or undefined but also due to issues with the test bench itself.

A fundamental rule is that no lab should use the test results if the experimental scatter remains inconsistent, unpredictable, and uncontained. To achieve reliable results, the labs must analyze the phenomena in-house and develop a methodology for applying the EN 14651 provisions to consistently achieve low and predictable scatter.

Most laboratories have not yet undertaken this effort, despite the EN 14651 standard being available for 20 years.

Table 10.1 shows typical results comparing the same mixes analyzed by four different labs.

Example of standard beam flexion test scatter

Each lab implements precisely all standard specifications for mixing, sampling, testing, and related procedures

Mix 1	same mix design and manufacture.		**LAB. X**	**LAB.Y**
		f_{R3m}	4,72 N/mm²	3,33 N/mm²
		C.O.V:	30%	17%
Mix 2	Idem		**LAB.V**	**LAB.W**
		f_{R3m}	5.26 N/mm²	8.59 N/mm²
		C.O.V:	25 %	13%

The standards EN 14651, DIN 1045-DAFSTB, JSC, and ASTM do not provide guidance on how to achieve consistent average values alongside consistent coefficients of variation (COV).
As a result, SFRC material is often wrongly perceived as erratic, unpredictable, and unsafe. Sampling SFRC by anyone unfamiliar with a consistent sampling method must be excluded to ensure reliability.

It is misleading and downplaying the issue to claim that the EN 14651 flexion test provides the material properties of SFRC in flexion. The EN 14651 test delivers only the standard flexural properties of small, statically determinate, notched beam specimens under center-point loading flexion above the notch.

10.1.2 The EN-SS 812310

Essentially, this design standard explains how to derive a design resistance in flexion or shear based on the strengths obtained from the EN 14651 test.

The EN SS 812310 introduces the important notion and influence of structural indeterminacy, which we discussed in an earlier chapter.

Therefore, a statistical indeterminacy factor η_{det} is defined. For statically indeterminate slabs, this factor is set at 2, reflecting the fact that at ULS, there are bi-directional yield lines in both the top and bottom, increasing the load to reach ULS by 100%. However, this increase is significantly less than what has been observed in the numerous full-scale test slabs we were part of.

The SS-EN 812310 also provides a method to calculate the design shear strength of SFRC slabs. We use it for solely fiber-reinforced slabs that do not include any primary flexural reinforcement.

A 5 N/mm² EN14651 flexion test strength becomes a design strength as 5 N/mm² x 2/1.5 = 6.66 N/mm² where the structural indeterminacy factor η_{det} = 2.

At SLS, it becomes 6.66/1.5 = 4.44N/mm².

10.1.3 The DIN 1045 DAFSTB

This document (also referred to as RICHTLinie/RILI) provides detailed guidance on how to determine the design resistance in flexion or shear for SFRC, based on results from small-sized, unnotched beam tests.

To address the issue of scatter, the characteristic value cannot exceed 70% of the average value. To obtain the design value, the 0.7 × average value is divided by a flat factor of 2 and then assigned to a category within a 0.2 MPa range, where the lowest value of the range is used at ULS.

In the following calculations, we present a typical calculation of the M_{Rd} for a 250 mm thick slab reinforced with only 45 kg/m³ of HE 75/50 steel fibers:

Moment resistance of SFRC based on performance classes:

Element thickness	h =	**250 mm**
Length or crack length of the element	l_R =	**3,5 m**
Concrete grade		**C 30/ 37**
SFRC performance classes	L1 =	**3,0**
	L2 =	**2,1**
	L2/L1 =	0.7
Steel section in tensile zone	a_3 =	**0 cm2/m**
Concrete cover	d_1 =	**50 mm**
Partial safety factor SFRC	γ^f_{ct} =	1,25
Partial safety factor concrete	γ_4 =	1,50
Partial safety factor steel	γ_5 =	1,15
Load duration factor	$\alpha^f_c = \alpha$ =	0,85

Moment resistance of SFRC based on performance classes:

Cracked area	$A_c = h \cdot l_R =$	0,88 m^2
Section of cracked area	$A^f_{ct} = 0,9 \cdot A_c =$	0,79 m^2
Element size factor	$\kappa^f_G = 1,0 + A^f_{ct} \cdot 0,5 < 1,70 =$	1,394
Fiber orientation factor	$\kappa^f_F =$	1.0

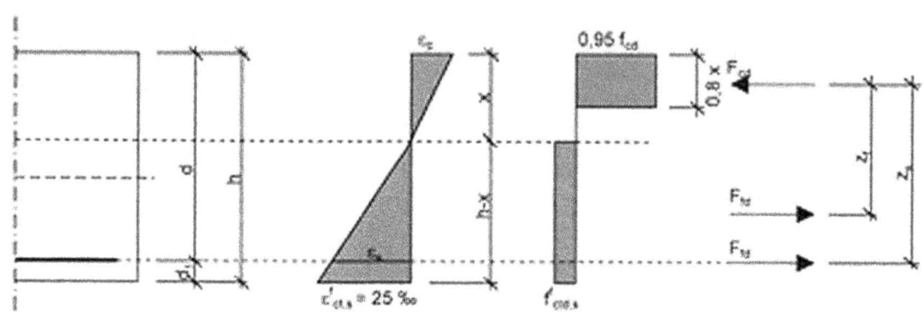

Depth of compressed zone x:	$F_{cd} = 0.8 \times 0.95 \cdot Fcd$	
	$F_{cd} = (h - x) \cdot f'_{ctd,s}$	
	$F_{cd} = a_s \cdot f_{yk} / \gamma_s =$	0.00 kN
	$=> F_{cd} = F_{td} = F_{sd}$	
	$=> 0.8 \times 0.95 \cdot f_{cd} = (h - x) \cdot f'_{ctd,w} + F_{cd}$	
	$=> x =$	13,4809 mm
Concrete strain	$c_t = c'_{ct,w} \, j(h - x) \cdot X$	1,42493 ‰
Applicable equilibrium forces	$F_{cd} = 0.8 \times 0.95 \cdot f_{cd} =$	174,17 kN
	$F_{cd} = (h - x) \cdot f'_{ctd,w}$	174,17 kN
	$F_{cd} + F_{sd} =$	174.17 kN = F_{cd}
Internal arm of forces SFRC	$z_f = h \cdot (h - x)/2 \cdot (0.8 \cdot x/2) =$	126.35 mm
Internal arm of forces reinforcement	$z_s = d - (0.8 \cdot x/2) =$	194.61 mm
Design moment of resistance according to Rili	$\mathbf{M_{R,d} = F_{td} \cdot z_f + F_{sd} \cdot z_s =}$	22,0064 kNm/m

Thus at ULS, M $_{Rd}$ = 22.00 kNm/m for a 250 mm thickness with 45 kg/m³ of HE 75/50 steel fibers in which there is 45 kg/m³ x 0,25m = 11.25 kg/m² steel fiber per square meter of slab.

Now we'll make the same calculation when a very light 6 mm diameter x 150 mm wire mesh of 3 kg/m² of slab is added together with the 45 kg/m³ steel fibers in the bottom of the slab:

Moment resistance of SFRC based on performance classes:

Element thickness	h =	**250 mm**
Length or crack length of the element	l_R =	**3,5 m**
Concrete grade		**C 30/ 37**
SFRC performance classes	LI =	**3,0**
	L2 =	**2,1**
	L2/LI =	**0,7**
Steel section in tensile zone	a_s =	**1,89 cm2/m**
Concrete cover	d_l =	**50 mm**
Partial safety factor SFRC	γ^f_{ct} =	1,25
Partial safety factor concrete	γ_c =	1,50
Partial safety factor steel	γ_s =	1,15
Load duration factor	$\alpha^f_c = \alpha$ =	0,85
Cracked area	$A_c = h \cdot l_R$ =	0,88 m²
Section of cracked area	$A^f_{ct} = 0,9 \cdot A_c$ =	0,79 m²
Element size factor	$\kappa^f_G = 1,0 + A^f_{ct} \cdot 0,5 < 1,70$ =	1,394
Fiber orientation factor	κ^f_F =	1.0

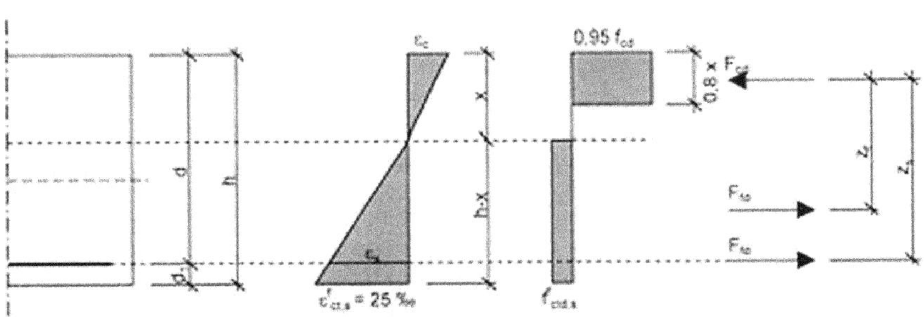

Depth of compressed zone x:	$F_{cd} = 0.8 \times 0.95 . Fcd$	
	$F_{cd} = (h - x) . f'_{ctd,s}$	
	$F_{cd} = a_s . f_{yk} / \gamma_s =$	82.17 kN
	$=> F_{cd} = F_{td} = F_{sd}$	
	$=> 0.8 \times 0.95 . f_{cd} = (h - x) . f'_{ctd,w} + F_{cd}$	
	$=> x =$	19,4981 mm
Concrete strain	$c_t = c'_{ct,w} j(h - x) . X$	2,11475‰
Applicable equlibrium forces	$F_{cd} = 0.8 \times 0.95 . f_{cd} =$	251,92 kN
	$F_{cd} = (h - x) . f'_{ctd,w}$	169,74 kN
	$F_{cd} + F_{sd} =$	251.92 kN $= F_{cd}$
Internal arm of forces SFRC	$z_f = h \cdot (h - x)/2 \cdot (0.8 . x/2) =$	126.95 mm
Internal arm of forces reinforcement	$z_s = d - (0.8 . x/2) =$	192.20 mm
Design moment of resistance according to Rili	$\mathbf{M_{R,d} = F_{td} \cdot z_f + F_{sd} \cdot z_s =}$	**37,3426 kNm/m**

Here, we see that the simple addition of something as minimal as a chicken mesh to the heavily steel fiber-reinforced section increases the M_{Rd} by 37/22=1.6937/22=1.69, or 69%, in the resisting moment of the hybrid section.

It shows how the German DAFSTB/RILI design standard leads to grossly overdesigned SFRC sections.

The equivalent elastic stress for 22 kNm/m is calculated as: $6 \times 22000/150^2$ = 2.11 N/mm² at ULS, which means that at SLS, the value becomes 2.11 N/mm²/1.5 = 1.41 N/mm², which is indeed a third of the first crack in flexion and one-tenth of the collapse load, as established by the numerous full-scale tests detailed in the earlier chapter.

Reviewing the numerous full-scale tests in this book, it is evident that such resisting moment limits confine the SFRC slab to only a small fraction of its actual capacity.

Both the Ternat and Townsville G-SFRS slabs, of exactly the same nature and size, exhibited first cracks at 100 kN under center-point loading, with collapse occurring at approximately 500 kN, a factor of 5 between the two.

According to the DAFSTB standard, these slabs could be used up to only 33 kN in-service loading, which is nearly the collapse load intensity divided by 17.

The domain of utilization for suspended slab tests, as shown in the diagram in Figure 10.2 according to the German RILI D, is represented by the light gray triangular area in the bottom left corner, starting from the origin.

In reality, under the service load SLS, the slab remains fully functional, while the ULS load, a plastic and stable case, involves cracking and deflections well beyond the SLS limit and is calculated as the SLS load multiplied by 1.5. The collapse load exceeds the ULS load by a significant margin.

At ULS, while the slab remains stable, it may exhibit excessive crack openings and deflections, making ULS a stable warning state. At ULS, the building should be evacuated but still stable.

Both the Ternat and Townsville test slabs, as well as the Bissen and the Talinn test slabs, far exceeded all these thresholds at collapse.

I must state that the DAFSTB design method represents a significant step backward and will hinder the development of SFRC structural application.

Once I discussed with a prominent international manufacturer of steel fibers that the German steel fiber DAFSTB design standard is a huge hinder to develop further the SFRC. His reply was that it sells a lot of steel fibers at good price. I would add it does it for slabs on grade with little responsibility as the document is of standard level in Germany.

It doesn't achieve the economical design nor the ecological design, as it results practically in a waste of resources and doesn't address the shrinkage issue.

It is amazing to see prominent engineers in Germany to provide dozens of pages of calculations with finite element calculations software including funny diagrams for a simple overdesigned slab on grade with sawn cuts shrinkage joints with 20 kg/m^3 dosage rates of steel fibers sometimes together with a chicken mesh-like reinforcing.

We should not forget to compare the DAFSTB's 1.41 N/mm^2 maximum SLS flexion stress to the 4.44 N/mm^2 maximum possible SLS flexion stress in service derived from the EN-SS 812310 Swedish design standard mentioned earlier.

It is clear that, in Germany, no structural application of SFRC is possible without a high quantity of rebars in combination. SFRC is mainly used to reduce standard crack openings, while the steel fiber dosage rate is kept low (25 kg/m^3) and combined with traditional rebar reinforcement (100–200 kg/m^3).

Otherwise, SFRC in Germany is not attractive, except for slabs on grade provided with sawn cuts every 6 m apart, as has been the standard practice in the past.

The DAFSTB standard emerged from a long-standing process that began with observations of standard beam specimens in flexion, which showed unpredictable behavior and high scatter. This led to the perception that SFRC is unreliable, often brittle, and prone to significant scatter, making it unsafe.

In my view, the concrete supply industry in Germany appears to be interested in providing SFRC but at very low-performance levels for design. Their responsibility often ends at the exit gate of the concrete plant, minimizing any "risk" associated with SFRC supply. Steel fibers are regarded and used merely as "spices" to traditional reinforced concrete applications.

10.1.4 The TR 34 4th Ed

The document serves as an excellent overall design guide for slabs on grade and piled slabs, addressing both traditional reinforced concrete and SFRC solutions.

However, regarding the design of G-SFRS, the document tends to underestimate the design M_{Rd} (resisting moments) as well as the V_{Rd} (shear/punching-out resistance). Because of the concerns about the adverse effects of shrinkage on joint openings, the recommended distance between joints is effectively limited to approximately 35 m. The focus is primarily on the type and installation of armored joints.

The design method for slabs on grade is based on Westergaard's K_w bearing coefficient, which assumes the ground behaves like an elastic liquid with no shear stresses and all pressures acting vertically, following the relation p = k x w, where w represents the elastic settlement. This approach is rooted in the Westergaard hypothesis.

However, the application of Westergaard's method often results in the overdesign of slabs on grade.

10.1.5 The ACI 544-6R15

The ACI 544-6R15 is a report on the design and construction of SFRC elevated suspended slabs. It is a "must read" document, as it is the most comprehensive resource on structural SFRC slab design.

We strongly recommend our readers to read it in full. The report is the result of a multidisciplinary team of contributors working diligently through a written contradictory process, involving numerous ballots and vetoes. It underwent 256 written vetoes and corresponding solutions before progressing further.

The present book you are reading does not replace or supersede the ACI 544-6R15. Instead, it shares my own experience of over 40 years in SFRC, specifically in slabs and foundations. This book is primarily based on the analysis of real E- and G-SFRS full-scale tests, or quasi-full-scale tests like the round indeterminate panel test, combined with over 30 years of field experience in G-SFRS.

10.1.6 The EuroCode 2 annex L (for SFRC)

The annex L is indeed, a draft that was supposed to be circulated during 2024 but is postponed to maybe 2027. Who knows?

- <u>In the EC 2 Annex L(2024) draft: we read for slabs design:</u> *The only partly replacement of minimum longitudinal tensile reinforcement is possible and limited to 50%.(!!!...)*
  In Reality: The first G-SFRS slab (SFRC slab on piles) was completed in 1993; since then ca. 20 million m² were completed and enjoy today a quite positive track record, all of them without any longitudinal nor shear reinforcement.
  The annex L of EC2-2024 as we know it today, and the TR 34 4th Ed. will continue to generate cost increases and excesses as well as a significantly larger CO_2 Carbon Footprint.
- <u>and for lightly reinforced structures, we read</u>:
  Without longitudinal rebars, the shear strength is limited to 0.37 f r3k/1.5.
  In our example of f_{r3k} = 4.50 N/mm², it shows a EC2 Annex L shear strength of 1.1 N/mm² so that a ULS punching-out load is underestimated (full-scale test experiences showed τ_{Rd} >= 1.95 N/mm²).
- <u>for fatigue, we read</u>: « *For fatigue verification of SFRC members, the contribution of fibers should be neglected* » *!!!*. This is unbelievable.
  According to a rich body of literature spanning 60 years, fatigue resistance has been one of the main advantages of using steel fiber reinforcing. Now, being forced to neglect it feels like a significant regression. This is more than just going backward.

10.2 EXAMPLES OF CALCULATIONS

The calculation of M_{Rd} here below is done according to the ACI 544-6R15.

$t := 200mm$ — Slab thickness

$f_{ck} := 30$ — Cylinder strength (characteristics)

$f_{ckc} := 37$ — Cube strength

Tensile strength of plain concrete

$$f_{ctm} := 0.3 \cdot \left(f_{ck}\right)^{\frac{2}{3}} = 2.896$$

Ratio

$$\omega := \frac{f_{ck}}{f_{ctm}} = 10.357$$

$f_{r3k} := 4.50$ — EN 14651

$\sigma_{cr} := f_{ctm}$

ACI 544 8R16

$$\mu := \frac{f_{r3k}}{3.104 \cdot f_{ctm}} = 0.501$$

$$u := 1 \frac{N}{mm^2}$$

(Eq. 6.4. 1b and H.8)
Structural resisting moment ULS

$$m_{Rdaci} := \frac{3 \cdot \omega \cdot \mu}{\omega + \mu} \cdot f_{ctm} \cdot \frac{t^2 \cdot u}{6} = 27.658 \cdot kN \cdot \frac{m}{m}$$

$t := 200mm$ — Slab thickness

$f_{ck} := 30$ — Cylinder strength (characteristics)

$f_{ckc} := 37$ — Cube strength

Tensile strength of plain concrete

$$f_{ctm} := 0.3 \cdot \left(f_{ck}\right)^{\frac{2}{3}} = 2.896$$

Compressive to tensile strength ratio

$$\omega := \frac{f_{ck}}{f_{ctm}} = 10.357$$

$f_{r3} := 4.50$

$\sigma_{cr} := f_{ctm}$

ACI 544 8R16

$$\mu := \frac{f_{r3}}{3.104 \cdot f_{ctm}} = 0.501$$

$$u := 1 \frac{N}{mm^2}$$

$D := 180mm$

SFRC slab shear capacity

$$v_{pc} := \left[2 \cdot \pi \cdot \left(\frac{3 \cdot t + D}{2}\right) \cdot t \cdot 0.66\right] \cdot \left(\mu \cdot \sigma_{cr}\right) \cdot u$$
$$= 468.931 \cdot kN$$

The same section is calculated here below according to the TR34 4th Edition

---

$H := 200mm$

$f_{R4} := 4.1 \dfrac{N}{mm^2}$

EN 14651

$f_{RI} := 4.2 \dfrac{N}{mm^2}$

$\sigma_{RI} := 0.45 \cdot f_{RI} = 1.89 \cdot \dfrac{N}{mm^2}$

In 6.3.4

$\sigma_{R4} := 0.37 \cdot f_{R4} = 1.517 \cdot \dfrac{N}{mm^2}$

$\gamma_f := 1.5$

$M_u := \dfrac{H^2}{\gamma_f} \cdot (0.29 \cdot \sigma_{R4} + 0.16 \cdot \sigma_{RI}) = 19.795 \cdot kN \cdot \dfrac{m}{m}$

TR 34 6.3.5, (Eq.6-p.23)

$d_t := 200mm \qquad u := 1mm$

Slab thickness

$d := \dfrac{d_t}{u} = 200$

$k_S := 1 + \left(\dfrac{200}{0.75 \cdot d}\right)^{0.5} = 2.155 \qquad k_{smax} := 2$

EC2 6.2.2

$f_{ck} := 30$

Compressive strength

$v_{Rdcmin} := 0.035 \cdot k_{smax}^{1.33} \cdot f_{ck}^{0.5} = 0.482$

Shear strength concrete

$f_{r1m} := 4.2 \cdot 1.2 = 5.04$

$f_{.r2m} := 4.3 \cdot 1.2 = 5.16$

When $f_{rim}/f_{rik} = 1.20$

$f_{r3m} := 4.5 \cdot 1.20 = 5.4$

EN 14651

$f_{r4m} := 4.1 \cdot 1.20 = 4.92$

$v_f := 0.015 \cdot (f_{r1m} + f_{.r2m} + f_{r3m} + f_{r4m}) = 0.308$

Shear strength steel fiber contribution

$D := 180mm$

Pile diameter

$\mu_1 := \pi \cdot (D + 4 \cdot 0.75 \cdot d_t) = 2.45\,m \qquad v := 1\dfrac{N}{mm^2}$

Critical shear perimeter

$P_d := (v_{Rdcmin} + v_f) \cdot v \cdot \mu_1 \cdot 0.75 \cdot d_t = 290.284\,kN$

Shear capacity of stab

The calculation is done according to the EC2-2024 Annex L for SFRC, as known on 26-08-2023

$h := 200\,mm$ — Thickness of slab

$L := 3\,m \qquad u := 1\,m^2$ — Span unit of area

$f_{flk} := 4.20\,\dfrac{N}{mm^2}$ — EN 14651 SLS

$f_{f3k} := 4.50\,\dfrac{N}{mm^2}$ — EN 14651 ULS

$f_{Fts} := 0.40 f_{flk} = 1.68 \cdot \dfrac{N}{mm^2}$ — EN 14651 SLS Tensile strength

$\gamma_f := 1.5$ — material factor

$$M_{Ud} := \left[\left(0.57 \cdot f_{r3k} - 0.26 \cdot f_{rlk}\right) \cdot \dfrac{h^2}{2} + \left(f_{Fts} - 0.57 \cdot f_{r3k} + 0.26 \cdot f_{rlk}\right) \cdot \dfrac{h^2}{6}\right] \cdot \dfrac{1}{\gamma_f} = 20.56 \cdot kN \cdot \dfrac{m}{m}$$

$$M_{Ud} := \left(1.14 \cdot f_{r3k} - 0.15 \cdot f_{rlk}\right) \cdot \dfrac{h^2}{6 \cdot \gamma_f} = 20 \cdot kN \cdot \dfrac{m}{m}$$

Ultimate Material Resisting Moment

$A_{ct} := 0.9 \cdot h \cdot L = 0.54\,m^2$

$\kappa_{rd} := 1.00 + 0.5 \cdot \dfrac{A_{ct}}{u} = 1.27$ — Scale factor $< 1.5 = 1 + A\,\underline{ct} \times .0.5$ $< 1.5\,A$ ct in m^2

$M_{Rd} := \kappa_{rd} \cdot M_{Ud} = 25.4 \cdot kN \cdot \dfrac{m}{m}$ — Structural Resisting Moment ULS including KRd factor

Definition: A_{ct} is the area of the tension zone (m^2) of the cross-section involved in the failure of an equilibrium system.

10.3 COMPARISONS

In Table 10.1, standard values of M_{Rd} and V_{Rd} are compared:

Table 10.1 Comparison of standard values

Comparison table at ULS:		
G-SFRS of 200 mm:	M_{Rd}	V_{Rd}
ACI 544-6R15:	28 kNm/m	469 kN
TR 34 4th Ed.:	20 kNm/m	290 kN
EC2, Ann. L 2024:	25 kNm/m	None
		rebars are needed (mandatory)
DIN 1045 DAFSTB:	13 kNm/m	**idem**
Full Scale Tests:	31 kNm/m	**> 600 kN**

According to the EC2 Annex L draft for SFRC (2024), the following is understood:

For slabs:

The partial replacement of minimum longitudinal tensile reinforcement is allowed but limited to 50%

Reality: The first G-SFRS slab (SFRC slab on piles) was completed in 1993. Since then, over 20 million m² of such slabs have been constructed, all without any longitudinal or shear reinforcement, and they maintain a positive track record to this day.

The Model Code 2010 is slightly less stringent than the Annex L draft for the next Eurocode but still results in significant overdesign as it does not focus adequately on specific applications.

In Reference 1 by Lin, de la Fuente, regarding the SFRC tunnel segments of the Barcelona metro, it states "The direct application of the requirement from the MC 2010 would render the use of fibers as the only reinforcement almost unviable... In contrast the new philosophy proposed here yields 42.3 kg/m³(45 kg/m³). This represents a reduction of approximately 31% in the fiber consumption and provides enough ductility with a safety margin for the design load considered...."

This design conclusion aligns well with Table 10.1, which shows that SFRC design standards result in varying degrees of overdesign—ranging from 10% for ACI 544 6′R15 to as high as 300% for the German Din 1045 DAFSTB.

10.4 CURVATURE OF SMALL SPECIMEN BEAM TEST IN FLEXION

We can compare the curvature of the round indeterminate panel test (ACI 544 6R15), which has a 1.5 m clear span and a 150 mm depth, to the curvature of an EN 14651 notched beam test.

The same curvature abscissa is achievable for specimens of significantly different sizes, allowing the superimposition of diagrams for small beams, round indeterminate slabs (essentially quasi-full-scale), and full-scale slabs. This comparison is illustrated in Figure 10.1.

We observe that at a 0.5 mm deflection (corresponding to the f_{r1} level, applicable for the service limit state), the EN 14651 small notched beam specimen undergoes a curvature of 8/1,000,000. In comparison, the round slab achieves the same curvature at a 4.5 mm deflection (approximately span/333, greater than span/500) under its maximal possible point-centric loading intensity of 180 kN.

At a 2.5 mm deflection (representing the f_{r3} level of ULS), the beam undergoes a curvature of 4/100,000, while the slab shows a 22 mm deflection (approximately span/68).

The f_{r4} level (3.5 mm deflection) corresponds to a beam curvature of 5.6/10,000, which does not even exist in indeterminate slabs as it would

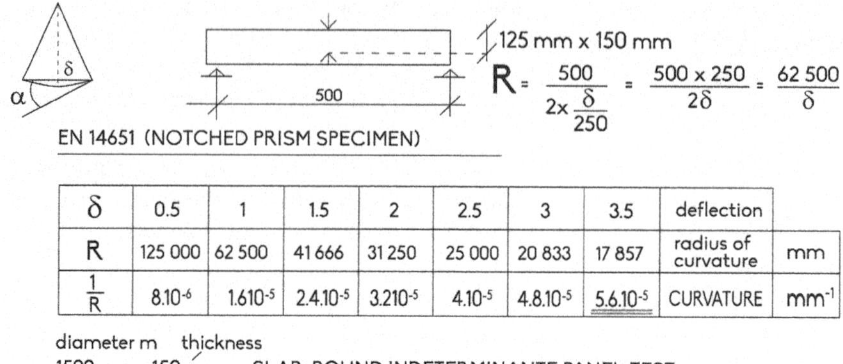

$$R = \dfrac{500}{2 \times \dfrac{\delta}{250}} = \dfrac{500 \times 250}{2\delta} = \dfrac{62\,500}{\delta}$$

δ	0.5	1	1.5	2	2.5	3	3.5	deflection	
R	125 000	62 500	41 666	31 250	25 000	20 833	17 857	radius of curvature	mm
$\dfrac{1}{R}$	8.10^{-6}	$1.6.10^{-5}$	$2.4.10^{-5}$	$3.2.10^{-5}$	4.10^{-5}	$4.8.10^{-5}$	$5.6.10^{-5}$	CURVATURE	mm^{-1}

diameter m   thickness
1500mm x 150 mm   SLAB: ROUND INDETERMINANTE PANEL TEST
ACI 544-6R15

$\dfrac{1}{R}$	8.10^{-6}	$1.6.10^{-5}$	$2.4.10^{-5}$	$3.2.10^{-5}$	4.10^{-5}	$4.8.10^{-5}$	$5.6.10^{-5}$	
δ	4.5	7.2	10.8	18	22.5	27	31.5	mm^{-1}

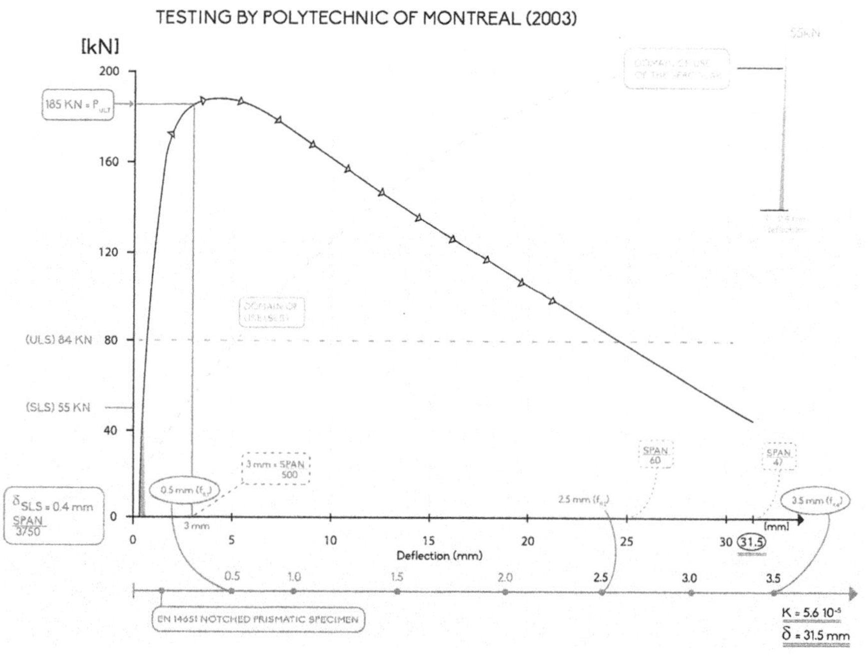

Figure 10.1 Loading vs. curvature diagrams.

imply a 31.5 mm deflection (span/48)—a condition unrealistic for such slabs.

The same analysis applied to full-scale elevated indeterminate slabs, as presented at the BEFIB 2008 conference and shown in Figure 10.2,

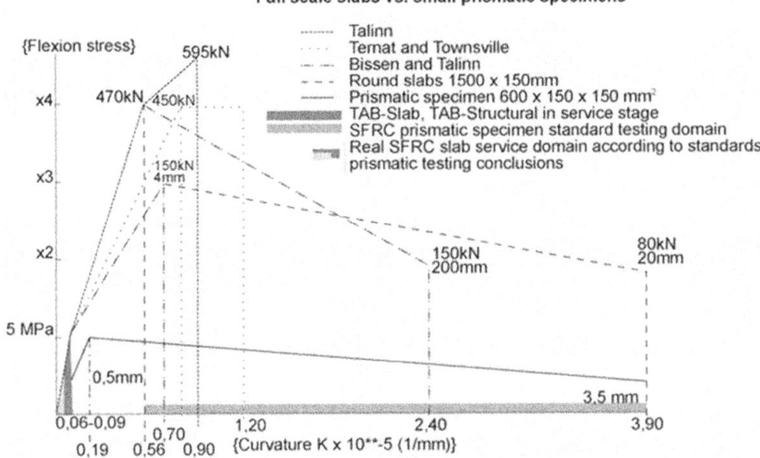

Figure 10.2 Flexion stress vs. curvature diagram of EN 14651 notched prismatic specimen and full-scale elevated suspended SFRC indeterminate slabs. The steel fiber dosage rate is 45 kg/m³.

confirms that the f_{r3}(SLS) and f_{r4} (ULS) levels, defined for small EN 14651 beam specimens, are irrelevant for real-world slab behavior.

The most frustrating aspect is that both f_{r3} and f_{r4} are divided by a factor 1.5 to 3 to calculate the resisting moment at ULS, and further divided by 1.5 to derive the SLS case.

In Figure 10.1, the small green triangular area represents the domain of use for the round indeterminate slab subjected to test loading, as defined by the 2024 draft of Eurocode 2, Annex L for fiber-reinforced concrete.

Such a slab capable of carrying 185 kN at collapse may, according to the draft of Annex L in the Eurocode draft, be used in service (SLS) for a static load of up to 25 kN at most. Other standards allow slightly varying limits: 27 kN by ACI 544-6R, 20 kN by TR34 4th Edition, and 11 kN by the German DAFSTB-RILI.

The practical flexion limit for a two-way suspended elevated SFRC slab is approximately 5 N/mm² to 6 N/mm² at SLS. This corresponds to a practical loading limit of 55 kN, as shown in Diagram 10.1.

Note that at a 55 kN SLS maximum load, the ULS load is 1.5×55 kN=83 kN1.5×55kN=83kN, which is significantly lower than the collapse load of 185 kN.

With the application of the structural indeterminacy factor η_{det} = 2, as per the EN-SS 14651 design standard for fiber-reinforced structures in the case of two-way indeterminate slabs, the load-bearing capacity of the plate shown in Figure 10.1 approaches 55 kN. In contrast, the German method lags far behind, allowing only 11 kN when applied.

The small green triangle area in the load vs. deflection diagram represents a very limited domain of use compared to the entire area under the curve up to the maximum load of 180 kN and beyond, at a deflection of 31 mm, where the residual load remains higher than the allowable load specified in Germany.

If this plate were a prefabricated concrete cover for a pit or manhole, it would not be allowed to carry more than 40 kN point loading in service.

According to the German standard DIN 1045 DAFSTB, the allowable loading is even more limited to 11 kN!

This results in loading intensities so low that the corresponding deflections are only a tiny fraction of a millimeter, such as 0.3 mm, or span/5000.

10.5 THE TECHNICAL REPORT NO. 34, 4TH EDITION

It is an exhaustive document, recommended for reading, on the design and construction of slabs on grade and pile-supported slabs, both in SFRC and with traditional reinforcement.

As noted in an earlier chapter of this book, the document underestimates the resistance and bearing capacity of SFRC jointless slabs and SFRC piled slabs, as demonstrated by numerous full-scale slab tests (see Chapter 3 of this book).

One reason for this underestimation is the reliance on the ground-bearing k-value of Westergaard in all design moment calculations. The Westergaard k factor is based on the simplifying hypothesis that the supporting ground behaves as an elastic liquid, offering no shear resistance. For example, while a finger can easily penetrate a glass of water, it cannot do so in a glass filled with compacted sand. This holds even more true for a well-graded granular base, which provides significant friction and further limits deformations.

In the case of pile-supported SFRC slab design, the document is overly conservative. It recommends a minimum slab thickness of 200 mm and a maximum center-to-center span of 3.5 m, regardless of loading intensity or the presence of pile heads.

However, we have successfully designed and completed hundreds of such slabs with spans of 4 m, 4.5 m, 5 m, and even up to 6 m, achieving complete customer satisfaction.

The TR 34 4th Edition does not account for the design advantages of SFRC slabs benefiting from chemical post-tensioning generated by restrained chemical expansion. Primekss' experience and full-scale tests have demonstrated that a restrained chemical expansion of 600 µm/m (as per ASTM C 157) can reduce slab thickness by 8–10%.

The TR 34 4th Edition, in its "Rigorous assessment of moment capacity of SFRC" (Appendix C), uses the EN 14651 small notched prism test, which is biased by artificial large variations, at 0.025 flexion strain, corresponding to f_{r4} of a single hinge-crack.

The reality from full-scale tests shows that multiple cracking and hinging occur at a post-ULS stage of 1.5 times the SLS loading intensity. The collapse

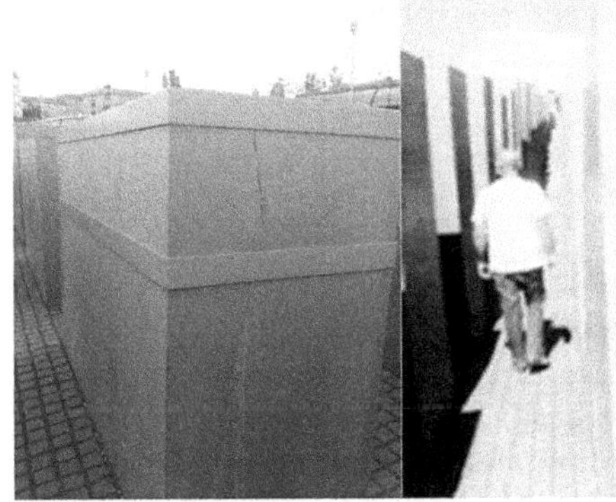
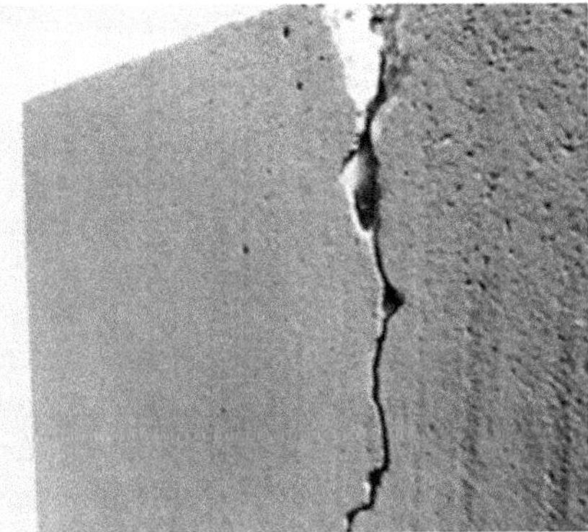

Figure 10.3 Overall view of the Monument of holocaust, and belts and cracks of blocks.

of a piled slab usually happens at about 4 times the maximum SLS load, or around 2.7 times the ULS load. This SFRC structural ductility is completely ignored in the TR 34.

The TR 34 limits the design shear capacity of a 200 mm SFRC suspended elevated slab (C35 concrete with 40 kg/m³ of HE=1/60 steel fibers on a 220 mm diameter pile) to 170 kN, which corresponds to a pile reaction or point load intensity of 113 kN in SLS. However, full-scale test results show that such a slab, even without pile heads, can support a pile reaction of 250 kN in service conditions.

Note that the TR 34 also defines the effective depth of SFRC as 0.75 of the total depth, regardless of fiber integration or slab installation methods. Chapter 7 shows how the Primekss method of blowing fibers at high speed over the entire surface of fresh concrete in the truck mixer significantly reduces deviations in fiber content and minimizes differences in fiber concentration between the top and bottom of the slab. This method gives results much better than any mixing method using truck mixers.

The 75 % thickness for effective depth, implies a bad segregation with 25% of the section without fiber reinforcing. The experience shows that segregation of steel fibres goes together with segregation of large size aggregates. This deficiency results only from a very bad mix design that produces an unstable mix with a much too high a fluidity.

The steel fiber reinforced concrete, even at high dosage rates, doesn't need to be like liquid on site., indeed a very bad practice with high water content and high superplasticizer content.

Such a concrete will show excessive shrinkage with time, like severe cracking, joint opening up to 20 mm, curling of edges and rocking of joints.

Both the Technical Report 63 of 2007 and the Eurocode recommend developing the design using a testing-assisted route method, which is unfortunately not followed in the TR 34.

Regarding the ACI 318 Code of traditional reinforcing design, it only briefly mentions the use of steel fibers for shear reinforcement in beams, specifying a prescriptive minimum dosage rate of 65 kg/m^3, along with a stringent upper limit for the most demanding cases. ACI 318 does little to support SFRC, to say the least.

10.6 CONCLUSIONS

Cost savings throughout the building process are more critical than ever, from the design stage to project completion, as cost reductions significantly improve the carbon footprint of the project.

This is particularly evident when using less material and labor to complete the project. Steel fibers on-site are installed at tons per hour, unlike rebars, which require hours per ton.

Moreover, the dosage rates of steel fibers used in suspended slabs (G-SFRS and E-SFRS) generally range between 40 and 70 kg/m^3, compared to steel rebars, which typically range between 80 and 160 kg/m^3 or even higher.

As a general rule, for slabs on piles (G-SFRS) and raft foundations, ASTM Type I steel fibers (0.9–1 mm diameter and 1500 MPa wire tensile strength) require only about 45% of the steel quantity needed for rebars. For example, 50 kg/m^3 of HE 1/60–1500 MPa steel fibers achieves the same performance as a traditional rebar solution at 100 kg/m^3, with the added advantage that SFRC does not require the concrete cover needed to protect rebars, leading to further savings.

We see that the ACI 544-6R15 method of design and construction aligns most closely with the reality of our 35 years of experience, as well as with the observations and conclusions from full-scale tests.

For example, an ACI 544 6R15 designed slab of 270 mm thickness at 50 kg/m^3 steel fibers dosage rate in a C30-37 mix, is equivalent to a traditional reinforced concrete slab of 300 mm thickness with 12 mm diameter wires at 150 mm spacing in top and bottom meshes (90 kg/m^3, including cuttings and overlaps).

The traditional piled slab solution uses 25 kg re-steel per m^2 while the SFRC solution only 13.5 kg/m^2 steel fibers.

Regarding labor, we estimate that placing wire mesh requires approximately 1 hour of manpower for every 5 m² installed.

For a 5,000 m² piled slab, this means placing meshes and rebars requires around 1,000 hours of work, equivalent to 12 men working for 10 days, at a cost of approximately $50,000 (2023). The SFRC solution eliminates this labor, saving 2 weeks on the critical path.

Running a medium-sized project site of 5,000 m² costs around $3,000 per day, resulting in savings of $30,000 for 10 days. This adds up to total savings of approximately $80,000 for 1,500 m³ of concrete or around $53/m³.

Additionally, the SFRC solution eliminates the rebar supply cost, which would amount to 25 kg/m² x 5000 m² x 0.75 USD = 94.000 USD.

Less construction joints are needed.

The EC2 Annex L draft solution provides approximately 20% less capacity, requiring 10% more thickness to meet flexion requirements, along with a minimum amount of traditional reinforcement combined with steel fiber reinforcement.

The TR 34 4th Edition follows with an additional 10% thickness but without the minimum traditional reinforcement required by the EC2 Annex L draft.

The German DIN 1045-DAFSTB trails far behind, requiring approximately 60% more thickness than the ACI 544-6R15 method, in addition to traditional rebars.

One could question why building regulations in Germany remain so backward regarding steel fiber reinforced concrete (SFRC) and why they lag so far behind.

A general manager of a major global steel fiber manufacturer recently told me that, while my opinion of the German design standard is correct, they still appreciate the standard because it allows them to sell large quantities of steel fibers in Germany.

It seems that the carbon footprint is either not considered a priority or is seen as irrelevant to the business interests of German academics and industry stakeholders.

Why is it so in Germany?

There are several reasons, including:

- Nobody wants to take responsibility beyond certain limits. The general contractor is only liable for two years for the standard project design and execution.
- All materials and designs must be certified according to established standards. If standards are not followed, liability increases to 30 years of civil liability, even if the design is based on a testing-assisted method.
- There are official checking engineers, licensed by the local government, who control and approve standard designs, ensuring strict adherence to regulations.

- Standards are largely influenced by material suppliers and a small group of specialized academics in concrete and related materials.
- The German State Institute for Building Technique in Berlin is a heavily bureaucratic body that oversees non-standard projects. Dealing with this institution is expensive, time-consuming, and often leads to years of discussions without guarantees of practical, advantageous, or economical outcomes.
- The building industry in Germany is tightly segmented between material supply, design, and execution, limiting integration and innovation.
- Corporatism still prevails in the building sector, maintaining traditional roles and practices.

A notable example of the outcome of such a system in Germany is the Memorial Building to the Murdered Jews of Europe (Holocaust Memorial) in Berlin, an open-air concrete monument consisting of 2,400 hollow concrete blocks, each measuring 1 m × 4 m × 1 m and up to 4 m in height. By design specification, it was intended to be highly durable, with a life expectancy of 300 years.

However, within 2–4 years, the blocks developed severe cracking due to concrete shrinkage and thermal variations. As a remedial measure, some of the blocks were fitted with stainless steel belts to prevent falling concrete that could pose a danger to pedestrian visitors. Unfortunately, this solution is neither aesthetically pleasing nor reassuring for the monument's long-term durability.

A team of 12 prominent German concrete academics was appointed to design the blocks for the Memorial to the Murdered Jews of Europe. In adherence to German standards, they developed high-strength plain (!) concrete with a low modulus of elasticity, intending to allow the blocks to swell or shrink without generating internal stress. A brilliant concept on paper! However, combining low modulus of elasticity and high-strength concrete is as contradictory as ice and fire. Even with my limited knowledge, the feasibility of such a solution seems highly questionable.

It appears that no one has taken responsibility for the outcome, and to my knowledge, no viable repair process has been identified. I have also been informed that composite concrete could not be used because it is not included in the standards.

If you ever travel to Berlin, I strongly recommend visiting the Holocaust Memorial, just a short walk behind the iconic Brandenburg Gate. From as far as 50 meters away, you can see the cracks in the blocks, a striking testament to the limitations of the standards-based approach.

Very unfortunately for Europe, a similarly bureaucratic administrative system prevails in France, making it increasingly difficult to implement innovative, game-changing solutions. If such an administrative system had existed between 1890 and 1930, pioneering contractors and engineers would likely have been unable to develop both reinforced concrete and post-tensioned concrete applications.

We think plain concrete was chosen because German standards did not include novel reinforced concrete types to meet durability requirements, and thus, cracks in the concrete were inevitable.

10.7 A FRUSTRATING REACTION

In 2009, the 17-floor Rocca Tower was constructed in Tallinn, Estonia. All suspended elevated slabs were cast in situ using the original patent by ArcelorMittal, which I invented and which later became the basis for the ACI 544-6R15 design and construction method for free-suspended elevated SFRC slabs. The details of this system are provided in Chapter 5 of this book.

Figure 10.4 Rocca tower in Tallinn, Estonia, with all elevated slabs in SFRC including the ACI 544-6R15 defined APC rebars.

The Dutch Concrete Society and the Cementrum Research Center organized a visit for a group of Dutch and German structural engineers to see the completed building, as it was the world's first high-rise structure to use SFRC elevated suspended slabs.

The visit was highly successful, and an article about the project was published in the Dutch Concrete Association magazine and the Cementrum Research Center. The cover of the magazine, published on April 19, 2011, prominently featured the iconic SFRC tower structure. Figure 10.4 shows the 17-story building with all slabs in SFRC, along with the APC rebars as defined in the ACI54-6R 15 document.

It was very frustrating that the German delegation chose to visit the structure only from the outside, refusing to enter the building. Such a reaction was quite disheartening and leaves little hope for the adoption of this invention in Germany, at least for the foreseeable future.

Following the visit, however, the Cementrum team decided to launch a research and development initiative on the topic. This effort culminated successfully with the full-scale testing of SFRC tunnel-formed slabs and walls conducted at TU Eindhoven, as detailed in Chapters 5 and 6 of this book.

REFERENCE

1. L. Lin, A. de la Fuente, S. Cavalario, A. Aguado. "Design of FRC tunnel segments considering the ductility requirements of the Model code 2010." Polytechnic University of Catalonia (UPC). Tunnel, Mining and underground Space Technology.

Chapter 11

SFRC metal decking

The traditional concept of metal deck systems considers the steel profile primarily as a permanent formwork during the casting process. Once the concrete hardens, it forms a composite section with the steel, providing sufficient moment capacity to carry the imposed loads.

11.1 SFRC COMPOSITE DECK SYSTEMS

The primary advantages of using SFRC in composite metal deck systems for contractors, professionals, and clients include the following:

- Safety: Eliminating mesh fabric reduces hazards such as lifting heavy loads to upper floors, minimizing the risk of safety rule violations during handling. It also avoids the manual handling of large mesh sheets from the delivery point at the steel deck level to the workplace and eliminates tripping hazards caused by workers moving across mesh fabric placed on irregular profile decking.
- Program Efficiency: The removal of the mesh handling and fixing process significantly reduces time, streamlining the production workflow. From a production perspective, SFRC simplifies the casting process for composite slabs.
- Quality Assurance: Removing mesh fabric minimizes congestion around openings and eliminates issues like incorrect wire mesh positioning, insufficient concrete cover, and related defects, ensuring higher-quality construction outcomes.
- Can provide a fire resistance up to 120 minutes

The benefits described above are directly attributable to the concept of adding steel fiber reinforcement to the concrete at the batching plant or on-site in the truck mixer, and then pumping the ready-reinforced concrete to the workplace. It is important to note, however, that in several regions of the world, metal decking concrete slabs are still made of plain concrete with minimal or no fire resistance. Evidently, 30 kg/m^3 of steel fiber reinforcement is more expensive than plain concrete. The investigation into the fatal progressive collapse of the 9/11 World Trade Center also highlighted the weaknesses of each floor—100 mm thick light metal decking with a chicken mesh "reinforcing," a standard design type in 1966.

The first application of SFRC on metal decking I was involved with was in Toronto in 1984, where I opened a pioneering sales office in Rexdale, Toronto, Canada, under Eurosteel Fibers Canada. The challenge was to introduce a new and completely novel market where steel fibers were virtually unknown.

Developers quickly became interested in the economic benefits combined with better control of concrete cracking. All traditionally reinforced slabs at the time were heavily cracked, even more so than those observed in Europe. The wire mesh used was typically at most 3 kg/m² or even 2 kg/m² in commercial and office metal decking applications, and in many cases, only plain concrete was used.

We started by selling a 20 kg/m³ dosage rate of Eurosteel 1 mm x 60 mm for slabs on grade with saw cuttings and down to 10 kg/m³ for light metal decking in 100 mm total thickness as in traditional applications.

It proved to be competitive and was appreciated. The cracking was somehow improved, leaving the clients satisfied. A floor covering was always used as a finishing method. The deflection under the own weight of the metal decking slabs was of no concern, and customers or engineers didn't mention any fire resistance requirements at that time.

The deflection equal to span/250, thus up to 12 mm for a 3 m span, was also of no concern. However, deflection of L/250 mm in slender metal decking slabs, as permitted by most standards, will always result in a crack line over each supporting beam at 3 m or 3.50 m distance apart. Cracking becomes more significant in industrial applications where no additional slab covering layer is used on top: despite cracks with potentially large openings, the slab must remain serviceable.

Therefore, in traditional industrial applications, most metal decking slabs are provided with a layer of light wire mesh of sections ranging from 141 mm²/m to 189 mm², 252 mm², 335 mm², or 393 mm². I can say that, in most cases, the regular basic steel fiber of the HE 1/50 type(hooked ends of 1 mm diameter, 50 mm length) at a 30 kg/m³ dosage rate in a C 30-37 mix is suitable to replace the traditional wire meshes.

11.2 FIRE RESISTANCE OF SFRC METAL DECKING SYSTEMS

11.2.1 Traditional wire mesh reinforced solutions

In traditional reinforcing solutions, fire resistance relies on the mesh fabric to ensure adequate plastic moment capacity, both positive and negative, in the event that fire spalls concrete from or up to the slab surface, reducing the section depth and, consequently, the moment capacity. Different mesh sizes and thicknesses of the metal decking slabs are commonly specified in the supplier's design tables to meet fire resistance requirements of 30 minutes, 1 hour, or 2 hours.

The fire rating of SFRC composite deck systems, as well as with mesh fabric, is reasonably well-documented. Designers typically use a simple set of tables provided by the manufacturer to correlate section depth, profile type, reinforcing type, and quantity.

11.2.2 SFRC composite decks

The technical advantages of SFRC in composite metal deck systems are perhaps best summarized by V. Kodur and T. Lie in Reference (1):

> The superior performance of fibre-reinforced concrete, under fire conditions, will compliment the other beneficial properties such as resistance to crack growth, resistance to thermal shock and energy absorption. This make steel fibre-reinforced an attractive material in many high temperature applications.
>
> By properly designing the fibre-reinforced concrete structural members, required fire resistances can be obtained without the need for additional fire protection measures.

From a fire rating perspective, understanding the behavior of the slab subjected to extremely high temperatures and the collapse mechanism is essential.

Bednar and Wald in Reference 2 wrote the following:

> These material properties of the fibre-concrete allow the slab to create a different load bearing mechanism, which increases the fire resistance of the slabs and further concluded:
>
> The material test of fibre concrete demonstrated that its behavior of fibre concrete is ductile enough to create a membrane action if the cracks are developed at yield lines.

The Steel Construction Institute in the UK concludes in its September 2006 report to ArcelorMittal, that *"30 kg/m³ dosage rate of HE1/50-1200MPa steel fibres (Hooked Ends of 1 mm diameter by 50 mm length) is, in most cases, sufficient to provide longitudinal shear resistance equivalent to an A 393 wire mesh in a 150 mm solid slab of f cu = 30 N/mm²."*

Economically, the 7 kg/m² (0.07kN/m²)of A 393 wire mesh is much heavier than the 4.5 kg/m²(0.045 kN/m²) of HE 1/50 steel fibers. Instead of using a crane to place all the meshes on each floor and then carrying them manually to the required positions, the 30 kg/m² steel fiber concrete is pumped directly from the ground level to the higher floors without any difficulty.

It results in savings in the very expensive crane time so that the crane becomes free to be used for more valuable tasks. Steel fibers speed up the installation process of the composite slabs considerably.

11.2.3 Fire tests and design of SFRC metal decking

For ArcelorMittal, we attended fire testing of SFRC metal decking slabs at the Warrington Fire Research Laboratory (UK) and the CTICM Federation (Centre Technique de l'Industrie de la Charpente Métallique) fire laboratory in Nancy-les-Metz, France. Although the CTICM fire lab held a Ministry-level certification, it was not accepted by the CSTB (Centre Scientifique et Technique du Bâtiment) for DTU (Documents Techniques Unifiés) standard-level approval. Testing in the UK proved to be more straightforward, effective, quicker, and practical, with lower costs for investigating the fire resistance of ArcelorMittal SFRC!

In all cases, the testing demonstrated fire resistance of REI > 30 minutes, 60 minutes, and 120 minutes for SFRC metal decking with trapezoidal and swallow-tail reentrant section shapes supplied by Structural Metal Decking Ltd, as shown in Figure 11.1. The image features the ArcelorMittal UK team alongside the Warringtonfire Lab Manager.

The ArcelorMittal Design-SMD (Figures 11.2, 11.3, and 11.4) summarize spans, loadings, metal sheet types, and fire resistance with fiber reinforcing. In the tables, numbers highlighted in italic black and white indicate spans limited by the critical fire resistance.

The steel fibers used in all cases are Hooked Ends, 1 mm in diameter, 50 mm in length, with a steel wire tensile strength of 1200 MPa.

Figure 11.1 Fire test result displayed on the screen by the testing team of ArcelorMittal UK.

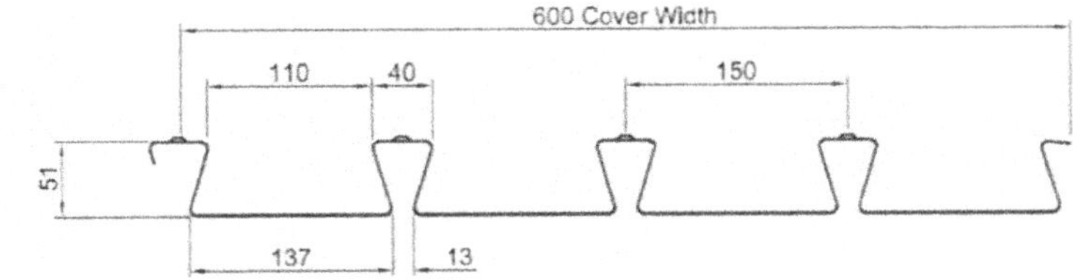

Normal Weight Concrete

Maximum Permissible Span (m)

Span Type	Fire Rating (hours)	Slab Depth (mm)	Steel Fibre	0.9mm Gauge				1.0mm Gauge				1.2mm Gauge			
				Total Unfactored Applied Load (kN/m^2)											
				3.5	5.0	7.5	10.0	3.5	5.0	7.5	10.0	3.5	5.0	7.5	10.0
Double Span	1.0	101	HE 1.0/50	3.23	3.23	3.23	3.18	3.47	3.47	3.47	3.34	3.78	3.78	3.78	3.46
		130	HE 1.0/50	3.05	3.05	3.05	3.05	3.30	3.30	3.30	3.30	3.51	3.51	3.51	3.51
		150	HE 1.0/50	2.89	2.89	2.89	2.89	3.14	3.14	3.14	3.14	3.44	3.44	3.44	3.44
	1.5	110	HE 1.0/50	3.22	3.21	2.82	2.54	3.39	3.34	2.93	2.64	3.69	3.58	3.14	2.83
		130	HE 1.0/50	3.05	3.05	3.05	2.79	3.30	3.30	3.19	2.89	3.51	3.51	3.42	3.10
		150	HE 1.0/50	2.89	2.89	2.89	2.89	3.14	3.14	3.14	3.14	3.44	3.44	3.44	3.35
	2.0	125	HE 1.0/50	3.09	2.85	2.60	2.35	3.33	3.07	2.70	2.44	3.45	3.27	2.88	2.61
		150	HE 1.0/50	2.89	2.89	2.89	2.71	3.14	3.14	3.09	2.81	3.44	3.44	3.28	2.98
		175	HE 1.0/50	2.72	2.72	2.72	2.72	2.98	2.98	2.98	2.98	3.26	3.26	3.26	3.19

Figure 11.2 R51 SFRC design table (ArcelorMittal/SMD Ltd.).

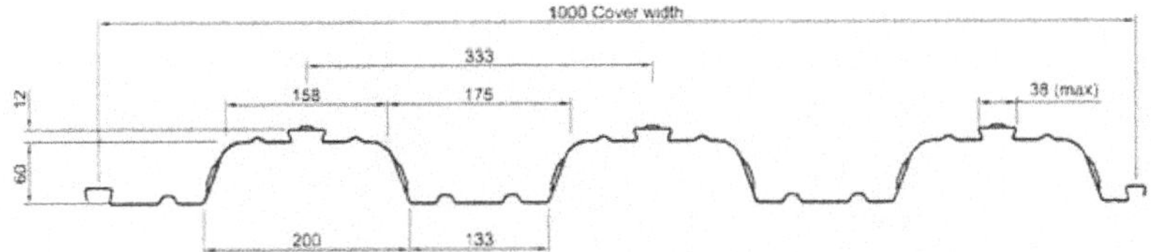

Normal Weight Concrete

Maximum Permissible Span (m)

Span Type	Fire Rating (hours)	Slab Depth (mm)	Steel Fiber	0.9mm Gauge				1.0mm Gauge				1.2mm Gauge			
				Total Unfactored Applied Load (kN/m^2)											
				3.5	5.0	7.5	10.0	3.5	5.0	7.5	10.0	3.5	5.0	7.5	10.0
Double Span	1.0	130	HE 1.0/50	3.59	3.42	2.99	2.68	3.83	3.54	3.09	2.78	4.19	3.80	3.33	2.99
		150	HE 1.0/50	3.38	3.38	3.30	2.98	3.70	3.70	3.42	3.10	4.12	4.10	3.63	3.28
		200	HE 1.0/50	2.99	2.99	2.99	2.99	3.20	3.20	3.20	3.20	3.78	3.78	3.78	3.78
	1.5	140	HE 1.0/50	3.21	2.90	2.54	2.12	3.29	2.96	2.61	2.18	3.47	3.14	2.75	2.48
		150	HE 1.0/50	3.37	3.06	2.69	2.26	3.46	3.15	2.77	2.50	3.63	3.30	2.90	2.62
		200	HE 1.0/50	2.99	2.99	2.99	2.92	3.20	3.20	3.20	3.20	3.78	3.78	3.62	3.32
	2.0	150	HE 1.0/50	3.21	3.01	2.58	2.14	3.31	3.01	2.64	2.20	3.46	3.14	2.76	2.32
		175	HE 1.0/50	3.17	3.17	2.64	2.40	3.45	3.28	2.71	2.48	3.72	3.42	2.83	2.57
		200	HE 1.0/50	2.99	2.99	2.99	2.79	3.20	3.20	3.20	2.86	3.78	3.78	3.46	2.96

Figure 11.3 TR60 SFRC design table (ArcelorMittal/SMD ltd.).

Figure 11.5 shows a typical flowing mix containing a 30 kg/m³ dosage rate of HE 1/50 steel fibers being pumped to the SMD TR60+ metal decking level during installation.

11.3 EXAMPLE OF APPLICATIONS

Figure 11.6, depicts the Centre Hospital in Luxembourg, showcasing, as in the reference (3) by Brauch and Wolperding, a parking structure project featuring 4

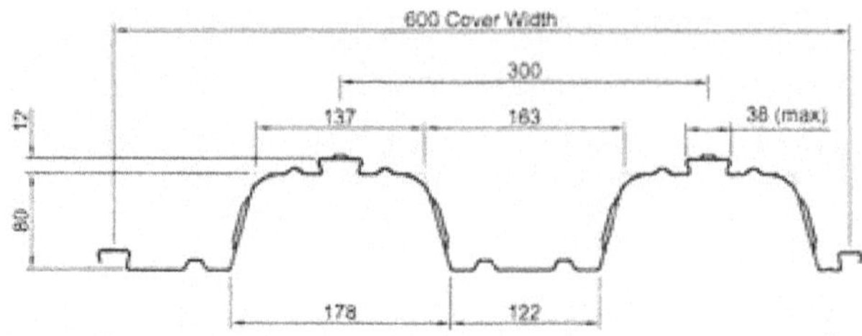

Normal Weight Concrete

Maximum Permissible Span (m)

Span Type	Fire Rating (hours)	Slab Depth (mm)	Steel Fiber	0.9mm Gauge				1.0mm Gauge				1.2mm Gauge			
				— Total Unfactored Applied Load (kN/m²)											
				3.5	5.0	7.5	10.0	3.5	5.0	7.5	10.0	3.5	5.0	7.5	10.0
Double Span	1.0	140	HE 1.0/50	4.35	3.93	3.43	3.09	4.50	4.07	3.56	3.20	4.84	4.39	3.84	3.45
		160	HE 1.0/50	4.19	4.18	3.68	3.33	4.49	4.35	3.84	3.47	5.02	4.66	4.12	3.72
		200	HE 1.0/50	3.78	3.78	3.78	3.78	4.13	4.13	4.13	4.03	4.63	4.63	4.63	4.31
	1.5	150	HE 1.0/50	3.57	3.23	2.82	2.54	3.72	3.37	2.94	2.65	3.98	3.60	3.15	2.84
		175	HE 1.0/50	3.88	3.54	3.13	2.63	3.99	3.64	3.22	2.71	4.22	3.85	3.40	3.08
		200	HE 1.0/50	3.78	3.78	3.54	3.22	4.13	4.07	3.63	3.31	4.63	4.27	3.81	3.47
	2.0	160	HE 1.0/50	3.58	3.24	2.85	2.38	3.69	3.35	2.94	2.46	3.91	3.55	3.12	2.62
		175	HE 1.0/50	3.88	3.35	2.96	2.50	3.79	3.46	3.06	2.58	4.02	3.67	3.24	2.73
		200	HE 1.0/50	3.78	3.78	3.41	2.88	4.13	3.92	3.50	2.96	4.45	4.10	3.66	3.33

Figure 11.4 TR80+ SFRC design table (ArcelorMittal/SMD Ltd).

Figure 11.5 Metal decking slab at 30 kg/m³ dosage rate of HE 1/50 steel fiber at installation in the UK.

floors of jointless SFRC metal decking slabs. The structure spans 125 m x 44 m, with a total area of 16,500 m² and a slab thickness of 130 mm. The slabs utilize ArcelorMittal Coffra Plus 60 sheets, reinforced with a 35 kg/m³ dosage rate of TABIX steel fibers (1.3 mm diameter, 50 mm length, with an undulating shape and 850 MPa wire tensile strength) in a C 30-37 concrete mix.

There is only one construction joint located at ca. mid-length of each composite floor level.

The project adheres to French standards and has received approval from the SOCOTEC technical control company. The composite deck spans 2.5 m and rests on IPE 400 beams with a 16.5 m span, fitted with top studs to

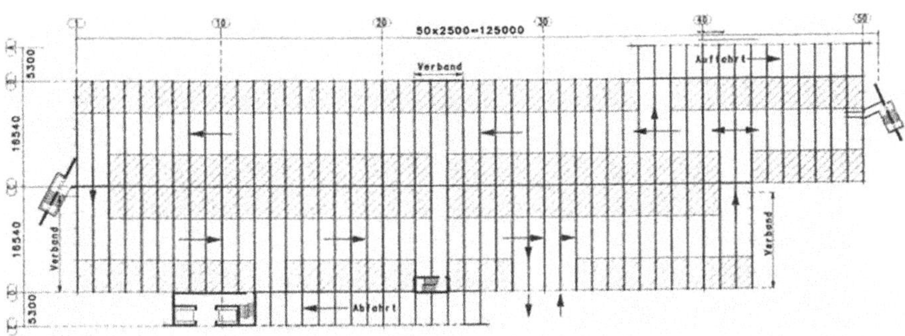

Bild 2. Grundriss Ebene 1
Fig. 2. Plan view floor 1

Figure 11.6 Overall view of a typical floor level 44.80 m x 135 m typical floor drawing of the Centre Hospitalier in Luxembourg.

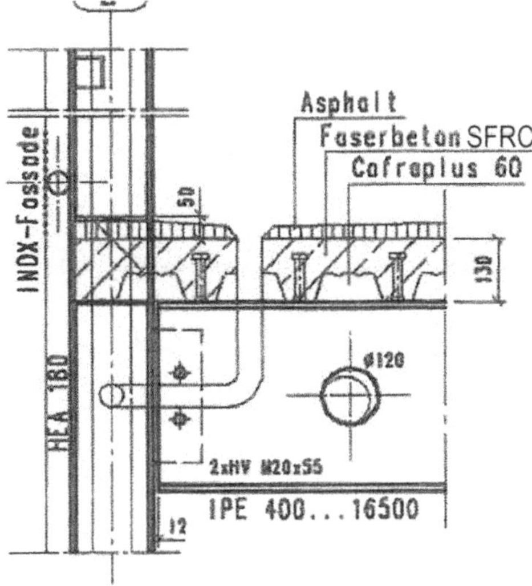

Figure 11.7 Edge column section composite slab section detail.

enable the beams to act as composite beams using the SFRC slab as a compression table.

The metal decking and supporting IPE400 beams were installed with a 140 mm camber to offset the deflection caused by the own weight of the freshly laid SFRC.

Figure 11.7 shows a typical section at an edge column support where the Coffra Plus 60 sheet, the shear studs, the fiber-reinforced concrete, the IPE 400beam and the HEA180 column are all visible.

Figure 11.8 shows the installation process of an SFRC slab reinforced with 35 kg/m³ of TABIX steel fibers (undulating fiber with1.3 mm diameter,

Figure 11.8 The installation of the SFRC on the ArcelorMittal COFFRA + metal decking slab.

Figure 11.9 The SFRC I decking structure of the Children Hospital parking in Luxembourg.

50 mm length- 850 MPa wire tensile strength). A notable aspect of the construction is the 140 mm camber of the ceiling, which corresponds to the subsequent floor above. This camber is designed to return to level under the self-weight of the steel fiber concrete slab after its installation, ensuring structural precision and uniformity.

An overall view of the building structure is shown in Figure 11.9.

After the year 2005, all ArcelorMittal SFRC metal decking applications were with a 30 kg/m³ dosage rate of HE1/50 steel fibers, proving to be more economical than the 35 kg/m³ dosage rate of TABIX1.3/50 and in compliance with the design conclusions of the UK Steel Construction Institute, based on the Warrington Fire Laboratory report.

11.4 CONCLUSION

A simple consideration of the safety risks involved in lifting and using mesh fabric on uneven surfaces in multi-story projects suggests that SFRC is a safer alternative to traditional methods.

When comparing the production techniques and technical merits of the material, it is clear that the use of SFRC for composite metal decks offers superior performance and is safer, easier, quicker, and more economically viable than traditional systems.[1]

NOTE

1 For the reader in need to study and understand in full the car park building structure under fire, I would recommend the following reference: "Fire Analysis of Car Park Building Structures", B. Fettah, Polytechnic Institute of Bragança, July 2016.

REFERENCES

1. Kodur and Lie. "Fire Resistance of Fibre-Concrete" by Kodur and Lie in "Fiber Reinforced Concrete, Present and Future" pp. 189, The Canadian Society for Civil Engineering. (1998) Editors: N. Banthia, A. Bentur.
2. F. Bednar, F. Wald, A. Kouhoutkova, J. Vodicka. "Membrane action of composite fibre concrete slab in fire." Czech Technical University in Prague, in Steel Structures and Bridges 2012, *Procedia Engineering*, 40, 49, 2012.
3. J. Brauch, G. Wolperding. "Parkhaus CHL Centre Hospitalier de Luxembourg:Verbunnddecken mit Faserbeton." *Stahlbau* 76, 2007. Heft 11, Ernst & Sohn Verlag für Architektur. und technische W References.

Chapter 12

The sustainability challenges

The concrete volume used today is about 30 billion tons per year worldwide.

Concrete, however, remains the most sustainable construction material, outperforming timber, steel, and other alternatives.

Nevertheless, the production of cement in about 4000 cement plants worldwide is a significant source of CO_2 emissions, averaging around 700 kg of CO_2 per ton of cement.

Between 2023 and 2030, the cement industry will reduce this to about 300 kg CO_2 per ton through optimization of processes in each plant, along with replacing clinker with other hydraulic binders and by-products with very low CO_2 content.

Each cement plant will need to invest approximately €60 to €80 million to achieve this target, a phenomenal amount globally within a few years.

After 2030–2035, further reduction to 0 kg CO_2 seems almost impossible but will depend on carbon capture and sequestration in underground cavities, with high costs for transporting the captured CO_2 through pipelines to these cavities.

The governments are expected to handle this massive task. However, as of today, no action has started, leaving uncertainty about how it will be addressed after 2030.

In my view, an essential goal to achieve as soon as possible is to drastically reduce the volume of concrete used through more effective design and a better concrete matrix that is ductile rather than brittle. This could significantly reduce cracking and creep, both of which lead to greater deformations and require thicker sections in traditional concrete as well as the structure life expectancy..

Cracks provide pathways for aggressive elements to penetrate concrete and traditional rebars, making the durability of concrete structures uncertain and limited. Thus, many cases exist where concrete structures have failed in service without ever reaching ULS (Ultimate Limit State) conditions.

Steel fiber reinforced concrete (SFRC) is already recognized today as significantly improving the durability and crack resistance of concrete. It can

DOI: 10.1201/9781003188315-12

prevent cracking in the cover thickness of rebars, thereby protecting the steel rebars from corrosion, which, once initiated, is irreversible.

Concrete shrinkage is a major cause of cracking in all concrete structures. It seems obvious to me that reducing shrinkage to zero should be a top priority.

It had been my idea since the 1980s to develop zero-shrinkage slabs on grade. While many attempts were made in Europe and the USA, none succeeded in the markets. Once again, let's revisit the Primekss case, where I found the right team and mindset to identify, research, develop, implement, and improve the zero-shrinkage concept. Interestingly, the chemical post-tensioning effect achieved, as documented in the existing literature, turned out to be serendipitous.

The Primekss chemical post-tensioning method stiffens the section to minimize flexural deformations, allowing further reduction in section thickness. Primekss has demonstrated over 15 years of experience that this approach works effectively for slabs on grade and other structural slabs, reducing their thickness by nearly half.

This goal is also realized on-site during slab installation, with quality control measures monitoring expansion in real time using embedded laser fiber optic sensors. This technique, inspired by Prof. Dr. Gregor Fischer at TU Riga (Latvia) and TU Denmark, enabled practical implementation in slabs. However, such a method requires highly skilled crews on-site and focus on the importance of training and education.

The laser optical sensors monitor stress and micro-cracking rates from the hardening phase through the full life of the slab.

These sensors show that the static mechanical stresses in slabs are significantly lower than those estimated by academic calculations, particularly when composite fiber reinforcement is used to arrest cracking at the micro-level.

Laser optical fiber sensors can precisely locate cracks anywhere in the section, in real-time, of a concrete structure, transmitting all data through WiFi from the structure site to Primekss' quality control management system. This allows the structure's owner or end-user to access accurate, real-time information about their structure's condition, rather than relying on theoretical values derived from academic models, which often leads to imprecise or incorrect conclusions.

Traditional concrete design standards significantly hinder the optimization of solutions like those achieved by Primekss in slab construction.

The actual standards for concrete promote the use of construction materials that comply with standard performance requirements for the material's design.

The challenges of sustainability could be addressed more effectively if innovative solutions proposed by pioneering engineers and contractors were accepted more rapidly by overcoming both technical and psychological barriers.

I observe every day that traditional structural engineers from international design firms tend to oversaturate industrial slabs on grade, metal decking, footings, rafts, piled slabs, and large machine foundations with high concentrations of rebars right beneath each face of the structure, leaving the interior completely unreinforced and brittle. This often makes it difficult to place and compact concrete between the intricate rebars and meshes.

For example, to address overdesign, we recently redesigned a typical 70,000 m² metal decking slab for a large battery plant. The original specification required a 200 mm thickness in C 35-45 concrete over a 51 mm x 1 mm steel decking with a reentrant section shape. It was reinforced with a 140 kg/m³ rate of heavy top and bottom wire meshes (10 mm and 12 mm wire diameter) along with additional rebars.

Our Primekss redesign utilized self-stressing SFRC at a dosage rate of 40 kg/m³ HE 90/60 hooked-end fibers, combined with a minimal 10 kg/m³ of 10 mm diameter rebars placed over the composite beams with studs, as shown in Figures 12.1 and 12.2.

These 10 kg/m³ rebars were, according to the engineer, necessary to address longitudinal shear along the beams. We did not challenge this perspective further, as the total quantity of residual rebars was relatively small, and the 150 mm spacing posed no obstacle to the proper compaction of the SFRC.

Figure 12.1 40 kg/m³ HE 90/60 in an SS-SFRC Primekss Concrete.

Figure 12.2 Overall view of a phase of the metal decking project.

In total, we achieved 50 kg/m³ of steel reinforcement compared to the original 140 kg/m³ in the traditional design. This resulted in an estimated CO_2 saving of approximately 1,400 tons for this 70,000 m² application.

Often, projects are not fully designed but are instead specified with high steel rebar concentrations in kg/m³ and concrete defined only by its standard compressive strength. In some cases, the structure is so stiff and undeformable that the rebars do not even engage unless cracking occurs due to shrinkage. By introducing ductility to the concrete mass and addressing shrinkage cracking, it is possible to save around 200 kg of CO_2 per cubic meter of concrete.

The most effective approach to sustainability at a low cost is to minimize material usage. This also reduces workforce requirements, leading to further CO_2 savings, as well as safer jobsites. Moreover, each tons of rebar saved on-site reduces the need for at least one worker-day on site, shortening the critical path of the project schedule. Keeping a large jobsite operational can cost up to €30,000 per day for utilities, security, insurance, fuel, crane rentals, and administration, making such savings significant.

At the 65,000-seat Friends Arena in Solna/Stockholm, the use of SFRC for the suspended raft foundation in 2011 saved over 2,000 tons of cut and bent rebars, as well as approximately six months in the project timeline, with a daily cost of €10,000 at the time.

Large-diameter rebars are ineffective against cracking and require thick coverage of up to 125 mm to protect them from corrosion in critical cases. This approach falls far short of the optimal solutions needed to address sustainability challenges.

The sustainability challenges of concrete can be met if optimal solutions are invented, developed, and widely adopted. This will require pioneering and responsible engineers and contractors who are freer to innovate, create, and make these solutions available.

Carola Edvardsen, Technical Director at COWI, a world-scale engineering firm in Denmark, stated in References (1, 2) the sustainability advantages of the SFRC as follows:

- *Approximately 5 to 10 times the resistance to chloride-induced corrosion.*
- *Significantly reduced construction cost and time.*
- *A drastic reduction in steel consumption and tunnel segment thickness and therefore a smaller volume of concrete is needed.*
- *Top structural performance with reduced cross-sectional dimensions ans inherent immunity to degradation mechanisms.*
- *An ideal solution for bored tunnels that require a 100-year service life and beyond and are located in areas of harsh environmental conditions*
- *A reduced demand for cement, as waste by-products from other industries such as slag, fly ash, silica fume are used within SFRC and inherently emit very low levels of carbon dioxyde.*
- *More importantly, carbon dioxyde emmissions from the production of SFRC with fly ash and/or slag as cementitious replacements are reduced by a remarkable 60-70 per cent compared to traditional reinforced concrete production.*

In Reference (2), the author demonstrated Figure 12.3:

As shown in Figure 12.3, prepared by Carola Evardsen, the CO_2 emissions decrease significantly when transitioning from standard traditional

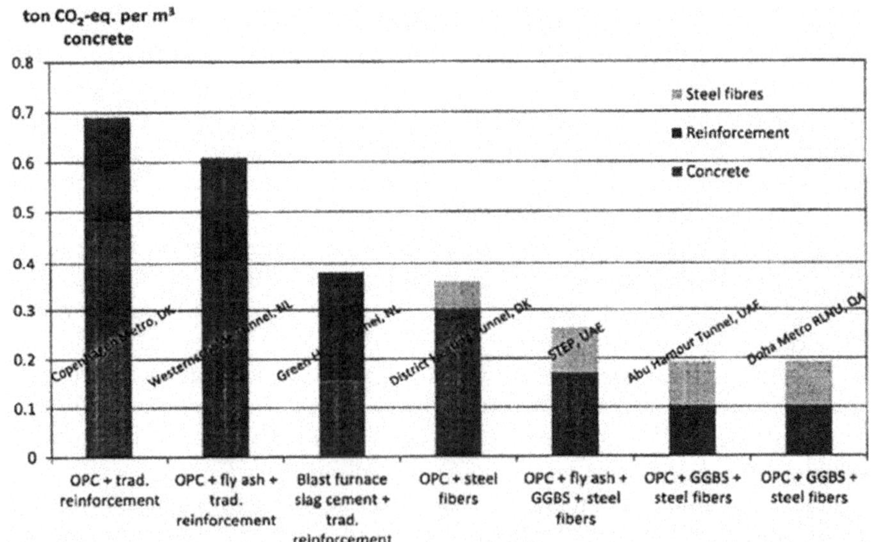

Figure 12.3 Reduction of CO_2 emissions from standard traditional reinforced concrete down to SFRC including GGBS.

reinforced concrete to steel fiber reinforced concrete, dropping from 700 kg/ m³ of concrete to 200 kg/m³ of concrete.

It can be assumed that approximately 20% of this reduction is due to the possible decrease in thickness enabled by the use of a thinner section, made feasible through fiber reinforcement.

Primekss self-stressing SFRC (denoted SS-SFRC), utilizing proprietary binders, achieves a self-stressing compression stress of up to 1.5 N/mm². Combined with its stiffening effect and zero shrinkage, it allows for a further 35% reduction in concrete volume.

The 200 kg CO_2 emission per cubic meter is further reduced by 70 kg CO_2 resulting in a final total of 130 kg CO_2 emission per cubic meter of concrete. which is approximately 20% of the initial traditional initial figure.

This remarkable performance is not limited to a single large-scale project reported in peer-reviewed literature but is part of everyday PRIMEKSS slab installation, occurring at a rate of 500 to 1000 m³ daily.

Through its R&D efforts in design, materials, and installation techniques, Primekss alone saves an estimated 700,000 tons of CO_2 annually—a figure equivalent to the emissions from ca. half a million European combustion engine cars in a year.

Primekss sustainability effort is explained in References (3, 4).

I would recommend reading Reference (5) to understand an original and innovative study of the SFRC suspended elevated slabs, not only for their better sustainability in general but also for the social, and economic, and environmental requirements, the reduction of labor and execution time, the reduction of the noise pollution, and the reduction in discomfort for the labor and third-party.

The performance detailed in the reference has been further improved today. The previously used 100 kg/m³ dosage rate of TABIX (1.3 mm diameter, 50 mm length, 850 MPa tensile strength, undulating shape) has been replaced with an equivalent 50–60 kg/m³ dosage rate of HE (1 mm diameter, 60 mm length, 1500 MPa tensile strength).

Moreover, extensive green steel fibers are now available, achieving a reduction of 300 kg CO_2 per ton and further down to 100 kg CO_2 per ton in ArcelorMittal plants powered entirely by renewable energy sources.

I would also recommend the reading of Reference (6), as it is a clear summary of the sustainability in numbers of the SFRC elevated suspended slab application: "*It is a simple rule of thumb established over almost 30 years of research in many countries that when using a suitable structural SFRC, the amount of steel required per cubic meter is rediced by a factor of up to 2.2.*"

I must say that I understand both Prof. A. Aguado and Prof. de la Fuente have been, for SFRC, true pioneering academics, promoting and developing the total omission of all rebars in SFRC tunneling prefabricated segments in Spain together with Mr. M. Van de Walle, and later promoting full-scale testing for SFRC elevated suspended slabs to analyze and design applications, and detailing the significant sustainability advantages.

For slabs like those Primekss can produce, the common barriers are the old traditional mentality still taught at universities, reflected in outdated standards or even in draft documents like Annex L draft of the next Eurocode 2 for fiber-reinforced concrete structures. It is indeed unfortunate, as these are mainly written by academics and traditional mentality engineers, and are still in the preparation stage.

I have little hope that such a draft can be amended, as the EC committee is a highly bureaucratic body, following a procedure with one or two members per country, most of whom have no practical experience in SFRC, and there is no effective public vetoing process in writing, unlike at ACI.

12.1 PRINCIPLES FOR SUSTAINABLE CONCRETE SLABS

The authors in Reference (7) detail seven general principles from P1 to P7 regarding the sustainability of the building process. I'll use them here with modifications to adapt it more for slab contractors, as it has been implemented with success at Primekss Sia.

P1: Get smart

Design, Materials processing, Constructions, Operations, Preservations/ rehabilitation, Reconstruction/Recycling are involved.
- Design what you need: Overdesign is wasteful.
- Design holistically: consider support conditions, concrete type, availability and properties, and **avoid joints.**
- Develop education programs for new and practicing professionals (no Black Box concepts or attitudes).
- Develop new materials and practices or support the development of these new materials and practices.
- Communicate fluently with all stakeholders.
- Analysis of CO_2 of all raw materials from "cradle to grave", transports, equipment, wastes, and demolitions.

P2: Design to serve

- Practice context-sensitive design to focus on the needs of the user.
- List and discuss the needs.
- Discard and replace all these being not sustainable and explain.
- Eventually create new needs.
- How the slab contractor will meet these needs? Indeed by a multidepartment action in the company.

- Meeting them will result in increasing the project efficiency and acceptance.
- The slab contractor and customers should be able to advertise and market for better sustainability of their products.

It becomes possible to charge more for a sustainable product as it sells better.

P3: Choose what you use

Don't reject out of hand an attractive material or process or machine, because it has never been used so far.

Think of the slab solutions to meet user needs, economically (including planning to shorten the duration of the jobsite), technically, and environmentally.

Examples:

- A low resistance to freezing and thawing aggregate could be excellent for inside.
- Supplementary Cementitious Materials and recycled aggregates are worth to look at (type and percentage).
- Tend to reduce down to zero the number and length of joints.
- Avoid all traditional reinforcing (not optimum at all environment and is labor-intensive, steel fibers are more efficient and preferable).

P4: Less is more sustainable

A design that uses fewer virgin materials is generally more sustainable.

Favor a simple solution that is easy to implement.

Examples:

- Use less Portland cement (CEM I).
- Less shrinkage is more sustainable, zero shrinkage with zero-joint slabs are today available at Primekss SIA.
- A tight tolerance base (obtained by specific laser robotized machines) allows a further possible reduction of the slab thickness.
- Get rid of pile heads and eliminate tasks (each task implies costs and generation of CO_2).
- Increase significantly the precision of the steel fiber dosage rate. Conveyor belts discharging in truck mixers are not quite precise and not optimal.
- Less number of phases of construction is generally more sustainable (think of a zero joint slab).
- Fiber reinforcing is "less" than rebars and meshes.
- Promote Self-Healing Concrete (crack self-hard filling techniques).

- Increase the productivity per Workman Day of production in flooring by using machines and robots.

P5: Minimize negative impacts

Less disruption to others and less noise on site (less or no poker vibrating, use flowing concretes).

Promote light reflexivity (use of suitable SCM) of slabs: better visual comfort and less artificial lights are needed.

Smooth and flat surfaces (tyer wear reduction, smoother and faster ride, less maintenance) are needed.

Less construction wastes (re-use or don't use at all) are required.

Improve at each level the efficiency of processes, equipment, and vehicles.

Reduce trip number and length.

P6: Take care of what you have produced

Maintain slabs in good conditions: Promote the DESIGN-BUILT-Monitor and MAINTAIN contracts.

Provide relevant and interesting slab contractor typical data to the client.

Drop the Black-Box attitude.

Develop rehabilitation and preservation techniques and products.

- When possible, develop and propose overlays (bonded or debonded depending on the specific case).

P7: Innovate. It is much more challenging than doing what is familiar

- The sustainability achievement will need development of new ways of thinking and doing by collaborating in depth together by all departments so that ... $1 + 1 + 1 = 5$ or 6.
- Promote two-lift construction (two-layered slabs with top and bottom functions), internal curing, and self-healing concrete.
- Develop further SFRC zero shrinkage. Structural products with the elimination of rebars are replaced by fiber reinforcing together with smart concrete mix designs. Note that in slabs and rafts 2.2 kg rebars/mesh are replaced by 1 kg of steel fibers.
- Enhance and implement chemical post-tensioning as Primekss SIA started successfully from the year 2009.
- Communicate constantly at all steps to all stakeholders (for $1 + 1 = 4$ unlike 1,5 sometimes).
- Evaluate, learn, and then teach so that $(1 + 1 = 4)$.

- Don't escape risks. Take risk and manage it.
- Understand in depth the slab products.
- Understand in depth relevant competitor's and stakeholders' products like so many carbon capture, and so many minerals and organic admixture types and propositions are a strong message that PRIMEKSS can grow quicker, more safely for all stakeholders, become much stronger and more robust, will develop and survive much further when we all will embrace fully the sustainability.

You understand that the seven Ps are an immediate concern at all levels from the top management, to sales and marketing, to R&D, to design engineers, to concrete engineers, to production and administration/accounting.

Sustainability is a great opportunity for a slab contractor to unleash all departments for the future. Don't let some big players like cement and concrete plants, large general contractors, business federations, developers and academics to lock us in their backyard. We have to be ahead of them.

REFERENCES

1. C. Edvardsen. Engineering sustainable and durable concrete to reduce our carbon footprint, in COWI report, circulated in public from 05.10.2018.
2. C. Edvardsen. Durability and sustainability aspects of steel fibre reinforced concrete (SFRC) under severe exposure conditions.
3. X. Destrée, B. Pease. "Reducing the CO_2 Emissions of the Concrete Slabs Construction with the PrimeComposite slab system." *ICCS 13, Tokyo, The first International Conference on Concretes: Sustainability.*
4. X. Destree, B. Pease, D. Martin. Reducing CO_2 emissions by design: the PrimX system. *Concrete*, March 2021. P.20-22 (UK).
5. A. Blanco, A. de la Fuente, A. Aguado. Sustainability of steel fiber reinforced concrete slabs; Second International on Concrete Sustainability. ICCS16, Madrid.
6. D. Martin, X. Destrée, M. Suta. "The inexorable rise of the steel-fibre-reinforced elevated slab construction." *Concrete (UK)*, 55, 27–29, June 2021.
7. T. Van Dam, P. Taylor. "Seven principles for sustainable concrete pavements." *Concrete International*, pp. 49–52, Volume 33, Nov. 2011.

Index

Pages in *italics* refer to figures and pages in **bold** refer to tables.